Geographic Information Systems and Science

A Concise Handbook of Spatial Data Handling, Representation, and Computation

Steven A. Roberts and
Colin Robertson

OXFORD
UNIVERSITY PRESS

Oxford University Press is a department of the University of Oxford. It furthers the University's objective of excellence in research, scholarship, and education by publishing worldwide. Oxford is a registered trade mark of Oxford University Press in the UK and in certain other countries.

Published in Canada by
Oxford University Press
8 Sampson Mews, Suite 204,
Don Mills, Ontario M3C 0H5 Canada

www.oupcanada.com

Library and Archives Canada Cataloguing in Publication
Roberts, Steven A., 1963–, author
Geographic information systems and science: a concise handbook of spatial data handling, representation, and computation / Steven A. Roberts and Colin Robertson.

Includes bibliographical references and index.
ISBN 978-0-19-900363-1 (paperback)

1. Geographic information systems—Textbooks. I. Robertson, Colin (Colin John), author II. Title.

G70.212.R62 2016 910.285 C2016-901535-1

Part/Chapter Openers/Cover image: © iStock/marigold_88

Oxford University Press is committed to our environment. Wherever possible, our books are printed on paper which comes from responsible sources.

Printed and bound in Canada

1 2 3 4 — 20 19 18 17

Contents

(A) indicates an advanced section that may be skipped on first reading or reserved for readers with greater background in GIS.

Preface

We collect here a few thoughts on this textbook for the student and teacher that we hope will provide some practical advice on the logistics of using the book. Here we also collect some insights on the authors' perceptions of the subject of the text. The last section describes our approach to the mathematical content of the book.

For the Instructor

We have used this text in both third year and fourth year geographic information systems (GIS) courses, and portions of it in a third year spatial analysis course. For an intermediate GIS course a suggested path through the text (see Figure 1) is as follows: Chapters 1 through 7, Chapter 12, and the appendices as needed. For an advanced GIS course a suggested path through the text is as follows: Chapter 1, Chapter 4, Chapter 5, Chapters 7 through 11, and Appendices I and II. For an introductory spatial analysis course selected topics from Chapters 4, 5, 7, 8, 9, 10, and 11, along with Appendices I and II, are suggested. The entire text might be comfortably covered in a first year graduate course in GIS or spatial data handling (SDH). In such a course, emphasis might be placed on the advanced sections and full derivations. The symbol Ⓐ in the

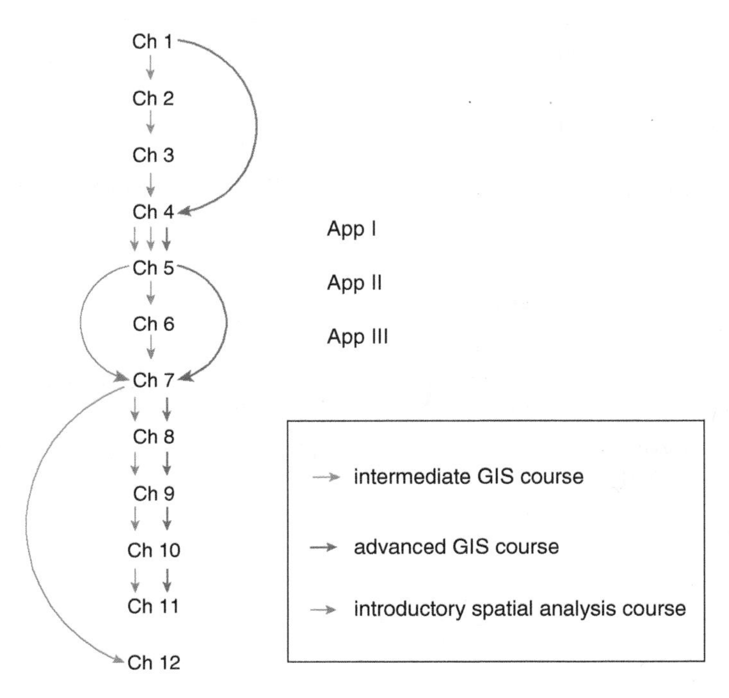

Figure 1 Potential pathways through the text.

margin indicates an advanced section. Finally, many figures and the datasets used throughout the pages that follow are available on the website associated with the text.

For the Student

This text is meant to provide a bridge from your introductory exposure to the ideas of geographic information systems/science (GIS/GISci) to more advanced work and a critical reading of the current literature. The symbol Ⓐ in the margin indicates an advanced section that may be skipped on first reading or reserved for readers with greater background in GIS. We expect and hope that a few of you will go on from such a course to create the new ideas and technologies for the next generation of GIScience. We hope this book will give you some of the background knowledge and tools for this journey. Our approach to building this bridge might be illustrated with the following analogy. If you commute to work on a bus you use the bus as a means to an end but you need not be very concerned with how the bus works. If, however, you are a professional race car driver, to be successful you need to know in detail how your car works. You must be able to interpret what your instruments and telemetry are telling you about the car's performance and your interaction with, and responses to, it. You may go on to be a researcher in GIScience or have a managerial position where you must evaluate GIS software or direct large projects or application development. In those instances, having awareness of the issues and ideas raised in this book should help you make better, appropriate, and critically informed decisions.

Levels of abstraction in GIS

When we work with geographic data in a GIS, remote sensing (RS), or other spatial data handling domain, we must be aware that we are often dealing with several levels of abstraction from "reality" (see Figure 2). We have long created geographic abstractions in language and maps. We have more recently formally added geometric abstractions in

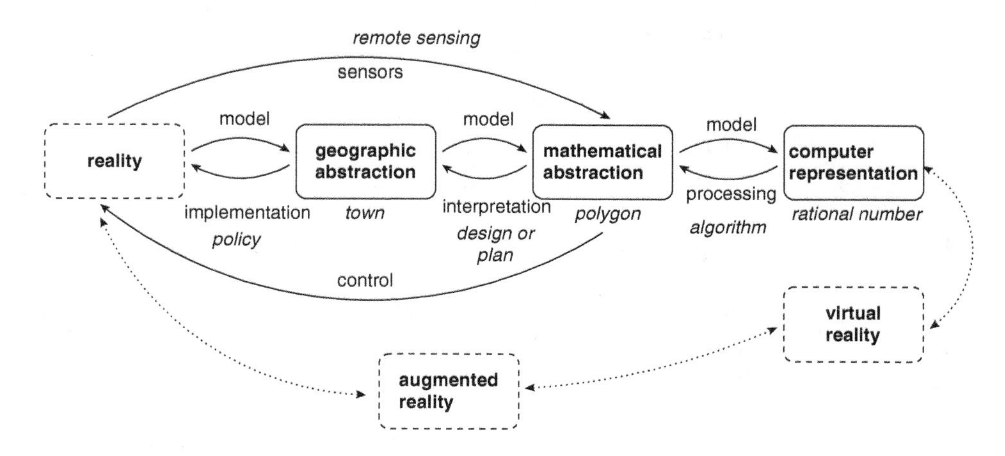

Figure 2 Levels of abstraction in GIS.

order to handle geographic data in a digital environment. These digital environments (computers) have their own abstractions, finite models of various types of numbers, and topological relationships. Note that in the figure some examples of the abstract categories are given in *italics*. We can also move backward through this hierarchy of models as shown by the labelled arrows in the figure. This represents actions of, for example, implementation or interpretation. We will say more about these ideas throughout the text. Also emerging are sensor networks with perhaps attached automated control systems, which are starting to become embedded in infrastructure and machinery. These systems can act directly from and to the outside world, often without an intervening geographic abstraction. We also include in the diagram the ideas of augmented reality and virtual reality. How these emerging notions fit into the current hierarchy of abstractions is not yet clear, so the dotted arrows in the diagram here are perhaps just a starting point for further discussion.

Why did we include derivations?

It might seem at first glance that we included so much detail in the derivations to appeal to math-literate students and faculty. On the contrary, these people will likely find this presentation somewhat pedantic although hopefully in places enlightening. We primarily included as much detail as we did to allow those with a spotty or incomplete mathematical background to be able to follow the technical arguments. Another way of saying this is that we hope to deconstruct the technical arguments to lay bare much of the mechanics that are usually assumed. This may show the student (or instructor) that a deeper level of understanding is indeed accessible. Further, the types of arguments and algorithms presented are typically and widely used in the literature, and a greater familiarity with these tools is invaluable to understanding and extending knowledge in our discipline. Finally, although we sometimes do, we often do not just leave a derivation to a literature reference, as this does not serve the need of timely explanation. The original source of a result may be in a different context, with different notation, and sometimes for older sources, hard-to-penetrate prose. Many of our derivations are original syntheses from several sources, and often many gaps in the explanations have been filled in to help the reader.

Acknowledgements

Steven A. Roberts would like to thank the following people for providing support, mentorship, and friendship: Brent Hall, Paul Calamai, Barry Boots, and the late Ferko Csillag and the late James Kay.

We also note that parts of section 9.4 were strongly influenced by the presentation in the excellent texts by O'Sullivan and Unwin [89] and Bailey and Gatrell [6]. As always we take full responsibility for any remaining errors or omissions in the text. We are greatly indebted to the software projects that were used in creation of this book, particularly LaTeX and the developers of R and its many packages.

— Steven A. Roberts and Colin Robertson, June 2016

S.A.R. dedicates this book to Lynne and Genevieve

C.R. dedicates this book to Christine.

Chapter 1

Introduction

1.1 Introduction

In this chapter, we introduce the approach to the study of geographic information systems (GIS) taken in this book. We discuss the reasons for decisions about coverage and depth of topics, and highlight two recurring themes in the book. The first is the idea of viewing GIS as enabling an abstraction process—from the geographic forms, patterns, and processes that encompass reality on earth to digital representations of that reality. We return to this theme often, as many of the fundamental concepts and issues in GIS relate to these abstractions and their imperfections. The second core idea is the notion that there is something special about spatial, geographic data. That is, data that describe processes and patterns of the geographic world require new methods, theories, and collectively embody the field of *GIScience* that underlies the technology of GIS. We explain methods and algorithms in enough detail for readers to get an intuitive grasp of their core underlying ideas while also demonstrating how these ideas play out when operationalized with geographic information. However, we start with a cursory overview of the development of GIS, a story that started in the universities and government laboratories of North America and is still unfolding today around the world.

1.2 Development of GIS: People and Institutions

The first GIS is credited to the Canadian Geographic Information System (CGIS), developed in Canada in the late 1960s and largely spearheaded by Roger Tomlinson, often called the "Father of GIS" (see Figure 1.1). In the early 1960s, Tomlinson had been involved in aerial surveys for collecting and analyzing forest information in Africa and was faced with the challenge of analyzing vast amounts of geographic information to identify the sites for new plantations. When manual methods of compiling the relevant data sources on natural and social factors were deemed too onerous, Tomlinson reasoned that a computer would be able to manage and analyze this amount of data much more efficiently.

Jana Chytilova/The Canadian Press

Figure 1.1 Roger Tomlinson (1933–2014), a key figure in the history of GIS in Canada.

Coincidentally, the Canadian Government (specifically the Department of Agriculture) was planning a Canadian Land Inventory (CLI) that would produce maps for agricultural suitability of populated regions across all of Canada. Again, the scale of data and questions to be answered vastly exceeded the capabilities of cartographers and analysts. It can be easily argued that the invention of GIS was simply the result of a big data problem. However this problem differs from what might conventionally be described as such today because the data—in digital form—did not exist yet. The *information need* exceeded the tools and technologies (mainly paper, but some automated maps) available at the time. More important than mapping was developing the ability to search and manipulate that information. The need was not to create an ultra-efficient computer-automated mapping system that could produce excellent paper maps, but to produce something entirely new that could manage information about the world in a way that made information accessible and useful.

The project to create the CLI led to many innovations critical for GIS, including the first rudimentary topological data model used for geographic data, and in 1971 the CGIS was fully operational [26]. The mix of a pressing problem (handling and querying large amounts of geographic information), ongoing public investments, passionate individuals, and technologies ripe for adapting to new uses led directly to the CGIS. Like many technology-oriented fields, GIS was invented to solve a problem, and its use and historical development have remained largely empirical and grounded in real-world problems.

While Tomlinson created the spark and motivation behind the CGIS, he left the project in 1969 and was also active in the early developments of GIS in the US. In 1965 the US Bureau of the Census faced a similar challenge, as had Canada's Department of Agriculture: how to compile and map growing volumes of census data from records that could only be linked to a postal address. Experiments in address matching, computer mapping, and areal data analysis between 1969 and 1973 under the auspices of the New Haven Census Use Study led to the development and distribution of address-matching software to support analysis of census data geographically. Again, the demand for information from data appears to be at the heart of the motivation leading to the development of GIS, as articulated by Voight of the US Census Bureau [118]:

> *As the American population living in urban areas continues to expand, the number and scale of programs concerned with improving the urban environment increases. Each program generates a demand for new and better data about life and activity in the area, data necessary to define factually both the nature and the scale of the problems to be dealt with, and data essential for planning future action.*

And noting later,

> *there is also a growing need to improve the system for relating census data with locally generated data at a finer scale than was ever required before. The linking of data from census and local sources becomes essential to a more penetrating analysis of various urban problems and trends of change.*

In these quotes we see two salient features relevant to why GIS emerged in the US government: the need to solve problems related to urban living standards (at the time, urban crime and poverty were major social issues in the US), and the technical need to fuse data from the census with other sources of geographic information. We see in this two of the principal functions of GIS—handling large amounts of geographic information and using geography as way to bind data from disparate sources. It is not surprising that one result of the growth of GIS has been the concurrent growth in methods and applications of spatial analysis and modelling.

However, the most active institutions in the development of GIS in the US were in universities. The most notable university institution for early research into GIS was Harvard University's Laboratory for Computer Graphics (LCG), which started in 1965. Led by Howard Fisher, the lab produced a computer mapping software package called SYMAP, which used a line printer to create areal and isoline maps. Throughout its lifespan until 1980, the lab served as a centre of gravity for individuals interested in computer mapping and spatial analysis, including many individuals who went on to be highly influential in establishing the field of GIS/GIScience, such as Nicholas Chrisman, Jack Dangermond, and Thomas Poiker. Notably, an extension to the lab's name was added in 1968, making it the Harvard Lab for Computer Graphics *and Spatial Analysis*.

By the early 1980s, the centre of gravity in the development of GIS technology had shifted to the private sector. Companies that had been active in the development of computer-assisted drawing (CAD) had begun to recognize the importance of geographic information. Intergraph was an early developer of vector-based GIS hardware and software, including mapping and GIS applications designed for civil engineering and emergency dispatch, initiated in part by LCG alumnus David Sinton. Another LCG alumnus who formed what was initially a non-profit organization in environmental consulting was Jack Dangermond, who started Environmental Systems Research Institute (ESRI) in 1969. By 1982 ESRI had launched ARC/INFO, which could run on multiple computer platforms and included a topological data model. The program was command-line driven, and implemented in FORTRAN. Additional functionality was added for analysis of triangulated irregular networks, network datasets, and survey data. As the

1980s wore on and the use of micro-computers became widespread throughout all sectors of the economy, the focus of GIS innovation became developing GIS applications for an increasing set of domain areas. While many early adopters in government departments of Canada and the US had developed in-house GIS for specific domains, these systems began to be replaced by commercial GIS packages.

As discussed by Klinkenberg [69], the role of the government in North America in the development of GIS shifted from that of leader in the development of GIS systems (software and hardware) to one of provider of data: an enabler of GIS applications, research, and theory. Reasons for this shift are varied, from restraints on government spending to the success of mainstream commercial GIS packages. While parallel developments occurred in Canada and in the US during this period, the data cultures in each country differed significantly, largely boiling down to Crown copyright in Canada, which preserves copyright for any federal government-produced spatial data. In the US, where this concept does not exist, models of data distribution emerged as either free or intended to recover the costs of distribution (not production), which eventually led to a much more *open* data culture. While the reasons behind Crown copyright and data cultures are largely historical, they have dramatically impacted the evolution of GIS in North America.

Where data are open and available, value-added economies of scale emerge, which benefits the industry through lower costs and more research and development, and through better and lower-cost services for citizens. An example is the comparative case study of Canada's RADARSAT with the US's Landsat earth observation satellites. The Landsat data are freely available to anyone to use for commercial or scientific use. The Landsat program has spawned economic benefits that are estimated at over $2 billion per year (not including secondary uses and value-added products), with its over 40-year archive of earth imagery. Conversely, RADARSAT data are only available to government agencies cost-free, and data distribution is handled by a commercial enterprise. RADARSAT data, while very important for specific types of mapping, have by comparison remained a niche data product. The differences in data distribution at the government level between the US and Canada are being highlighted by the emergent open data movement and the developing data economy. As the history of GIS is still being written, the future of GIS development is likely to be one where data will play a major role (see Chapter 12 for a discussion of emerging issues). Interestingly, municipalities in Canada are not subject to Crown copyright, and municipalities have become the leaders of open data in Canada, often opening up their GIS datasets for public use first. Good examples of this are the City of Nanaimo, the City of Toronto, and the City of Surrey. Open source GIS software projects such as Quantum GIS, GeoServer, and GRASS GIS are also increasingly involved in advancing the state of GIS practice.

1.3 Abstracting and Formalizing Geographic Information

A GIS can be defined as *an information system tasked with representing geographic objects, relationships, and processes.* As with any information system, this implies that a GIS must be able to store data, create information products such as maps and data visualizations, interact with human users, and provide query processing and other analytical capabilities. The **G** in GIS means that these functions are uniquely developed for handling geographic information. GIS differ from many other information systems in that they are concerned with representing phenomena that are complex hybrids of physical reality, our imposition of order on this reality (i.e., coordinate space), human categorization and filtering of this reality (i.e., place), and a digital encoding of these phenomena. While other information systems may model purely abstract processes, such as financial constructs like stocks and bonds or business concepts such as employees and customers that have application-specific definitions, GIS are usually tasked with modelling phenomena that are thought to be shared-concepts—which can vary from person to person. Whereas the concepts of employee and customer are well defined in the sense that most people can agree on their meaning in a given setting, the concepts of forest and wetland, or labelling an area of higher elevation a hill or a mountain, are fuzzy and ambiguous. This ambiguity is inherent to many geographic objects and phenomena, complicating how they are digitally represented in an information system.

Maps are perceived as representations of reality, and the filters and biases embedded within them are well known by cartographers. These same biases carry over to spatial data models and GIS generally. This is not only true of naturally occurring phenomena, but also of anthropogenic structures. A very simple example is the choice of spatial representation for a building and parking lot in a GIS, which could be modelled as distinct or linked point objects, collections of various geometries, three-dimensional geometries, attributes of other objects, etc. The selection and definition of possible attributes on each of these is also infinite and necessarily context/application-dependent. The abstraction of these phenomena in a GIS is riddled with implicit and explicit decisions at conceptual and implementation levels, which, across individuals, are likely to differ.

We believe that these ambiguities are frequently forgotten or glossed over in GIS and GIS-based analysis, and that as more data are referenced to geographic space, explicitly considering these processes of abstraction should be standard practice for practitioners, students, and researchers in the field. For this reason we emphasize the explicit details of various data models used for GIS data throughout the text, and, where appropriate, show how manipulation of geographic data can produce results that may not coincide with our expectations, due to aspects of the data model, details of the algorithm, implementation of the algorithm, or other context-specific factors. While social processes, cultural processes, economic processes can all be modelled in GIS, it is the embedding of the relevant features of these processes in geographic space that makes these analyses suitable for GIS. An alternative definition of a GIS might be *a geographic representation for a particular purpose*, following Kaplan's definition of statistical models [67].

There are many unique properties of geographic information that make the functions of GIS particularly challenging, and throughout the rest of this chapter we will review some of these properties and challenges as an entry into the more detailed treatment that follows in later chapters. We'll briefly review how both data models and algorithms commonly used in GIS affect, and are affected by, the abstraction process of geographic information.

1.3.1 Data Models

GIS data models are how we operationalize the representation of geographic information in the computer. We later define in detail the different data models, but at the highest level of abstraction we can think of the first stage of data modelling as the filtering process where we decide what objects and relationships are included and excluded in our geographic model. The semantics of geographic information, as discussed above, are complex. And while simple modelling constructs became commonplace in traditional information systems, such as entity-relationship modelling for relational databases, it is surprising that a geographic analogue to this has never been normalized into GIS practice. The development of ontologies for geographic knowledge has become popular in the GIScience research community, but have not been widely adopted as a way to organize information and data for GIS representation. An ontology can be defined as a domain-specific representation of knowledge. This codified representation removes ambiguities in natural language that would be obvious for humans to recognize (e.g., a software piracy domain ontology differentiating software piracy from marine piracy), but difficult for a computer program to recognize. Typical components of an ontology include individuals, events, concepts, relations, rules, and axioms. While in the early days of GIS semantics were defined locally, within organizations and communities of users, this model is no longer tenable. The widespread use of the Internet to distribute both GIS data and services has created the need for *semantic interoperability.*

Almost all GIS represent reality formally as sets of things, which can have some interaction with each other. Model constructs such as entities, relationships, and attributes are taken to *be* the concepts they represent in the real world. We then can interrogate the quality of the abstraction by comparing how the modelled reality compares to the observed reality. However, Kuhn [71] illustrates the inherent limitations of this approach: humans understand meaning through behaviours, functions, and actions rather than by sets of entities and their properties. A lake is a lake not because it is polygonal and blue but because it holds standing water, can be used for swimming, sailing, a source of drinking water, etc. Kuhn highlights some of the special considerations for geographic information that relate to the challenge of semantic interoperability:

- Unlike most information that is abstracted for computer representation, geographic information is grounded in physical reality, which is linked to observations and measurements.
- Many of the geographic constructs we aim to model are based on social agreements and perceptions, with a special case being geographic place names (which can be contested in certain circumstances).
- Geographic information is understood as it relates to processes in space and time, yet computer representations tend to be static and discrete.
- Vagueness, uncertainty, and variable scales are essential for formulating the semantics of geographic information.

The above points illustrate why GIS data representations are laden with uncertainty and ambiguity. That these issues exist at such a foundational level is important for understanding the limitations of GIS-based solutions using today's technologies and methods in any given context. That poorly defined or crude representations of geographic information underly GIS does not negate its utility in solving problems, but rather identifies that important decisions are made when applying GIS to a certain context; and these decisions can and do vary by individuals, objective of the study or application, organizational issues, and technical and other factors. A solid grasp of the underlying data models and algorithms in common use in GIS today is therefore critical for making informed decisions in practical uses of GIS.

1.3.2 Algorithms

Algorithms are best thought of as a series of instructions to do a job. That job may be to minimize a function, measure a distance, or simply check whether the value of a variable is above some threshold. GIScience and GIS are heavily indebted to computational geometry for the algorithms that underly many of the tools required to identify and measure spatial relationships. GIScience researchers have also contributed to this knowledge base, adapting and creating new approaches for handling geographic information. Often, what works in 2D Euclidean space does not work for lumpy three-dimensional ellipsoids, so designers and users of GIS are faced with a choice: settle for a (sometimes poor) approximation, or employ empirical methods to adapt a mathematical approach to the irregularities of the earth's surface. Throughout the text, we examine both approaches and hope to give the reader a feel for the computational machinery underlying modern GIS, and to expose some of the assumptions and approximations that may not be obvious from a click of the button on a graphical user interface.

A 1.4 What Is Special about Spatial?

Inherent in the above discussions is that there is something fundamentally different about geographic data. We identified some of Kuhn's challenges for semantics for geographic information, which hint at some of the ways in which spatial *is* special, or, to be precise, *geographic is special.* We now examine this question in further detail for two important aspects of geographic information handling: the links between spatial processes and geographic patterns and the way that temporal change is represented in current geographic data structures.

1.4.1 Spatial versus Geographic—Patterns and Processes

When we talk about the unique properties of spatial data in the context of GIS, we are typically actually talking about *geographic data.* Concepts such as spatial autocorrelation and spatial heterogeneity, which relate to the way that values of a spatial variable are distributed across space, and described by Anselin [4] as the two fundamental properties of geographic data, are implicit in GIS algorithms, data structures, and analytical methods. Spatial autocorrelation, often summarized by Tobler's First Law of Geography as *all things are related, but near things are more related than far things*, captures an intuitive description of the patterning commonly associated with geographic phenomena. Spatial heterogeneity, also known as spatial non-stationarity, is when the expected value of a spatial variable varies over space. The world is both patchy and autocorrelated. Goodchild [48] describes how such general law-like statements about geographic data guide the development of GIS technologies. Both of these properties pertain to maps as static outcomes of stochastic spatial processes. Inference or representation of the dynamic parts of these processes is often ignored. Before exploring temporal dimensions of GIS in more detail, we will briefly review how spatial patterns and processes are connected.

The idea that spatial processes are the spatial aspects of complex processes—be they economic, cultural, or ecological in nature—that operate in the world, and which geographers are interested in learning about, has a long history in geography. Getis and Boots [43] use the metaphor of a mapped pattern as a single frame from a longer movie to describe how spatial patterns and processes are related. The objective of spatial analysis is to formulate a model for the movie as a whole (i.e., the spatial process) such that for any frame within the movie, adjacent frames could be predicted from the model. For movies with long periods of inaction followed by rapid change, obtaining maps for the representative frames are sufficient to describe the process as as whole. However, where adjacent scenes change frequently or in ways that are unrelated to previous frames, the model will be very difficult or impossible to specify. Fortunately, many spatial processes can be modelled using only a handful of core spatial models, such as clustering or diffusion. GIS have emerged as an important hybrid technology that supports the representation and specification of spatial models and spatial analysis.

Geographic information systems are also special in terms of scale. The scales of interest for GIS typically vary from millimetres to thousands of kilometres, but rarely beyond that, and usually within the 1 m to 1000 km range that corresponds to the distribution and range of sizes of most geographic processes and objects. Many patterns are scale dependent and are observed only at specific spatial scales, while others can be scale invariant and persist across many scales. The notion of scale, fundamental to almost all geographic studies, is uniquely important for understanding, representing, and analyzing geographic information because humans understand the geography of the world through both abstraction and scaling, provided by map representations. GIS enable scale-sensitive and multi-scale study of spatial patterns and processes. One particular scale issue has to do with aggregation of geographic information over areal units, as is common with administrative data, whereby the observed causal relationships or measures obtained over spatial units can change when the boundaries of the spatial units change. Since most of these boundaries are arbitrary, reflecting convex polygons of similar size or population rather than following contours of the variable of interest, these changes are potentially problematic. This issue is known in the literature as the modifiable areal unit problem (MAUP). At a minimum, GIS can be used to assess the sensitivity of a dataset to MAUP by enabling repeated analysis at multiple scales.

The unique properties of GIS data are also given meaning by their geographic manifestations—related to the scale and patterning of features on the surface of the earth. While spatial patterns and processes are abstractions of geographic patterns and processes, they are inextricably linked. We reason about spatial patterns and processes that are defined mathematically to better understand the geographic patterns and processes that shape and form the world around us. Similarly, given new empirical problems identified in the geographic world, we often refine and extend the theory and methods for spatial analysis. While the map-based history of GIS has led to under-treatment of time in GIS research and technology, that is starting to change, as space-time integration has become a core focus of GIS research (see Richardson [96]). Representation of dynamic processes, as we will see, still remains a challenging task after over 45 years of continual development of geographic-information software and systems.

1.4.2 Time and Geographic Information

As recounted at the beginning of this chapter, GIS evolved from computer mapping and has therefore inherited many characteristics of mapping. For one, the "layer view" of GIS is predominant, where different phenomena or processes are presented as distinct layers, usually encoded as a single geometric representation or data model, which can be viewed independently or simultaneously with other layers. This view is so prevalent within GIS as to have become synonymous with GIS itself. We will show in later chapters how the orthodoxy of this view limits development of GIS in some ways, if the ultimate goal of GIS is to be an information system for the geographic world. Perhaps more generally, one of the more obvious limitations is the frame of reference on geography taken by maps and GIS data structures and databases: a top-down view of the world, at a specific point in time. GIS are notoriously very poor at handling the temporal dimension, as they are inherently static data constructs; yet as noted above, we understand the world in relation to processes and functions, which are dynamic.

Sinton [104] lays out the basic context for the *analysis* and *representation* of spatio-temporal GIS data; that is, data within a GIS for which the analysis of how those data change over time in some way is required. As discussed above, GIS data are embedded in physical reality, in geographic space, and thus depend on observations and measurements. Observations for geographic information have three key properties: space, time, and theme(s). Each of these properties can be held constant, controlled, or measured. Most discussion of GIS in introductory textbooks relates to the storing and modelling (i.e., controlling) of space, such that the analysis (i.e., measurement) of theme can be carried out. As such, time is usually held constant. An example of this is given in Figure 1.2, whereby the control of space is determined to be the neighbourhoods of the City of Toronto. The second step is the varying of some theme of interest over this decided-upon delineation of space. In this case, this represents the ratio of positive to negative geotagged tweets in each neighbourhood, based on a sentiment analysis of tweet content (darker positive sentiment, lighter negative sentiment). We might then start to explore the spatial patterns of Twitter sentiment expressed in the map.

Considering common types of maps presented in Table 1.1, it is evident that this view of geographic information exposes how maps constrain our ability to represent time. It is clear that most forms of representation for geographic information inherently keep time constant. It may be that the cartographic-inspired "layer view" of GIS is responsible for perpetuating this issue into GIS data structures over 35 years after it was first articulated by Sinton [104].

Many models have been proposed to extend the representation of time in GIS; however, challenges persist. A summary model by Ferreira et al. [37] builds on Sinton's observations to propose three spatiotemporal data types required for GIS to represent the six possible combinations of constraints to geographic information: a time series, a trajectory, and a coverage. The *time series* data type holds space fixed, controls time, and measures theme. The *trajectory* data type shows spatial evolution over time: a GPS-tagged animal, a moving wildfire, or an automatic identification system (AIS) message tracking locations of ships at sea (see Figure 1.3 for an example). The *coverage* data type is the traditional snapshot model of change, representing spatial variation in a theme at a specific instance in time.

Figure 1.2 Traditional map view of geographic information: space is controlled, theme is measured.

Table 1.1 Representation of geographic information.

Dataset	Fixed	Controlled	Measured
Topographic map	Time	Theme	Location
Airline schedules	Location	Theme	Time
Nanaimo road centrelines	Time	Location	Theme
Mountain pine beetle point data	Time	Theme	Location
Victoria Landsat image	Time	Location	Theme

Source: Adapted from Sinton [104]; Langran and Chrisman [73].

Previous approaches to analysis of change using sequences of spatial snapshots are given in Sadahiro and Umemura [100] and Robertson et al. [99]. They aim to infer the change between two temporal snapshots by examining changes in spatial and topological properties. Ferreira et al. [37] infer events through observation of interactions of objects defined by the three spatiotemporal data types represented above. For example, Hurricane Katrina in New Orleans as an event might not just be represented as a signature spatial pattern and magnitude in a spatial coverage derived from microwave remote sensing, but might include a collection of objects including wind speed, rainfall, trajectory speed, and storm surge, as well as flooded area, emergency services,

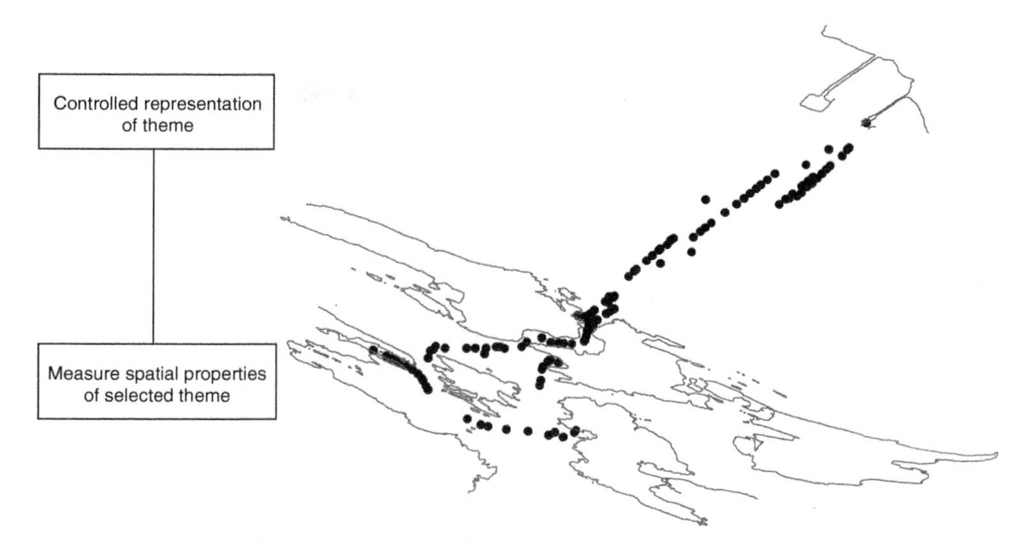

Figure 1.3 Spatial trajectory data, the Queen of Nanaimo in the Gulf Islands (BC) as recorded by satellite AIS.

etc. Representing such a complex event through separate, independent layers makes an integrated representation and analysis of the hurricane event much more difficult.

We highlight these issues, which we see as research challenges that continue to face the current state of the art of GIS, in order to invite the reader to think broadly about the assumptions embedded in GIS practice today. We stress this critical approach throughout the text by describing selected aspects of the field in sufficient depth to scrutinize the formalisms underlying the simple user interfaces that greet new GIS users. Other areas of the text, specifically those well covered by most introductory texts, are covered in lesser detail. We envision this text as a gateway to higher study and reading in the GIS literature by readers interested in the fundamental research challenges facing the field today.

Chapter 2

Geographic Data Acquisition

2.1 Introduction

Geographic data describing features of the earth's surface can be obtained from an increasingly diverse set of technologies. This chapter will highlight the major modes of geographic data acquisition used in geographic information analysis and processing today. While the production of geographic information was traditionally the domain of professional surveyors and cartographers, a major shift in the production of digital geographic information has occurred over the last three decades, marked by the emergence and maturation of two data acquisition technologies: satellite-based remote sensing and global positioning systems (GPS). In this chapter we review the basic principles of these two technologies, followed by a brief discussion of approaches for converting analogue geographic information into digital representations.

A general model of geographic data acquisition, which forms the basis for the topics of this chapter, can be formulated as in Figure 2.1. GIS-based geographic information is the result of a series of processing steps that transform analogue signals into discretized digital representations describing locations on the earth's surface. These discrete representations are first recorded by a sensor, which senses a noise-contaminated signal from a target object. In the case of remote sensing, the target object is the earth's surface and the sensor can be a handheld camera, a satellite-borne earth observation sensor, or an aircraft-mounted camera. With GPS, the sensor is a GPS receiver located on the earth, which receives radio signals from space-borne satellites to determine its geographic location. In all instances of geographic data acquisition, challenges are encountered during the processing stages between these states: filtering signal from noise, discretizing continuous signals into digital data, and transforming digital representations into GIS data models, structures, and file formats.

2.2 Remotely Sensed Data

Remote sensing is often defined as obtaining information about an object or surface without being in direct contact with the object. This expansive definition therefore includes a vast array of activities, from airborne mapping with active sensors to photography with handheld cameras, to viewing with the human eye. In terms of the

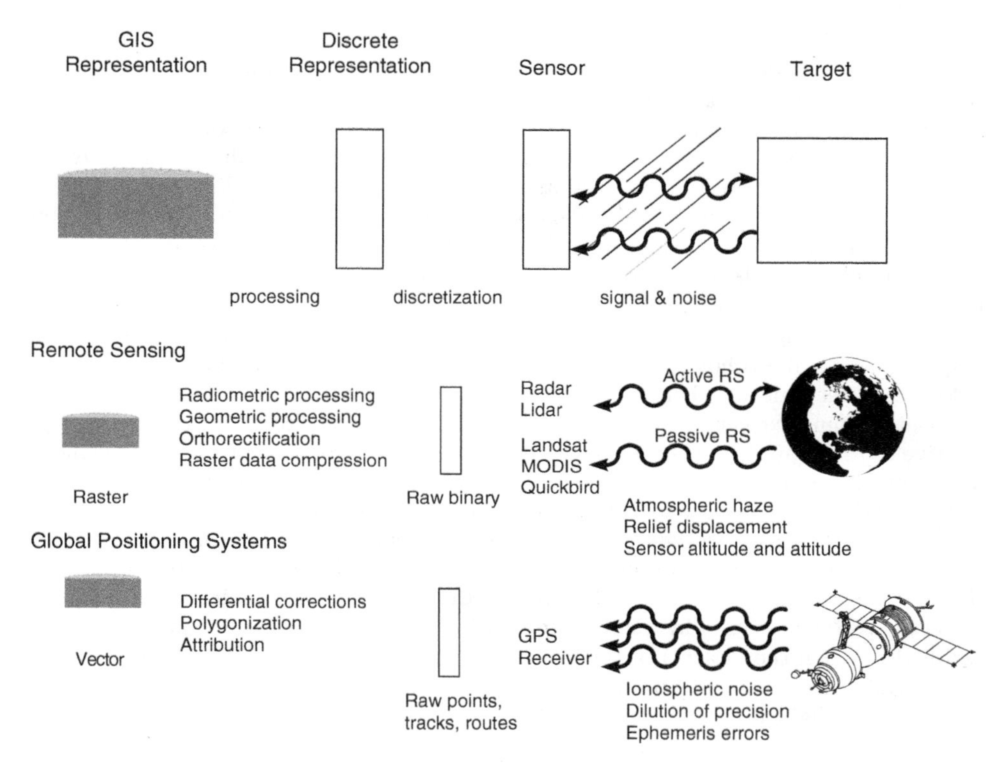

Figure 2.1 General model of geographic data acquisition with GIS representation on the left, and the processes from which these data are derived on the right. These categories are not mutually exclusive, as lidar data for example, are processed first as vector point clouds, then typically into raster formats.

types of data used in today's GIS, we can limit the scope of remote sensing to those methods that characterize portions of the earth's surface or atmosphere. The methods differ with respect to the sensing platform, the type of information that is recorded, the forms of data output, and the mechanism of data translation from continuous signals representing the variability in the target surface to a discretized digital representation of that surface using a GIS.

2.2.1 Aerial Photography

Film-based aerial photography has been and remains the primary source of GIS data in many fields. An aerial mapping camera installed on the floor of an aircraft (or as a payload on an unmanned aerial vehicle, UAV) is used to capture photographs of the earth's surface. Photo acquisition can be contaminated by noise introduced by the atmosphere (e.g., haze), which tends to increase with flying height. Lens filters can be used to filter out different types of light at the time of photo capture. Blue light filters are commonly used to reduce the effect of atmospheric haze because of the way the wavelengths in this part of the spectrum interact with the atmosphere. Filters can also be used to filter out both blue and green light when using films sensitive to near-infrared light, which is often used in vegetation- and crop-monitoring applications. As will be discussed later, the ability to tailor a sensor to a specific part of the electromagnetic spectrum is one of the most powerful features of using remote sensing for earth-observation applications.

Airphotos can be taken from different perspectives. Photographs taken where the tilt of the lens is within 3° of vertical are called vertical airphotos. Photographs taken at tilt angles greater than this are called oblique airphotos. Oblique airphotos are more difficult to use as as source of geographic information because they have a varying scale across the photograph such that distortions are introduced into measurements of distance and direction. Oblique photographs cover more area due to the tilted perspective but are usually used for visualization purposes rather than as a source material for mapping. Vertical airphotos can be used with photogrammetric methods for measuring distances, directions, and heights of features on the ground.

Determining the centre of an airphoto is required in order to orient the photograph relative to the flight line of the airplane. There are actually three centres on an airphoto. The simplest to find is the geometric centre of photograph, obtained by connecting the vertical and horizontal lines from the fiducial marks on the photograph (see Figure 2.2); this is called the *principal point.* On a perfectly vertical photograph, the principal point will coincide with the point observed by the intersection of a line connecting the centre of the camera and the ground, called *nadir.* However, in reality, airphotos are never perfectly vertical, so usually there is some displacement between the principal point and nadir. When the location of the principal point is marked on an adjacent airphoto, this is called the conjugate principal point, which connects to the principal point to represent the flight line.

A first step toward using a vertical airphoto as a source of geographic information is to determine the scale. The scale of an airphoto is, like the scale of a map, a ratio of the distance measured on the airphoto to the corresponding distance on the earth's surface, often standardized to one unit on the left-hand side (e.g., 1:25,000). Scale is therefore easily determined by measuring these distances. Scale is also a function of the flying height and the focal length of the lens that captured the photograph such that

$$\text{scale} = \frac{f}{(H - h)} \quad ,$$

where f is focal length, H is flying height, and h is the elevation above a known vertical datum such as mean sea level (MSL). Note that this scale will be generalized for the entire airphoto, which may be a poor estimate of the true map scale for many areas in the scene where elevation changes rapidly, and as such may not provide accurate distance measurements.

To illustrate, consider an aerial photograph taken at 7500 m with a 152 mm focal length lens (see Figure 2.3). A forest transect at location A is located at 44 m and a second transect at location B is at 119 m. Assume the transects at A and B both correspond to a 33.1 mm photo distance. The average height of the terrain in the whole scene is 62 m. Using the average terrain height to calculate the photo scale yields the following distance for the transect lengths at A and B:

$$\frac{0.152 \text{ m}}{7500 \text{ m} - 62 \text{ m}} = \frac{1}{48{,}934} \rightarrow 0.0331 \text{ m} \cdot 48{,}934 = 1619.7 \text{ m} \quad .$$

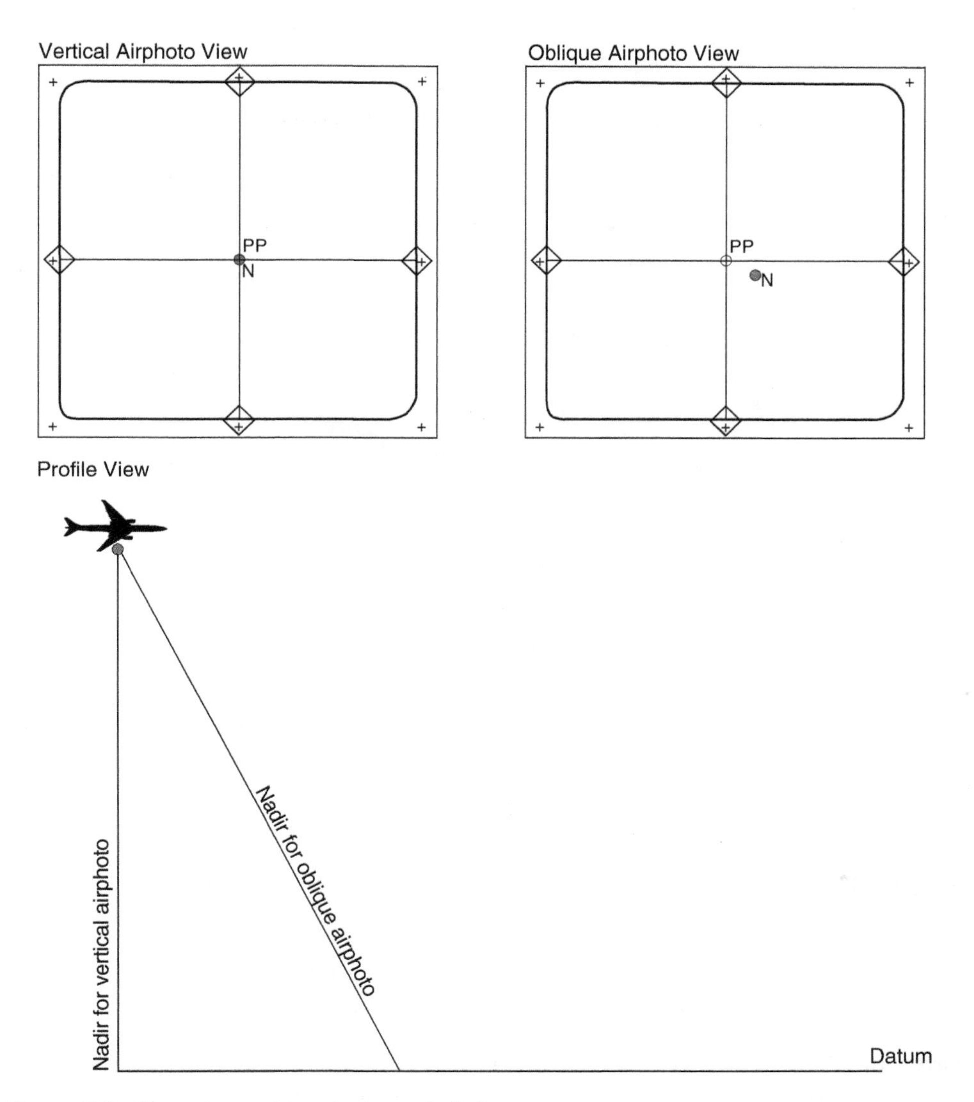

Figure 2.2 Geometry and terminology of airphotos.

But if we were to use the actual terrain elevation at locations A and B, the respective transect lengths would change to

$$\frac{0.152 \text{ m}}{7500 \text{ m} - 44 \text{ m}} = \frac{1}{49{,}053} \rightarrow 0.0331 \text{ m} \cdot 49{,}053 = 1623.6 \text{ m}$$

and

$$\frac{0.152 \text{ m}}{7500 \text{ m} - 119 \text{ m}} = \frac{1}{48{,}599} \rightarrow 0.0331 \text{ m} \cdot 48{,}599 = 1608.6 \text{ m} \quad ,$$

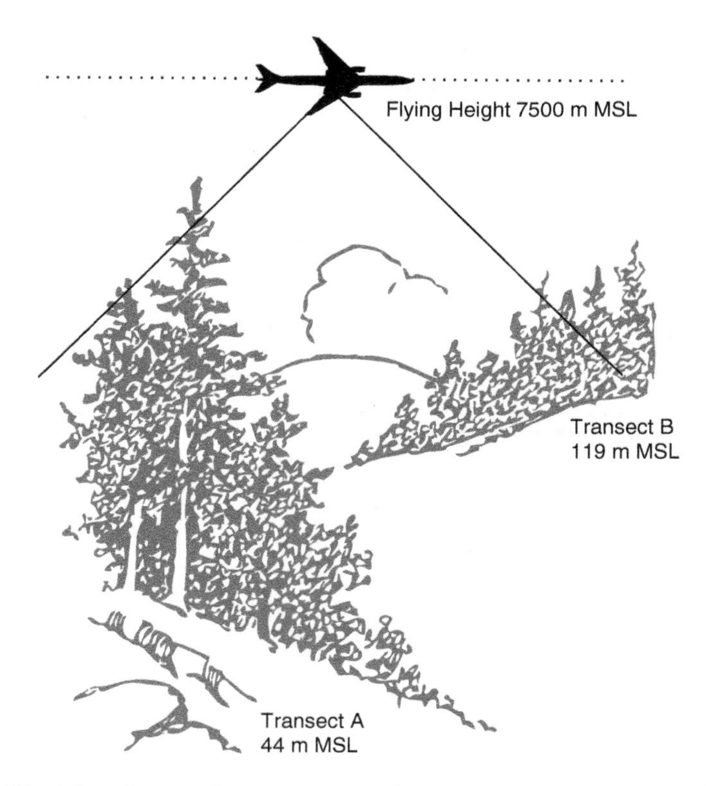

Figure 2.3 Obtaining photo scale measurements from two transects in one airphoto.

assuming heights are constant along the transects at both A and B. As the height of the terrain increases, the same photo distance (33.1 mm) corresponds to a shorter ground distance.

The chance of terrain variability being an important factor also changes with geographic location and the size of the scene in the photograph. For these and other reasons, such as the tilt of the aircraft and relief displacement, airphotos require many corrections before they can be used as a valid source of geographic information.

Relief displacement occurs when elevated features in an airphoto (e.g., mountain, tall building) are displaced radially away from the nadir point on the photo. This occurs because an airphoto is sensed from a single perspective point (like a polar projection). The degree of relief displacement of an object in an airphoto is a function of distance r between the object and nadir, the flying height H, and the height of the object h, with heights referenced to a common datum at the base of the object. Generally, the lower the flying height, the further the distance from nadir, and the higher the object, the greater the relief displacement will be. Individual objects on a single airphoto can be corrected to their true location by exploiting the relationship

$$d = \frac{r \cdot h}{H}$$

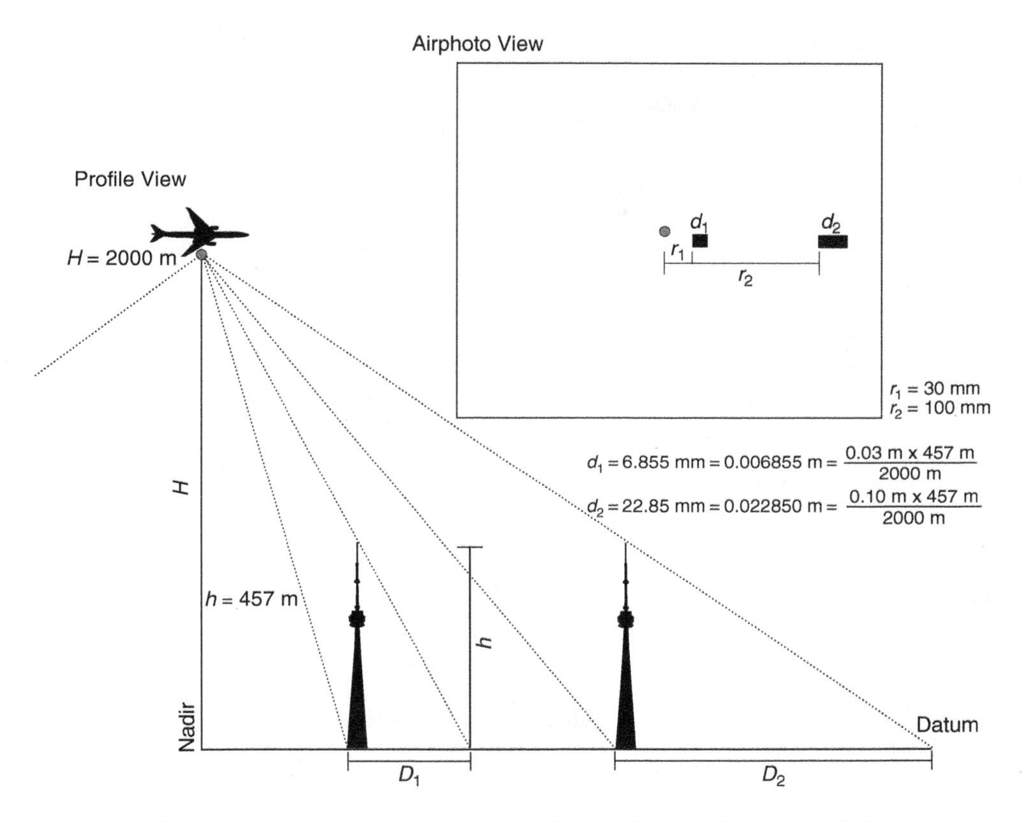

Figure 2.4 Relationships between object height, distance from nadir, and relief displacement.

to recalculate their true positions relative to nadir. Similarly, the relations can be rearranged to determine the height of an object on an airphoto.

Figure 2.4 gives a fictional example of relief displacement for two objects of identical height that are located different distances from the photo's principal point (assuming a perfectly vertical airphoto). We see that the displacement distance for the closer object (d_1) is smaller than that for the object closer to the edge of the photograph (d_2). Solving for object height for both relief displacement and radial displacement would give us

$$h_1 = 0.00685 \cdot 2000/0.03 = 457 \text{ m}$$

and

$$h_2 = 0.02285 \cdot 2000/0.10 = 457 \text{ m} \quad ,$$

giving the equivalent heights for both objects.

Entire airphotos can also be corrected for relief displacement and other geometric distortions. A process called *orthorectification* can account for distortions due to terrain relief and perspective of the camera at the time of acquisition. The resultant orthorectified airphotos can be treated as a planimetric map and used for spatial analysis and integrated with other geographic datasets within GIS. The science of airphoto measurement and interpretation is called *photogrammetry*.

Principles of photogrammetry

When subsequent airphotos are taken at an interval that ensures at least 50% overlap along the flight line, 3D reconstruction of surface scenes can be attained by viewing the photos through a stereoscope. A stereoscope is a mechanical device that enables each eye to view one of the photos independently, as each eye assumes the relative position of the camera lens at the time of exposure. Similar views can be attained using a soft-copy photogrammetric workstation and electronic 3D stereo-vision glasses connected to a computer where images are digitally positioned for stereo viewing. When independent views are perceived simultaneously (called *stereo vision*), a 3D stereo model is created. Photogrammetric methods that take advantage of overlapping airphotos and stereo vision are using what is known as parallax. *Parallax* is the apparent displacement of features on the earth caused by a change in the viewer's location. The displacement of elevated features described above is an instance of *parallax displacement*. Identifying the same point on two overlapping airphotos can allow us to calculate this displacement and use it to determine heights of objects in the scene.

The parallax at any point on an airphoto is measured with respect to the flight line, taken to be the x-axis of the photo coordinate system. The difference between the x-coordinate of the point on photograph 1 and the location of the same point on photograph 2 (for a stereo pair) gives the parallax measurement (P) for that point (sometimes called the absolute parallax). If parallax is measured at the top and base of an object in the scene and the difference between these parallax measurements is calculated dP, a simple formula can be use to determine the height of the object:

$$height = \frac{H \cdot dP}{dP + P} \quad .$$

Photogrammetry and airphoto interpretation were the main methods for creating geographic basemap information during the early evolution of GIS. These techniques continue to develop as unmanned aerial vehicles become more widely used as platforms for aerial photography. Additionally, earth observation sensors are increasingly being used as sources for geographic information, coupled with image interpretation as well as automated and semi-automated algorithm-based approaches for mapping features on the surface of the earth.

2.2.2 Earth Observation Sensors

Earth observation sensors record electromagnetic radiation (EMR) that is reflected by or emitted from the earth's surface and encode the EMR as digital data geographically referenced to locations on the surface. Geographic data produced from sensors is almost always represented as a raster data model, as sensors record spatial variation in continuously varying EMR for each location the sensor passes over. Sensors can be mounted on aircraft or UAVs, where the location and timing of data capture is dictated by the flight planning details, or on space-borne satellites, where location and timing of data capture is determined by the orbit of the satellite.

Satellites that carry earth observation sensors have two types of orbits: geostationary and sun-synchronous. Geostationary orbits are designed to orbit in the same direction and at the same speed as the earth's rotation; the satellite and sensor stay fixed over one location. Satellites in geostationary orbits tend to be at high altitudes, and imagery from sensors will have very large swath widths (ground track perpendicular to orbit path), low spatial resolution (cell size), and very high temporal resolution. Sun-synchronous orbits describe paths whereby a sensor passes over any given latitude at the same local time each day. Same-time-of-day imaging is useful for remote sensing so that images from different days can more easily be compared (solar illumination will be more or less constant day-to-day on anniversary dates). These orbits are achieved by rotating the orbital plane about the earth's axis in a way that counteracts asymmetry in gravitational pull of the earth created by flat poles and a bulging equator (i.e., the earth's shape). Sun-synchronous orbits have typical altitudes of 700–1000 km, 96–100-minute periods (time taken for one full revolution around the earth), and inclinations that put them around 8 degrees off-polar. Satellite altitude and orbital period can be related through

$$T = 2 \cdot \pi \cdot \sqrt{\frac{(R_p + H')^3}{G \cdot M_e}} \quad ,$$

where T is orbital period in seconds, R_p is the planet radius in metres, for earth 6,380,000 m, H' is the satellite altitude in metres, G is gravitational constant, taken to be 6.673×10^{-11} N·m^2/kg^2, and M_e is the mass of the earth, taken as 5.98×10^{24} kg.

For example, the most recent satellite launched by NASA, called Landsat 8, has an altitude of 705 km, so the orbital period works out to 98.9 min, or about 14.5 orbits per day. Since the earth is rotating about its axis (about 1670 km/hr at the equator), approximately 2752 km, or 24.7 degrees at the equator, pass unobserved between successive satellite tracks. This means that adjacent satellite scenes are actually sensed on different days. The entire earth between ±80° latitude is covered every 16 days. Landsat scenes are organized into a notation system called the World Reference System, which indexes each scene by a path and row number, with 233 paths, numbered 001 to 233, and 248 rows.

Satellite-borne sensors in sun-synchronous orbits cross the equator at an angle called the *orbital inclination.* The inclination determines how much of the earth's surface is covered by an orbit, with inclinations of −90° to +90° crossing all latitudes, and lower/higher inclinations covering less-polar areas. A geostationary orbit has an inclination of 0°, whereas the inclination of Landsat 8 is 98.2°. The combination of a satellite's orbital properties and the sensor's ground swath determines the revisit time, or temporal resolution, of the sensor.

Electromagnetic radiation and the spectrum

In order to understand how EMR is sensed and encoded into geographic information, the basic properties of EMR must be described, and these are best expressed through wave theory. The wave model of EMR posits that EMR behaves as an oscillating wave travelling at the speed of light. Each EMR wave is actually composed of two perpendicular fluctuating fields (electric and magnetic), which can be characterized by their fundamental wave properties: wavelength and frequency. *Wavelength* is the distance between successive wave peaks. *Frequency* is the number of times a wave peak passes over a specific point in a given period of time (see Figure 2.5).

Wavelength is measured in distance units, ranging from micrometres ($1\ \mu m = 10^{-6}$ m) to larger wavelengths of 20–40 m (see Figure 2.6). The other important property of EMR waves is their speed, which in a vacuum is a constant, the speed of light (3×10^8 m/sec), denoted c. Wavelength and frequency are inversely related to each other, and both can be used to characterize EMR. Wavelength is denoted λ, measured in metric units, and typically used for GIS and remote sensing,

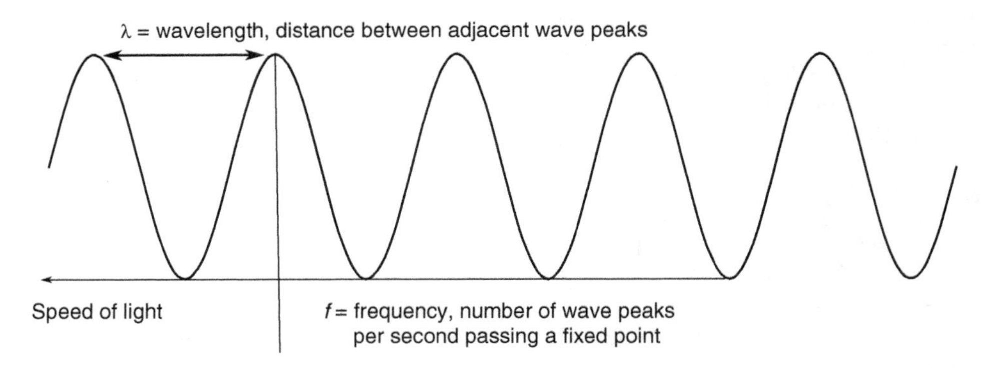

Figure 2.5 Properties of electromagnetic radiation waves.

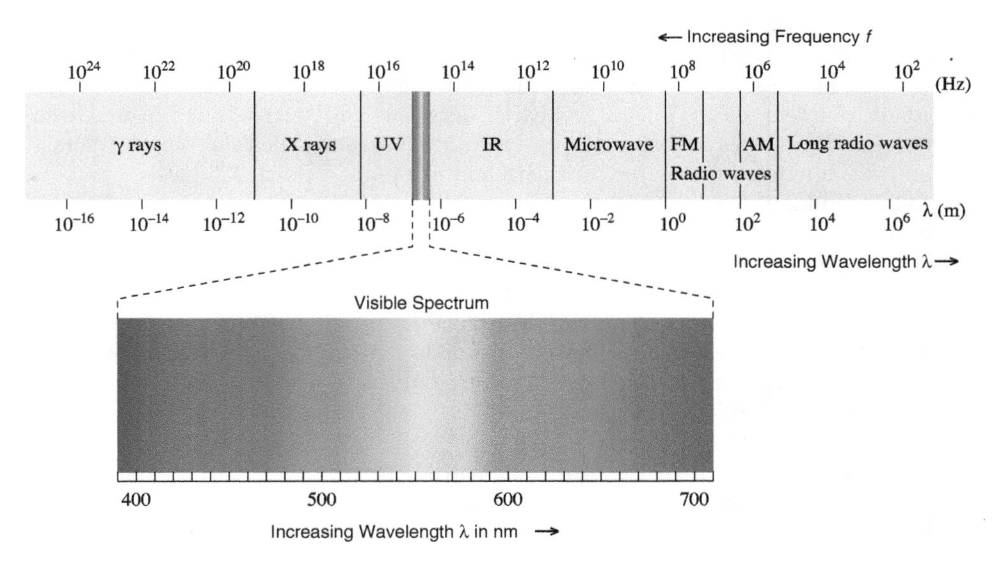

Figure 2.6 The electromagnetic radiation spectrum.

Image source: Wikipedia.

while frequency is denoted f (or sometimes v), measured in hertz, and typically used by engineers. The entire range of wavelength and frequencies is called the EMR spectrum, and discrete intervals of the spectrum are called *bands.* Earth observation sensors are designed to be sensitive to reflected or emitted EMR in specific bands of the EMR spectrum that are useful for characterizing different aspects of the earth's surface. While most film-based aerial photography is sensitive only to visible light (parts of the EMR visible to the human eye), earth observation sensors can be designed specifically for many bands in the EMR, such as near-infrared (NIR), thermal-infrared (TIR), and microwave. This gives great flexibility to the types of geographic information that can be sensed and recorded by earth-observation sensors.

EMR can be linked to physical properties of surface features through the relationship between an object's temperature and its energy spectrum. An *energy spectrum* is the distribution of *emitted* energy across all wavelengths for an object at a given temperature. This relationship is best demonstrated through a theoretical construct called a *blackbody,* which is a perfect absorber and emitter of incident EMR. The emitted EMR from a blackbody is proportional to $\frac{1}{\lambda^4}$. However, the wavelength of peak emittance is a function of the blackbody's temperature T, such that with increasing T the wavelength of peak emittance gets smaller, a relation that was discovered by Max Planck in 1900. Why this relation works wasn't known until Einstein's photon theory of light, which showed that EMR is delivered in discrete packets, called photons, rather than in a continuous wave. The energy content of a photon is proportional to its frequency $\times h$; thus, longer wavelengths, with lower frequencies, have lower energy content. This is important for remote sensing because we can construct EMR curves for blackbodies of different temperature T (Figure 2.7) and then measure the deviation from these theoretical curves for actual surface features. Distinct EMR curves characteristic of different surface types can be useful in interpreting and identifying features on the earth's surface based on its energy spectrum, as well as generalizing relationships between surface materials, their energy spectrum, and their temperature. This is especially important for targets that do not have substantial variation in reflectance properties. Low-frequency emitted thermal radiation therefore contains lower energy, and in order to detect these signals using remote sensing, they must be collected over large areas (i.e., larger pixels). For these reasons, remote sensing of polar areas covered in ice typically use emitted EMR and large pixel sizes or active sensors.

Before EMR is detected by a sensor, the energy interacts with both the atmosphere and the surface of the earth. A major objective of most remote sensing studies is to maximize the geographic information coming from the earth's surface while minimizing the information coming from the interaction of EMR with the atmosphere. Two of the most important effects of the atmosphere on EMR received by the sensor are *atmospheric scattering* and *atmospheric absorption.* Scattering occurs when EMR is diffused by particles in the atmosphere. Importantly, the scattering that occurs is often not evenly distributed across wavelengths; some wavelengths are scattered more than others. This selective atmospheric scattering causes many of the atmospheric phenomena we observe, such as blue skies and red sunsets. If the particles in the atmosphere are large relative to the incident EMR, all wavelengths can be scattered equally, commonly observed when large water droplets (10–15 μm) scatter all visible wavelengths, therefore appearing white (e.g., fog and clouds). Atmospheric scattering can greatly affect how

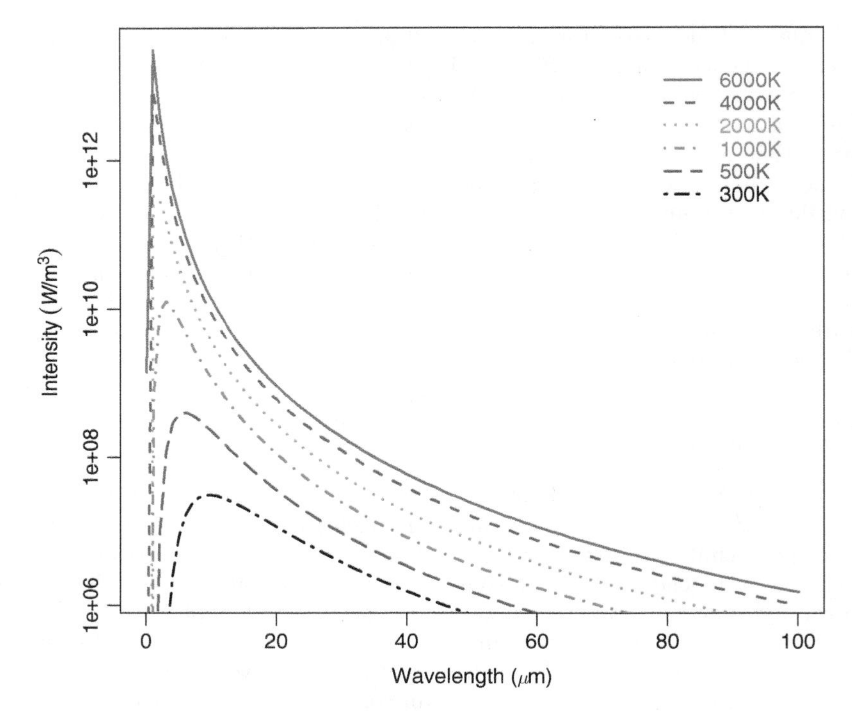

Figure 2.7 Energy spectrum for blackbodies of different temperatures.

much useable geographic information can be obtained from a remotely sensed image. Atmospheric correction methods are often employed during preprocessing of remotely sensed imagery prior to use within GIS.

Atmospheric absorption occurs in a more predictable manner than scattering, involving the absorption of specific wavelengths by atmospheric constituents. Generally, the makeup of the atmosphere can be considered constant, and therefore the major gases that absorb EMR (water vapour, carbon dioxide, and ozone) can be handled by "looking" in parts of the spectrum where wavelengths are unaffected by these absorption processes. These areas of the spectrum that are largely free of atmospheric absorption are called *atmospheric windows*. Where atmospheric absorption is low, atmospheric transmission is high, and these wavelengths are best suited to remote sensing of the earth's surface.

When incident EMR arrives at the surface of the earth, three types of interaction can take place: energy can be reflected, absorbed, and transmitted. Often, the relative importance of each of these processes varies by wavelength and with the type and condition of surface material. This variability allows the spectral response to be mapped to surface types for different wavelengths, in what are called *spectral signatures*. Figure 2.8 shows spectral curves for a variety of surface types. A spectral signature for a particular type might be found where there is maximum spectral separability between the surface type of interest and other features in the scene. For example, to map the grasslands

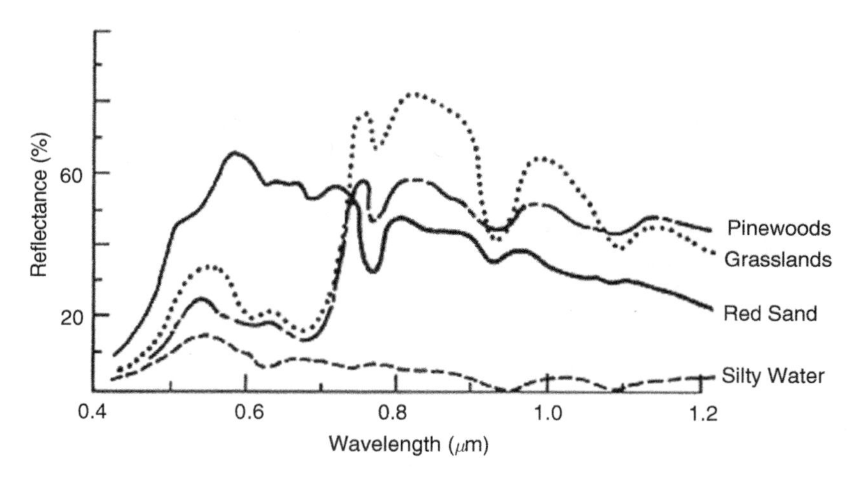

Figure 2.8 Spectral signatures of different hypothetical surfaces.

Source: Adapted from *The Remote Sensing Tutorial*, Dr Nicholas Short.

area for the scene from which these curves were extracted, we might look in the NIR range of the spectrum, where the reflectance of grassland areas is much greater than other surface types. Much remote sensing research is devoted to the development of algorithms for extracting spectral signatures for use in automated landcover and landuse mapping.

The two main types of earth-observation sensors are *active* sensors, which both send and receive EMR, and *passive* sensors, which record naturally reflected and emitted EMR. The majority of satellite-based earth-observation sensors are passive sensors that operate in the visible and infrared parts of the EMR spectrum. Tables 2.1 and 2.2 outline some common sources of remote sensing (RS) data. Active sensors have their own energy source and send out pulses of EMR at specific wavelengths, recording the intensity of reflected EMR to build up an image of the earth's surface. Active remote sensing is often used to look in low-energy parts of the EMR spectrum at longer wavelengths.

Radio detection and ranging—radar

Radio detection and ranging (radar) is a type of active remote sensing that utilizes EMR in the microwave part of the spectrum (1 mm–1 m). The misnomer is due to the historical evolution of radar as a military technology for determining the presence and distance of enemy vessels, which did originally use radio signals. As radar has its own energy source and uses longer wavelengths, it is able to operate in daytime and night. Radar waves can pass through cloud cover, fog, and rain, and are able to characterize low-energy surface features such as ice, snow, and ocean waves much better than passive remote sensing. In a radar system, a signal is transmitted toward a target surface/object, and a portion of the signal is returned (backscattered) to the sensor. The radar sensor observes the intensity and timing of the backscatter to build up an image of the surface. A number of bands are used in radar remote sensing, and they are designated with letters, originally a security feature when radar was a secret military technology. The major radar bands are listed in Table 2.3.

Table 2.1 Landsat earth observation sensors.

Satellite	Sensor	Spectral Resolution (μm)	Spatial Resolution (m)	Temporal Resolution (days)	Swath (km)
Landsat 8	OLI	B1 0.43–0.45	30	16	185
		B2 0.45–0.51			
		B3 0.53–0.59			
		B4 0.64–0.67			
		B5 0.85–0.88			
		B6 1.57–1.65			
		B7 2.11–2.29			
		B8 0.50–0.68	15		
		B9 1.36–1.38	30		
		B10 10.60–11.19	100		
		B11 11.50–12.51	100		
Landsat 7	ETM+	B1 0.45–0.52	30	16	185
		B2 0.52–0.60			
		B3 0.63–0.69			
		B4 0.77–0.90			
		B5 1.55–1.75			
		B6 10.40–12.50	60		
		B7 2.09–2.35	30		
		B8 0.52–0.90	15		
Landsat 5	TM	B1 0.45–0.51	30	16	185
		B2 0.52–0.60			
		B3 0.63–69			
		B4 0.76–0.90			
		B5 1.55–1.75			
		B6 10.40–12.50	120		
		B7 2.08–2.35			

Signals used in radar remote sensing are polarized. This means that the vibration of EMR occurs in one plane only, horizontal or vertical, perpendicular to the direction of transmission (unpolarized EMR vibrates in all directions perpendicular to the path). The polarizations used in remote sensing are vertical (V) and horizontal (H), and systems can send and receive in different polarizations (designated VV for send vertical, receive vertical, VH for send vertical, receive horizontal, etc.). Polarization can greatly impact image quality and mapping capability because surfaces also modify the polarization of backscattered EMR, so multiple polarizations are increasingly used in applied earth remote sensing. New radar sensors, such as RADARSAT II, are capable of obtaining images in all four polarizations, termed quad-polarized imagery.

Light detection and ranging—lidar

Light detection and ranging (lidar or LiDAR) is an active remote sensing approach that uses EMR in the visible/NIR range of the spectrum. Similar to radar, pulses of EMR are transmitted and the reflected response is used to determine the distance to the surface

Table 2.2 Earth observation sensors (* indicates active sensor; all others are passive).

Satellite	Sensor	Spectral Resolution (μm)	Spatial Resolution (m)	Temporal Resolution	Swath (km)
Terra	ASTER	V, NIR, SIR, TIR	15–90	16 days	60
	CERES	B1 0.3–100.0	20,000	hourly	whole earth viewable from satellite
		B2 0.2–5.0			
		B3 8.0–12.0			
	MISR	V, NIR	3 modes: 250; 550; 1100	2–9 days	360
	MODIS	V, NIR, SIR, MIR, TIR (36 bands)	250–1000	1–2 days	2330
	MOPITT	SIR, TIR	> 22,000		640
Aqua	AIRS	V, NIR, IR	2300; 13,500	16 days	1650
	AMSR-E	K-band, X-band	5400–56,000		1445
	AMSU	MIR, TIR	40,000	1 day	1690
	CERES	same as Terra	20,000	hourly	whole earth viewable from satellite
	HSB	K-band	13,500	1–2 days	1650
	MODIS	V, NIR, SIR, MIR, TIR (36 bands)	250–1000	1–2 days	2330
EO-1	ALI	V, NIR, SIR	30 m	16 days	185
	Hyperion	220 bands	30 m	16 days	7.7
Radarsat I*	SAR	C-band, HH polarization	7 modes, 8–100	6 days, 4 days above 48°, daily above 70°	45–500
Radarsat II*	SAR	C-band, selective quad polarization	11 modes, 3–100	6 days, 4 days above 48°, daily above 70°	
GOES-15	Imager	B1 0.53–0.75	1000	30 sec	hemispheric
		B2 3.8–4.0	4000		
		B3 5.8–7.3	4000		
		B4 10.2–11.2	4000		
		B6 12.9–13.7	4000		
SPOT-6/7		B1 0.455–0.525	6	daily	60
		B2 0.530–0.590	6		
		B3 0.625–0.695	6		
		B4 0.760–0.890	6		

Table 2.3 Radar bands used in earth remote sensing.

Band	Wavelength λ (cm)	Example Applications
K	0.75–2.4	Ground based systems, precipitation vertical profiling
X	2.4–3.75	Military reconnaissance, terrain mapping
C	3.75–7.5	Earth observation applications, precipitation
S	7.5–15	Soil moisture
L	15–30	Forest mapping, wetland mapping
P	30–100	Biomass, experimental research

being imaged. In lidar systems, the pulses are emitted with frequencies often greater than 100,000 per second. The returns (i.e., backscattered pulses) are converted into distance measurements based on the time of response (i.e., $\frac{1}{2} \cdot t \cdot c$, where t is response time and c the speed of light) and are integrated to build up an elevation profile. Lidar has thus been mostly widely used for high-resolution terrain mapping. The rapid pulse rate of lidar creates returns that have ground distances of less than 1 m. The actual ground distance between pulses depends on the flying height of the sensor, the forward speed of the sensor (i.e., along track), and scanning angle, among other factors. The geometry of the light pulse is also an important component of the data obtained from a lidar sensor. As the light beam expands as it is transmitted toward the surface, by the time it reaches the surface the beam covers a circular area rather than a single point. The beam can therefore come into contact with different features when there is some variation in relief within the pulse's footprint. In this way, a single pulse can generate multiple returns (often differentiated as first, intermediate, and last), as each surface of a different elevation produces its own return. Differences in return elevations can be used to map the structure of surface features such as the height of a forest canopy or building, while last returns might be used to map bare-earth elevation.

Lidar data are usually processed into a series of large ASCII files, which might represent first returns, intermediate returns, last returns, and intensity. A typical lidar file includes the time, x-coordinate, y-coordinate, z-coordinate, and intensity, which in a GIS would form what is called a *lidar point cloud.* Massive point-cloud datasets represent a challenge to many GIS data structures and software packages, although optimized binary lidar file formats and database extensions for handling lidar data are now available.

2.2.3 Properties of Remotely Sensed Imagery

Remotely sensed imagery can take many forms. It is helpful to distinguish and evaluate different sources of geographic information obtained from earth-observation sensors, using characteristics common to all types of remotely sensed data. For example, satellite imagery has been widely used for monitoring the health status of crops. Crops are susceptible to pests, environmental contaminants, and natural environmental variability, which impact yield and thus overall productivity for farmers. Remote sensing can provide a synoptic view of a field in order to determine the overall health of a given crop and to forecast the expected yield early in the growing season. Additionally, changes in crop health that are not visible to the human eye are sometimes evident in other parts of the EMR spectrum.

Reflected radiation varies with wavelength in distinct ways for different surface features. In agricultural applications, reflected radiation is usually a mixture of spectral responses from crop canopies and ground soil. Reflected radiation from plants has a slight peak of reflectivity in the green band, due to chlorophyll absorption of EMR in the blue (4.3 μm) and red (6.6 μm) parts of the spectrum. In the near-IR band (7–13 μm), reflectivity spikes due to internal scattering of EMR within air spaces of the cellular structure of plants. This difference in reflectance for healthy vegetation (low red, high NIR) is the basis for many "band ratio" approaches to monitoring vegetation.

Spatial resolution refers to the smallest discernible distance resolvable in the image. In practice, spatial resolution is described by the ground distance associated with one side of a pixel in an image. Where one square pixel in a remotely sensed image covers a ground area of 1 ha (100 m $\times$ 100 m), the image is said to have a spatial resolution of 100 metres. *Temporal resolution* refers to the frequency with which the sensing platform passes over the same location. Temporal resolution generally only applies to space-borne satellite sensors operating in a fixed orbit. Temporal resolution is also sometimes called the revisit time of the sensor. While spatial and temporal resolution can be properties of all types of geographic data, *spectral resolution* is unique to remotely sensed imagery; it describes the number and dimension of the bands of EMR represented in the data. Sensors can be highly tailored to one specific band of the EMR spectrum (e.g., radar), or general environmental monitoring sensors can have detectors in many visible and infrared bands (e.g., Landsat). *Radiometric resolution* refers to the sensitivity of a spectral sensor to changes in spectral response (EMR). Radiometric resolution is described by the number of possible unique values for a given dataset, often reported in base-2 notation. For example, many RS data sources represent geographic variability in EMR as whole integers within the range 0–255, which corresponds to 256 unique values. This is the outcome of an 8-bit storage level, or 8-bit radiometric resolution ($2^8 = 256$).

2.2.4 Handling Remotely Sensed Geographic Data in GIS

Remotely sensed geographic data are almost always represented using the raster data model. A variety of data storage formats are used in today's GIS software packages. However, remotely sensed imagery from satellite-borne sensors (via the Internet or on disc/memory storage) usually require some processing before being used in a GIS. Sensors such as the Thematic Mapper on Landsat 5 or the Enhanced Thematic Mapper + on Landsat 7 (available freely at USGS websites) produce massive amounts of data. To make them useful for the wider scientific community, the data are downlinked to an international network of ground stations, which process the raw sensor data. For ETM + 7, different processing levels (e.g., level 0R, level 1), which correct for atmospheric, radiometric, and geometric distortions, are available. To facilitate distribution to a wide user community using varied GIS and RS software packages, often with proprietary data formats, data are delivered in standardized binary compressed formats that are usually unreadable natively to GIS software. Translation to more common GIS data formats usually requires some RS data processing software.

2.2.5 Radiometric Processing

The discretization of radiance (L_λ, the signal detected by the sensor) to calibrated binary integers (called digital numbers, abbreviated DN) occurs at the sensor level and depends on the sensitivity of the band-specific detectors onboard the satellite. This information is included in radiometric calibration coefficients, and is published in metadata files or headers included with imagery data files. When using remotely sensed imagery, it is sometimes necessary to know the radiance values received at sensor, and the calibration coefficients can be used to back-calculate radiance for input into, for example, biophysical models. To recalculate radiance from DNs, we need the scaling factors $LMIN_\lambda$ and $LMAX_\lambda$, which were used during the original rescaling process and proceed as follows:

$$L_\lambda = \frac{LMAX_\lambda - LMIN_\lambda}{DN_{\max} - DN_{\min}} \cdot (DN_{\text{cal}} - DN_{\min}) + LMIN_\lambda \quad ,$$

where L_λ is spectral radiance at sensor, DN_{cal} is the calibrated DN value, $DN_{\min}$ is the minimum DN value, $DN_{\max}$ is the maximum DN value, $LMIN_\lambda$ is spectral at-sensor radiance scaled to $DN_{\min}$, $LMAX_\lambda$ is spectral at-sensor radiance scaled to $DN_{\max}$. One way to think about the calibration is as a linear model where $(DN_{\text{cal}} - DN_{\min})$ is the independent variable, $LMIN_\lambda$ is the y-intercept, called bias (B), and

$$\frac{LMAX_\lambda - LMIN_\lambda}{DN_{\max} - DN_{\min}}$$

is the slope of the model, called gain (G), and L_λ the dependent variable, or rather

$$L_\lambda = B + G \cdot DN \quad .$$

The results of this calibration might then be used to derive empirical relationships between spectral responses and radiance measured on the ground using in-situ spectroradiometry [19]. More commonly, L_λ is converted into reflectance or radiance received at the top of the atmosphere (TOA) which accounts for variability in radiance due to non-surface factors such as sun-earth distance, sun angle, and differences in spectral band sensitivity.

2.2.6 Geometric Processing

Remote sensing (RS) data are fundamentally different from non-geographic digital image data, as each pixel maps to a physical location in geographic space. A required processing step to represent RS data within a GIS is assigning image pixels to real-world geographic locations, a series of techniques called *georeferencing* (see Figure 2.9). There are two steps to georeferencing image data: determining geographic coordinates for each pixel and then assigning image data to the appropriate pixels. These steps are sometimes called spatial interpolation and intensity interpolation, and taken together they

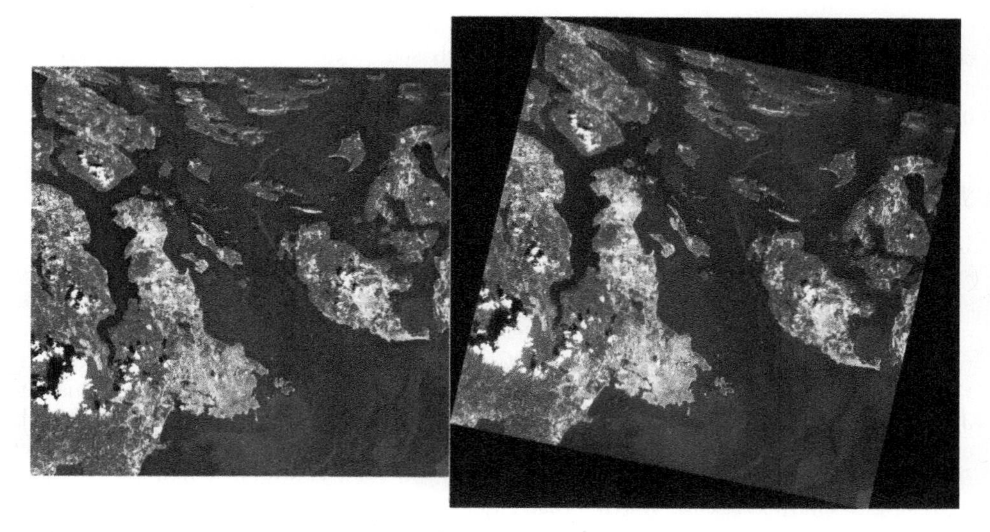

Figure 2.9 Geometrically uncorrected and corrected versions of a Landsat 7 scene from Vancouver Island, BC, Canada.

are a major component of geometric corrections needed to make images planimetric. While most RS data today are produced with systematic geometric corrections already done and a coordinate reference system assigned, random distortions in the image often require additional geometric corrections. Distortions in the geometry of imagery are introduced through random variation in attitude and altitude of the sensor, relief displacement, and off-nadir scale distortions. The general method to correct for these errors is through the collection of a set of ground control points (GCPs), which are identified in the imagery and on an already georeferenced image or basemap, and modelling the geometric distortion by mapping coordinates in one domain to coordinates in the other domain. Different modelling approaches (linear, quadratic, cubic) are available to account for different types of geometric distortions. A linear rectification might proceed as follows:

$$x' = a \cdot x + b \cdot y + c$$
$$y' = d \cdot x + e \cdot y + f \quad ,$$

where x and y are the input RS pixels in image space, and x' and y' are the output coordinates. The downside of this approach is that x' and y' may be in non-integer units, which can result in gaps and overlaps in the resulting grid structure, so to have more control over the structure of the output grid the equation can be modified to include the output grid structure in x and y (called inverse mapping). The six parameters (a, b, c, d, e, f) of the two linear models control the translation, scale, and rotation of the mapping between input and output space. Assignment of RS image data to the appropriate output grid locations requires intensity interpolation methods. There are two types of interpolation methods employed for this, those that maintain the exact values of the original input data (e.g., nearest-neighbour assignment) and those that create new values for unsampled locations based on distance-weighted averages of surrounding pixels (e.g., bilinear interpolation, cubic convolution). Further discussion of interpolation methods is provided in Chapter 10.

2.2.7 Remote Sensing Data Compression

Data compression is an integral part of remotely sensed data, which are by definition spatially continuous measurements. For sensors with high spatial resolutions, compression is important for providing data that can be distributed and processed easily. For example, an uncompressed Landsat 7 ETM+ image with an 8-bit depth in uncompressed format is 183 km by 170 km at 30 m grid cell size for Bands 1–5 and 7, 60 m Band 6, and 15 m Band 8, plus auxiliary files, resulting in a total of approximately 3.8 gigabytes of data. One of the simplest raster compression approaches is run-length encoding (RLE). RLE takes advantage of the fact that the earth's surface is patchy: forests, deserts, lakes, oceans, and urban areas are each contiguous patches of earth with similar characteristics. When a raster image has first been classified into landcover classes (i.e., assigning DNs to landcover classes), RLE can significantly reduce storage space by transforming the image DNs into RLE codes. RLE codes store data only when adjacent pixels change value. The RLE codes for the image in Figure 2.10 illustrate the cost savings in terms of data storage.

If the image in Figure 2.10 had not already been classified into landcover classes, RLE would be less useful for compression because we would expect fewer contiguous patches of identical DNs. Other approaches to compression useful for calibrated DN data (or any image) are commonly-used algorithms such as GIF, JPEG, and TIFF, some of which incur some loss of information (termed lossy) and others that do not (lossless).

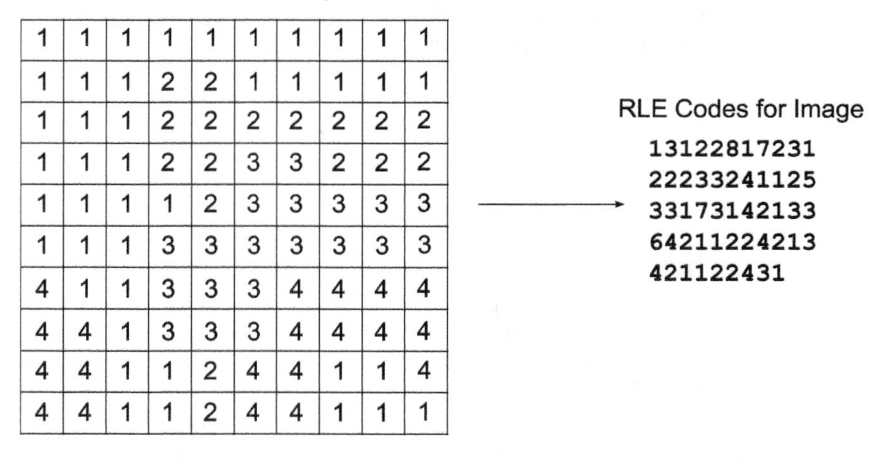

Figure 2.10 Run-length encoding example showing original image (left) and the corresponding run length codes (right).

Some compression algorithms are in the public domain (e.g., JPEG, a discrete cosine transform) while others are proprietary (e.g., the Lempel-Ziv-Welch algorithm used in GIF images was patented until 2004).

One widely used compression technique from signal processing is Fourier transforms. The basic idea of a Fourier transform is to represent any continuous, periodic signal by a sum of sinusoidal functions of different frequencies. In particular, the transformation approximates the signal with a finite series representation and, as with all series expansions, we can include as many or as few terms of the series necessary to provide the appropriate level of approximation. An inverse transform can be used to reconstruct an approximation of the original signal.

Given a 1-dimensional signal (e.g., a time series), which is a continuous function $x(t)$ with period T, we can decompose it into a spectrum of frequency components based on the angular frequency ω_0, where

$$\omega_0 = \frac{2 \cdot \pi}{T}$$

and can represent the function $x(t)$ as the weighted sum of complex sinusoids,

$$x(t) = \sum_{k=-\infty}^{\infty} a_k \cdot e^{i \cdot k \cdot \omega_0 \cdot t} \quad ,$$

which is called the signal's *Fourier series.* This represents the signal in the frequency domain, where interest centres on the Fourier coefficients,

$$a_k = \frac{1}{T} \int_{-\frac{T}{2}}^{\frac{T}{2}} x(t) e^{-i \cdot k \cdot \omega_0 \cdot t} dt \quad .$$

However, in practice the spectrum is limited to values of a_k that span a discrete range of k. As k increases, the signal reconstructed from these coefficients becomes increasingly similar to the original. The coefficients defined above are called Fourier coefficients, which allow a Fourier transform of $x(t)$ into a linear combination of frequency components. This is due to Euler's formula,

$$e^{ix} = \cos(x) + i \cdot \sin(x) \quad ,$$

where i is $\sqrt{-1}$. This provides a definition of sine and cosine functions as weighted sums of exponentials, which allow us to model a sine wave in the Fourier series. So filling in from above,

$$e^{i \cdot k \cdot \omega_0 \cdot t} = \cos(k \cdot \omega_0 \cdot t) + i \cdot \sin(k \cdot \omega_0 \cdot t) \quad .$$

This is just one possible method of signal decomposition. In place of the exponential form $e^{-i \cdot k \cdot \omega_0 \cdot t}$, alternate functions can be used as long as another orthogonal function

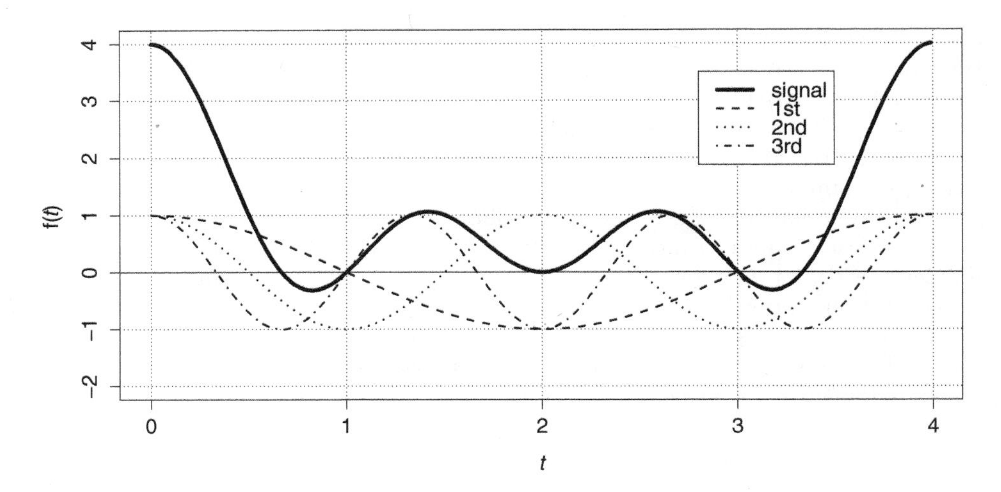

Figure 2.11 Fourier decomposition of a 1D signal into first-, second-, and third-order harmonics.

exists to define expansion coefficients. An example of a signal and its first three harmonics is given in Figure 2.11 This means that many signal forms can be decomposed into a linear combination of a basis function and its expansion coefficients.

The above examples illustrate signal decomposition in a continuous, one-dimensional case. However, in image compression we have two dimensions and a discrete signal. The 2D discrete Fourier transform is given by

$$X[n,p] = \sum_{k=0}^{N-1}\sum_{l=0}^{N-1} x[k,l] \cdot e^{-i\cdot 2\cdot \pi\left(\frac{nj}{N}+\frac{p_l}{N}\right)} \quad ,$$

where the value of the DN in the image is given by $x[k,l]$, and in frequency (Fourier) space is $X[n,p]$, and to convert back from frequency to spatial domain, the inverse Fourier transform is

$$x[k,l] = \frac{1}{N^2} \cdot \sum_{n=0}^{N-1}\sum_{p=0}^{N-1} X[n,p] \cdot e^{i\cdot 2\cdot \pi\left(\frac{n_k}{N}+\frac{p_l}{N}\right)} \quad ,$$

which is computed with two separate one-dimension transforms of length N (i.e., sums of sines and cosine functions). This method of 2D signal representation has many uses in image analysis, including image compression. The JPEG compression algorithm uses a similar method called a discrete cosine transform to express a signal as a sum of cosine functions with different frequency components.

A limitation of the frequency-domain transforms is that while the frequency of the signal is preserved, the actual temporal locations of frequencies are not. Computing the

Fourier series along a moving window (i.e., convolution) is often a poor solution because the window size is arbitrary and cannot adapt to non-stationary signal changes with time. *Wavelets* address this problem by examining frequencies in portions of the signal *at various scales* (i.e., windows) smoothed by a zero-mean function called the wavelet function (or the mother wavelet). While Fourier analysis deals with decomposing a signal or *wave* into the frequency domain, wavelet analysis decomposes a signal by separating out discrete parts of the wave (i.e., wavelets). So in wavelet compression, we have an arbitrary set of functions, which are localized in the time domain to capture high-frequency and low-frequency patterns, to model all of the important variation in a signal. The two important parameters for wavelet analysis are the form of smoothing (the wavelet function) and the scales at which the wavelet will be evaluated (the scaling function). In many cases, the 1D wavelet can be applied in the x and y directions so long as the scaling and wavelet functions are separable (similar to the DFT above).

The JPEG2000 compression algorithm is a recently developed approach that uses the discrete wavelet transform to develop a multiple resolution representation of spatial frequencies of an image. Wavelet compression is increasingly used in raster image compression formats (e.g., ECW, Mr Sid). The basic design of a wavelet algorithm is iterative filtering of high-frequency and low-frequency components of the image pattern. High-frequency components would be where cell values change rapidly over short distances. Low-frequency components would be where gradual spatial trends occur across the image. After each filter, the signal is downsampled in order to compress the signal. The compression is multi-resolution in the sense that each level of the coefficients corresponds to a finer scale of the signal. The number of scales at which the wavelet is applied determines how much of the signal is "lost," and in fact many wavelet algorithms include a lossless mode for applications that do not tolerate any level of image degradation (e.g., medical imaging). The wavelet functions used in the JPEG2000 standard are the Daubechies 5/3 (which is restricted to integer-based expansion coefficients and thus no rounding/loss of information during quantization) for lossless compression and the Daubechies 9/7 for lossy compression (numbers refer to dimensions of low- and high-pass filter sizes).

Compression algorithms are evaluated in remote sensing on the basis of the *compression ratio*, which denotes the storage size of the uncompressed to the compressed as a ratio (e.g., 4:1, 10:1). At low compression ratios, the difference between the standard JPEG algorithm, which uses a discrete cosine transform, and the JPEG2000, which uses wavelet transforms, is negligible; however, at high compression ratios, JPEG2000 is much better, as there is no characteristic "blocking" created by the fixed-scale 8×8 blocks used in the JPEG method. This can have implications for remote sensing analysis. Zabala and Pons [126] recently compared JPEG and JPEG2000 compression and decompression on subsequent classification of Landsat scenes for a variety of compression ratios. At compression ratios of less than 10:1, the percentage of correctly classified was roughly similar for the JPEG and JPEG 2000 data; however, at compression ratios of greater than 10:1, the performance of JPEG2000 was much better than JPEG. This effect was also mediated by the type of landscape, where more homogeneous forest scenes were less impacted by compression algorithm when compared to fragmented forest scenes, which had significant classification errors with JPEG compression.

How RS data are actually stored on disk in file formats varies as well. As RS imagery typically contains spectral information from numerous bands, the way the data are stored in a file or memory can impact how easily it is accessed by GIS and image processing applications. The three most commonly implemented ways of storing RS data in files are band sequential (BSQ), band interleaved by line (BIL), and band interleaved by pixel (BIP). BSQ files store the binary pixel values as as separate block of data for each band in the image. BSQ data are optimal for spatial access, as all of the spatial data are stored together. BIP stores all spectral data for an individual pixel together in the file, and as such is optimized for spectral access. BIL stores all spectral data for individual rows of data, and represents a compromise between spatial and spectral optimization.

2.2.8 Selecting Remotely Sensed Imagery

In addition to preprocessing RS data, consideration needs to be given to the four types of resolution for a specific geographic analysis need. There is usually a trade-off between the resolution and costs (both financial and computational), making selection of appropriate RS data an important step in many geographic analyses. The first decision should be determining the spectral resolution. The objective of analysis and the phenomena under study will dictate which parts of the EMR spectrum need to be sensed. Sometimes, the selection of RS data is impacted by availability, and multiple sources are integrated for multi-temporal geographic studies. For example, Pope et al. [93] mapped the coastal sea ice extent at Yelverton Bay on the northwest coast of Ellesmere Island (Nunavut, Canada) with a suite of RS data sources ranging from the early 1950s to 2010. Using a combination of aerial photography, satellite-borne radar imagery, and imagery from MODIS satellites, the authors were able to show that significant loss (90% over a 50 year period) of coastal sea ice occured, and by 2010 the area had become completely free of coastal sea ice for the first time in the historical record. Such studies represent a careful balance between the object of study (sea ice, which is best detected with sensors capable of imaging in long-wave parts of the EMR spectrum) and the availability of RS data for specific geographic and temporal constraints (e.g., Yelverton Bay, 1950–present).

The other major factor to consider when deciding on a RS data source is the spatial scale of imagery. RS data can be obtained with pixels that are sub-centimetre to many kilometres in length. The appropriate spatial resolution is generally the largest that can adequately represent the process under study. Temporal resolution is an important consideration for multi-temporal studies, or when building image mosaics for study areas that cover more than one scene. Multi-temporal remote sensing is increasingly common as the RS record of the earth's surface increases, allowing long-term studies of environmental change. Radiometric resolution is usually not considered as a factor, as back transformation to raw EMR radiance values can usually be done to a level of precision required by most environmental applications. While radiometric resolution has been fixed at 8-bit for most environmental sensors, Landsat 8 has recently introduced 12-bit imagery (actually 16-bit after resampling and interpolation is applied).

2.3 Global Positioning Systems

Global positioning systems (GPS) refer to a group of satellite technologies used for determining locations on the earth's surface. Originally a military technology, GPS is now widely used in many industries and technologies ranging from airplane navigation to personal location sensors and smartphone location-based services, as well as traditional uses such as surveying and mapping. GPS is one of the principal acquisition methods for vector GIS data.

2.3.1 History of GPS Technology

GPS technology was developed and continues to be operated and maintained by the United States Department of Defense. The first GPS satellite was launched in 1978, and the system was completed with the launch of the 24th satellite in 1993, enabling global coverage. Satellites fly at an altitude of approximately 20,200 km. The system is composed of a space segment and a control segment. The GPS space segment is composed of a constellation of 24 satellites arranged into six orbital planes, each with four satellites. In practice, often more than 24 satellites are operational to account for service disruptions to individual satellites. Prior to May 1, 2001, GPS signals were intentionally degraded by the US military—to ensure, for example, that enemies could not use GPS against the US to locate targets. This purposeful signal distortion, termed *selective availability*, reduced positional accuracy to around 100 m, greatly limiting the application of GPS for civilian uses such as car navigation. Even today, in some countries GPS is considered a military technology, and civilian uses are strictly regulated. The GPS control segment is composed of a master control station, 12 command and control antennas, and 16 monitoring sites located around the world, which coordinate to monitor and analyze transmissions and send commands and data to the space segment.

The system of GPS satellites, formally called NAVSTAR GPS, was the first global navigation satellite system (GNSS). Other GNSS have been initiated and are in various stages of development. Russia launched the first GLONASS satellites into orbit in the early 1980s, but completion and maintenance of the system was interrupted due to economic and political events associated with the disintegration of the USSR in the 1990s. In 2010, GLONASS was restored to full global operation with 24 satellites. Europe has a GNSS in development named Galileo. China has also begun initiation of a GNSS with its Compass, the first satellite of which launched in 2007. Many believe that the future of GPS will be in integration of multiple GNSS to further improve accuracy, especially in areas currently poorly served, such as under heavy forest canopy or in urban canyons.

2.3.2 How GPS Works

GPS satellites transmit signals that are used to determine distances (i.e., ranges) between senders (space-borne satellites) and receivers (ground-based GPS receivers), which are subsequently used to triangulate position on the earth's surface. The first step of determining distance between the sender and receiver of one signal requires that the time the signal is sent and received is known to very high precision. The level of precision in timekeeping requires the use of atomic clocks on board GPS satellites to communicate with GPS receivers on the ground via time-stamped digital radio signals. GPS satellites transmit data on two signals, which are called carrier waves: the primary L1 (1574.42 MHz) signal and the secondary L2 (1227.60 MHz) signal. Information—such as satellite clock corrections, orbit, location, ionospheric information, satellite health status, and, most importantly, pseudorandom noise (PRN) ranging codes—are modulated onto the carrier waves. This information is part of a data package called the Navigation Message, which provides all the information needed to determine the receiver's position. These signals are continuously broadcast by GPS satellites, so whenever a GPS receiver is turned on the signals can be picked up by its antenna. The PRN range codes are of three types: the precision (P) code, the Y-code, which is an encrypted P code for redundancy/security purposes, and the coarse/acquisition (C/A) code, which is used as the primary range code for non-military-grade GPS receivers. The L1 signal is modulated by the C/A code, and the P code. The L2 signal is modulated by the Y code. Recent modernization plans for the system include the introduction a new civilian range code on the L2 signal called L2C, a third civilian signal L5 (on frequency L5, 1176 MHz) designed for high-performance civil applications such as aircraft navigation, and a separate military signal called the M code. A fourth civilian signal, L1C, has been designed to enable interoperability between NAVSTAR GPS satellites and other GNSS, specifically Europe's Galileo. These new developments will be part of next-generation GPS satellites, called GPS III, which have begun launching and replacing previous-generation satellites.

The atomic clocks are only needed on the satellites themselves to determine the distance between the receiver and the satellite. When a GPS receiver obtains a broadcasted signal, it can read the exact time the signal was sent by the satellite as well as the time the signal was received by the GPS receiver. Combining the time difference with the speed the signal has travelled, which can be approximated as a constant c (speed of EMR through space, 299,792,458 m/s), gives the distance. Determining the distance of the receiver from multiple satellites allows the location of the receiver to be triangulated. Theoretically, two-dimensional geographic location can be triangulated with three satellites (Figure 2.12). However, this requires perfect information, when in reality the parameters used in the calculations (for example, the constant speed of EMR at c) are estimates. One major source of error is time difference introduced by the inexpensive clocks onboard GPS receivers. To correct for these unavoidable artefacts that can greatly impact accuracy, a minimum of four satellites is required to determine location (four or more are needed for three-dimensional positioning). In practice, the number of "fixes" a receiver obtains, the higher the accuracy of location.

Differential GPS

Timekeeping is one of the most fundamental technologies of GPS. When a handheld receiver receives a signal from a GPS satellite, the position of that satellite in its orbit has to be known exactly. While theoretically satellites follow fixed orbits, in practice slight irregularities occur due to gravitational forces that can affect the accuracy of its own locational information sent to the receiver. One way of improving this is to introduce locational information of a known location on earth. In ordinary GPS, base stations play this role. However, the distance between the receiver and the base station can affect the accuracy that can possibly be obtained (there are only 24 base stations globally).

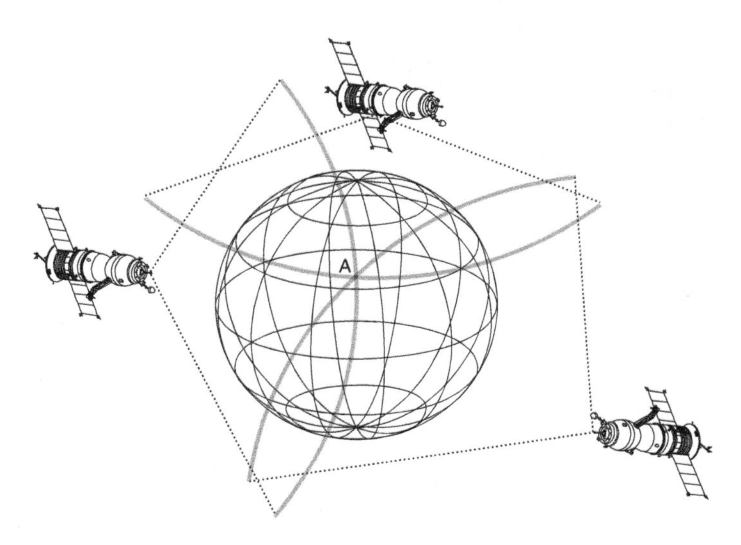

Figure 2.12 GPS triangulation, showing the intersection of three paths, each defined by the timing of a signal received by a GPS receiver at position A. With three paths, the exact location of the position can be determined.

An alternate configuration is to introduce another receiver at a known location, such as a survey marker, compute the location with GPS, and use the difference between the computed location and the true known location to correct for location measurements at unknown locations.

This type of configuration is called *differential GPS* (Figure 2.13) and can produce locational accuracies much higher than traditional GPS. Correction with differential measurements can be done after GPS measurements are taken, which is called post-processing, or in some systems this correction can be done at the time of data capture by encoding corrections onto carrier waves, systems called real-time kinematic differential GPS. Differential correction can greatly increase accuracy of location information. For a single-band (i.e., L1, C/A-code) receiver, accuracy in spatial location can be expected to shift from 5–10 m to 0.7–3 m.

Accuracy and GPS

Many factors influence the accuracy of GPS data. In order to outline how we can measure and evaluate the accuracy of GPS location information, we need some understanding of how errors are introduced into GPS data. One source of distortion is the orbital location information passed along by the satellite. This information can contain errors when the orbit has shifted slightly and onboard updates have not been received from a reference station (i.e., the location of the satellite is reported with error). Another source of error is atmospheric effects (ionospheric and tropospheric), which attenuate the signals and distort the true velocity of the transmitted signal (Figure 2.14). While the known effects of atmospheric attenuation can be controlled for, when periodic events such as solar winds occur, they are generally not corrected for in most receivers.

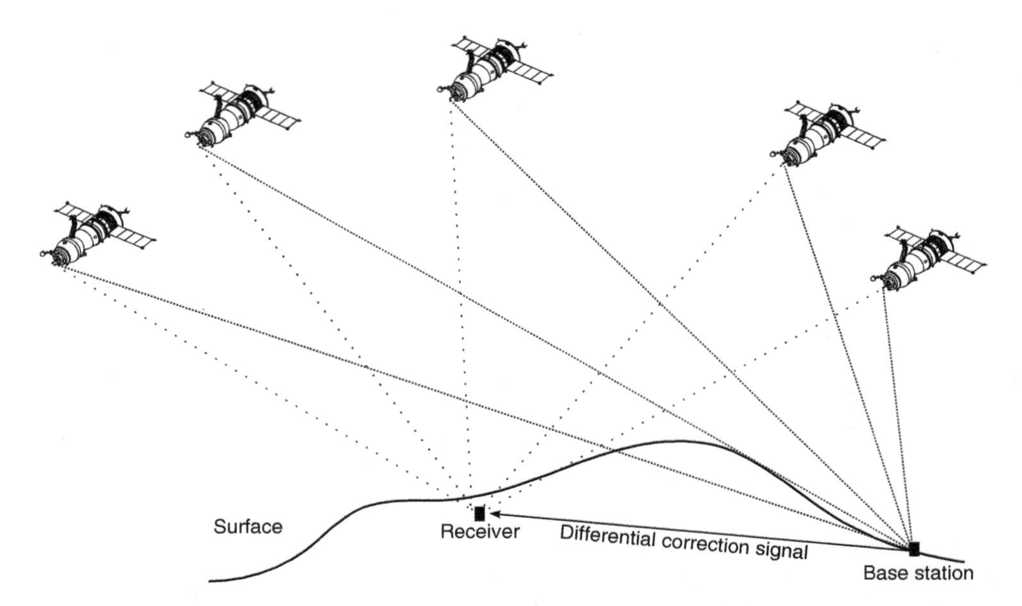

Figure 2.13 Differential GPS.

Satellite clock corrections can also contain errors. Signals can be affected by surface features in the field of view of the receiver's antenna, a phenomenon called multipath distortion. A final source of error in GPS location information is due to the positional arrangement of satellites in the sky that are contributing to the fix, sometimes called the dilution of precision (DOP). A fix consisting of satellites relatively close together in the sky is said to have "poor geometry" because the uncertainty in the area of intersection, based on the ranges of the satellites, is greater (i.e., greater portion of intersecting circles overlap rather than intersect at an exact point). While many of these effects are known and are corrected through modelling, some effects may remain. Newer GPS receivers are able to receive additional signals for error corrections (e.g., atmospheric and orbit errors) through wide-area differential GPS networks such as the Wide Area Augmentation System (WAAS).

2.3.3 Handling GPS Data in GIS

GPS data are increasingly becoming the easiest type of geographic information to collect. Due to the shrinking costs of and physical size of GPS sensors, the technology is increasingly integrated into automobiles, laptop computers, mobile phones, and other handheld devices. In some ways, all computing is becoming "location-aware" through the deployment of GPS, allowing the development of a new brand of location-based services. Some important characteristics of GPS data are worth noting, as the geographic information captured by GPS sensors is becoming ubiquitous representations of the real world.

Surveying and mapping applications were some of the first adopters of civilian GPS technology. The representation of geographic information took three forms: points,

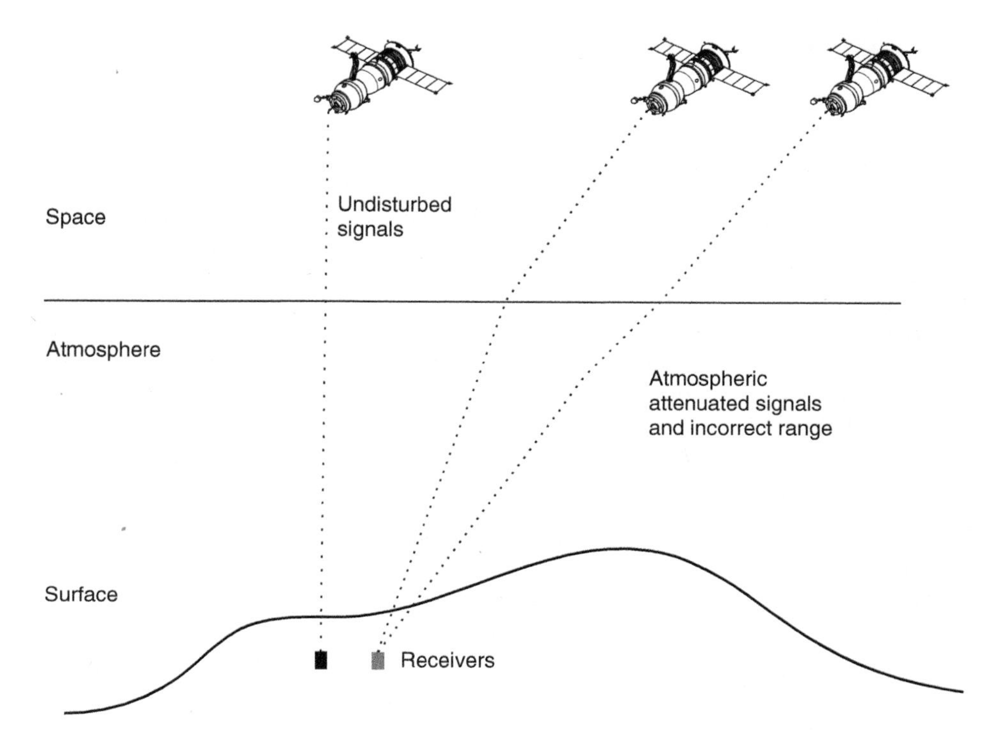

Figure 2.14 Atmospheric attenuation of GPS signals causing incorrect distance measurement.

routes, and tracks. Points (sometimes called points of interest or waypoints) are the basic fundamental element of geographic information in GPS. This consists of a coordinate pair representing location in Euclidean (projected) space or as latitude and longitude coordinates in an ellipsoidal reference system. Usually each fix also records the elevation of the receiver relative to some vertical datum and a timestamp at the point of data capture. A route is a sequence or collection of location points. A track is a recording of all location fixes during an interval of time. Tracks have attributes, such as time and elevation, stored as part of the path data, but the details of individual points that make up the track are usually unidentifiable.

Most GPS data formats are text-based, as they are designed to be translated easily between computing platforms and applications. The current standard file format is an XML schema called GPS exchange (GPX). The GPX schema is open and free to use by GIS developers. There are many possible elements in a GPX file (date, time, latitude, longitude, name, elevation), and they can be organized as waypoints, tracks, or routes. At a minimum, an entry must have latitude and longitude information. Many GPS receivers collect data natively in their own proprietary format but also provide export-to-GPX functions. Many collaborative mapping projects, where contributors collect GPS data on their own devices and contribute to a wider mapping project, depend on translation tools that can convert between GPS data formats, usually standardizing to GPX files as the common standard.

2.4 Converting Analogue Geographic Information

One of the most common methods of converting geographic information from analogue to digital form is via manual digitization. Digitization involves the use of a handheld control, usually a mouse or a digitizer, over an analogue data source (i.e., a paper map) or a digital data source (e.g., a remotely sensed image) to map individual features into vector format. The process of digitizing is by its nature labor-intensive. The classical form of digitizing requires the use of a large tablet on which a paper map is placed. The digitizer is registered to a coordinate system in the GIS software. There are two modes of data input: point and stream digitizing. In *point digitizing*, the user clicks along features, defining coordinate locations, which become vertices of linear features or polygon boundaries. *Stream digitizing* records coordinates at regular intervals, and the digitizing device is moved along the features being mapped. The accuracy of digitized data is affected by the resolution of the digitizing system, the skill of the person operating the device, and most importantly, the scale of original analogue data. Another common approach to inputting analogue sources of geographic information is via scanning. Usually, maps are scanned into a GIS using a document scanner. The digital raster image is then georeferenced, and digitizing is used to extract digital vector features from the raster dataset. Some details of conversion of binary raster data, as might be produced from a scanned map, to vector data are given in Chapter 4.

2.5 Synthesizing Existing Geographic Information

The synthesis of existing geographic information is one of the most important stages of many GIS projects. Increasingly, data are shared among researchers, corporations, governments and non-governmental organizations, and citizens involved in the production of geographic information. The sources of available geographic information are increasing and are easier to access than in the past. This requires greater emphasis on scrutiny of geographic information and descriptors of its lineage, collection specifications, coordinate system information, and processing history. The standard method of storing such information is via the metadata attached to files storing geographic information. However, metadata are only useful if they are a complete and accurate record. Metadata standards have been developed in the US, Canada, and Europe, as well as by data providers and distributors. This has led to confusion and reduced interoperability.

2.5.1 Metadata and Geographic Information

The US Federal Geographic Data Committee defines metadata as follows:

> *A metadata record is a file of information, usually presented as an XML (extensible Markup Language) document, which captures the basic characteristics of a data or information resource. It represents the who, what, when, where, why and how of the resource. Geospatial metadata are used to document geographic digital resources such as Geographic Information System (GIS) files, geospatial databases, and earth imagery. A geospatial metadata record includes core library catalogue elements such as Title, Abstract, and*

Publication Data; geographic elements such as Geographic Extent and Projection Information; and database elements such as Attribute Label Definitions and Attribute Domain Values.

The basic purpose of a metadata standard is to support the interoperability of GIS systems and the appropriate use and sharing of geospatial data. Metadata enable applications and users to describe, query, manage, share, and integrate geographic data.

Geographic data lacking metadata carry significant risks. Undocumented geographic data require extensive time and effort for users to familiarize themselves with the data. Previous data processing operations, attribute descriptions, or spatial reference information can be difficult or impossible to figure out with insufficient metadata. There is also greater chance that poor or missing metadata will lead to data not being widely disseminated, as metadata records can be easily published, queried, and reported. A final risk is that considerable time and money will be wasted on creation of duplicate data, often even within projects or organizations.

The first major metadata standard to gain traction was the FGDC Content Standard for Digital Geospatial Metadata (CSDGM), released in 1994. The CSDGM, updated to version 2.0 in 1998, was organized as a hierarchical structure of compound elements (containers) and data elements (properties). In 1995, the Canadian General Standard Board released the Directory Information Describing Geo-referenced Datasets, which described a geospatial metadata content for Canadian organizations. As noted above, the proliferation of standards reduces the likelihood of any specific metadata standard fulfilling its objectives. More recently, standards at the international level have been developed, culminating in the International Standards Organization (ISO) standard for Geographic Information—Metadata, ISO 19115. Different "profiles" of the ISO standard are developed for specific regions, countries, and industries, all compliant with the global standard but representing specific norms and processes. For example, the North American Profile (NAP) of the ISO 19115, jointly authored by the US and Canada, define updates to their previous independent standards that are ISO 19115 compliant. The NAP includes the addition of mandatory thematic categories, called Topic Categories, to facilitate sorting and characterization of geographic data. One way organizations are forced into adopting standards is through access to federal funds for geospatial projects or contracts with federal agencies, which often require that any geographic data produced in a project be documented according to the NAP standard. Table 2.4 identifies the 19 topic categories that will be mandatory for all NAP metadata records.

With greater production of geographic information in society in general, the need for metadata, and machine-readable metadata in particular, becomes increasingly important. Researchers looking at ways to link together disparate sources of information on the Internet have also begun to develop approaches for linking of geographic information. Figure 2.15 highlights a subset of semantic web projects that deal with geographic information. The purpose of these projects is to provide the tools to expose geographic information openly on the web so that data can be automatically linked together. In the semantic geo-web, geographic information is coded in a way that standardizes and exposes meaning inherent in that information such that it can be appropriately linked and consumed by other applications, datasets, and users. In practice, this means that geographic information databases will store links to universal resource indicators (URIs)

Table 2.4 Topic categories of the North American Profile, ISO 19115 metadata standard.

Topic Keyword(s)	Description
farming	rearing of animals and/or cultivation of plants (e.g., agriculture, crops, livestock)
biota	flora and/or fauna in natural environments (e.g., flora and fauna, ecology, wetlands, habitat)
boundaries	legal land descriptions (e.g., political and administrative boundaries)
climatologyMeteorologyAtmosphere	processes and phenomena of the atmosphere (e.g., processes and phenomena of the atmosphere)
economy	economic activities, conditions, and employment (e.g., business and economics)
elevation	height above or below the earth's surface (e.g., altitude, bathymetry, dems, slope, derived products)
environment	environmental resources, protection, and conservation (e.g., natural resources, pollution, impact assessment, monitoring, land analysis)
geoscientificinformation	information pertaining to the earth sciences (e.g., geology, minerals, earthquakes, landslides, volcanoes, soils, gravity, permafrost, hydrogeology, erosion)
health	health, health services, human ecology, and safety (e.g., disease, illness, factors affecting health, hygiene, substance abuse)
imageryBaseMapsEarthCover	base maps (e.g., land cover, topographic maps, imagery, annotations)
intelligenceMilitary	military bases, structures, activities
inlandWaters	inland water features, drainage systems and characteristics (e.g., rivers, glaciers, lakes, water use plans, dams, currents, floods, water quality, hydrographic charts)
location	positional information and services (e.g., addresses, geodetic networks, control points, postal zones, place names)
oceans	features and characteristics of salt water bodies (e.g., tides, tidal waves, coastal information, reefs)
planningCadastre	information used for appropriate actions for future use of the land (e.g., land use maps, zoning maps, cadastral surveys, land ownership)
society	characteristics of society and culture (e.g., anthropology, archaeology, religion, demographics, crime and justice)
structure	man-made construction (e.g., architecture, buildings, museums, churches, factories, housing, monuments, shops, towers)
transportation	means and aids for conveying persons and/or goods (e.g., roads, airports, airstrips, shipping routes, tunnels, nautical charts, vehicle and vessel locations, aeronautical charts, railways, trails)
utilitiesCommunications	energy, water and waste systems, and communications infrastructure (e.g., hydroelectricity, geothermal, solar, and nuclear sources of energy, water purification and distribution, sewage collection and disposal, electrical and gas distribution, data communication, telecommunication, radio, communication networks)

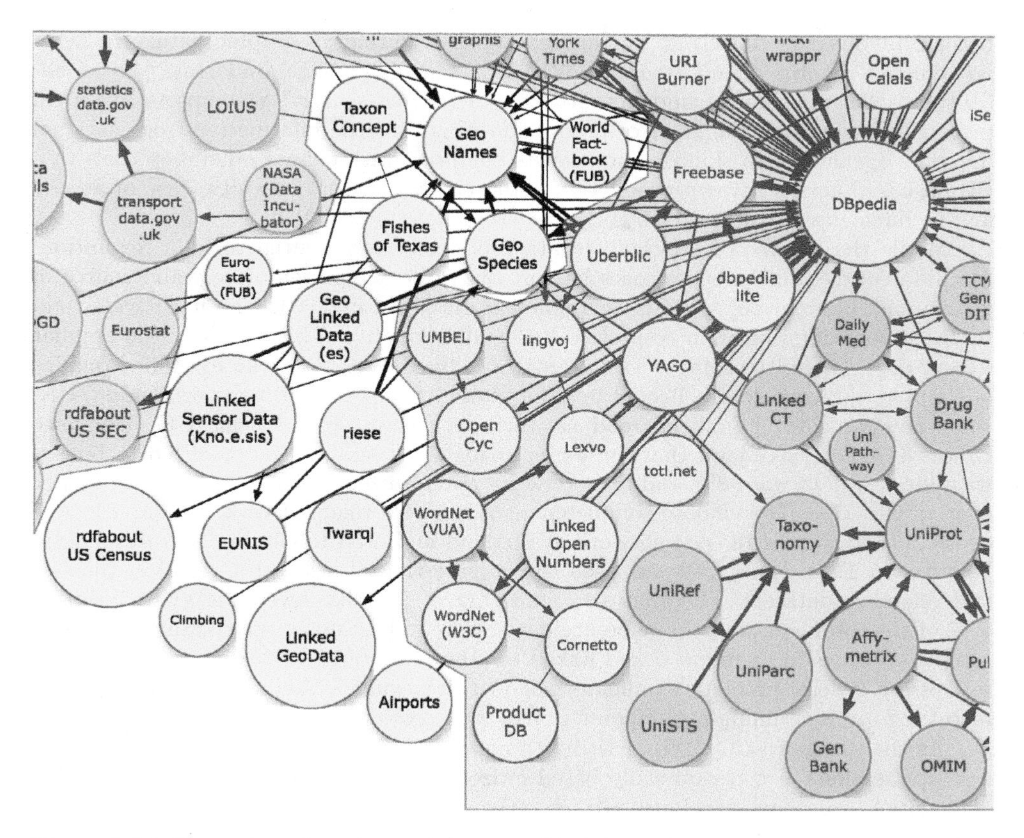

Figure 2.15 Semantic web projects with geographic information.
Source: Linking Open Data Project.

that link to online repositories of geographic information semantics. For example, instead of storing ID = 198221, Province = "Ontario," City = "Toronto," a relation in a semantic data model would store links to the GeoNames repository, which is building a database of all place names in the world. The geometry of the information itself could also be stored openly on the Internet, being linked, for example, to publicly curated data uploaded to OpenStreetMaps. Thus, differences in spelling and data type are overcome, enabling linking to other datasets using the same repository. The development of semantic standards for geographic information will also improve the reproducibility of scientific research using open geographic databases.

A 2.6 Volunteered Geographic Information

User-generated content via digital technologies, social networks, and ubiquitous low-cost location sensors have given rise to an explosion in the production of geographic data. This trend has been of great interest to geographers as more people become involved in collecting, managing, and distributing geographic data. The term *volunteered geographic information* (VGI) was coined by M.F. Goodchild [49] to encapsulate the phenomena and potential of "citizen as sensor." VGI upsets historical precedent by shifting the locus of production of geographic information from national mapping agencies with skilled practitioners equipped with calibrated survey tools to a citizen-based activity where production is driven by interest, curiosity, ease of access, and technologies of convenience.

While the volume of geographic data produced has potential to be a revolutionary data source, and in some cases has proven to be a representative, valid source of geographic information, there is also inherent risk in using VGI. The most widely researched consideration with respect to VGI is data quality. This aspect of the popular VGI project, OpenStreetMaps, was investigated by Haklay [59], who examined quality of the road network data collected by a motivated group of volunteers in London, UK, when compared to the official road network data maintained by the UK Ordinance Office. The analysis found that for major roads the average overlap of 5.6 m buffers with the test data was 88%, and for secondary roads with a buffer of 3.75 m the average overlap was 77%. To evaluate completeness of the OSM road network, a 1 km grid was overlaid onto the entire OSM network in England and used to clip line features at grid boundaries. Each 1 km^2 cell was used to compare the total length of road in the test and reference datasets. The analysis found that at a global level, the VGI comprised 69% of the total length of the reference data. More importantly, the spatial patterns in completeness could be mapped and linked to other variables. Such analyses might help to develop general predictive indicators of VGI data quality and suggest strategies for recruiting and sustaining participants in new VGI projects.

As the place of VGI within existing GI collection frameworks is better understood, new applications of geographically based citizen science and crowdsourced geographic information are being developed. In an early example [66], volunteers were recruited to identify and classify craters on Mars on images with previously identified craters. The goal of their study was to determine if volunteers could perform crater identification with sufficient accuracy. For large craters, 85% were found by at least two volunteers. These data were used to generate an age map of Mars based on crater visibility, which was found to agree with the the map produced by scientists. More recently, volunteer-driven analysis has been used in emergency response applications with some success [50]. The future development of frameworks and methods that can accommodate VGI is a rapidly developing area of geographic research [108].

A 2.7 UAVs and DIY Geography

Data described as VGI are typically and paradoxically coordinated by an expert in geographic information such as an academic researcher, government worker, or data specialist. These actors coordinate activities of participants through projects dedicated to collecting information that is then aggregated and integrated with additional datasets within GIS. An alternate trend in collection of geographic information is the purposeful collection of data using unmanned aerial vehicles (UAV) (i.e., drones). UAV technologies vary from low-cost quadcopter designs with custom mounts that hold regular digital cameras—capable of oblique and vertical images in the visible spectrum—to

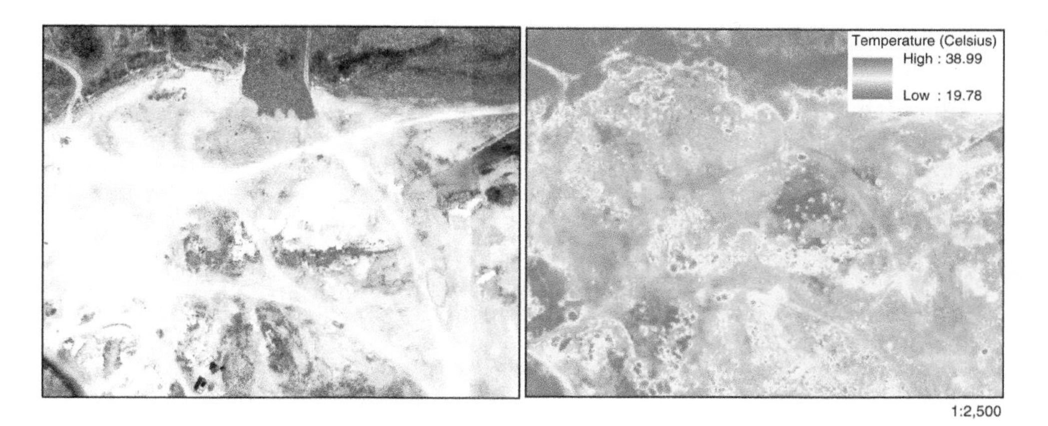

Figure 2.16 Orthophoto (*left*) and thermal image (*right*) obtained from UAV near Waterloo, Ontario.

autonomous-flight fixed-wing mapping drones equipped with multispectral and thermal sensors. Applications of UAV imaging include traditional GIS industries such as agricultural mapping and forestry, but also things like surveillance, traffic monitoring, and event management. Figure 2.16 is an example of UAV-obtained imagery in the visible and thermal spectrums. The imagery is at a ground resolution of 30 cm pixels, providing clear view of temperature differences between vegetation (cooler) and sandy and gravel areas (warmer).

The use of UAVs for collection of geographic information is exposing many more GIS users to the collection of raw data and the data quality issues that arise from airborne collection of images for mapping. Geometric distortions such as roll, pitch, and yaw on fixed-wing drones, and random instabilities on quadcopters have to be corrected in order for data in the scene to be correctly located. Additional distortions due to camera motion, which are typically controlled for in aerial mapping systems, are also increasingly important. Image blur distortions occur due to the forward and lateral motion of the camera during flight as well as the way pixels are encoded by the imaging microchip. These distortions can be corrected using an inertial measurement unit (IMU), which uses accelerometers and gyroscopes to record the UAV's speed and attitude and can therefore correct for these properties. IMUs can also be used to aid in georeferencing when GPS signals are lost (e.g., in tunnels). Citizen use of UAV-collected geographic information is creating a new generation of users that are interacting with GIS technologies in new and creative ways, and opening up new research questions for GISci researchers. Several challenges related to avian safety and privacy have emerged that highlight this trend. As with the dawn of many new technologies, the legal and societal regimes required to manage them tend to lag behind their uptake. UAV regulations continue to evolve in Canada and the US to ensure safe use for collection of high-resolution geographic information.

Problems

1. Using the Victoria Landsat data, identify three landcover types, and select three areas of each type. Use a digital globe or online map such as Google Earth to inspect high-resolution imagery of these areas. Discuss what characteristics of each site are obvious and what are hidden in the Landsat imagery compared to the higher-resolution imagery.
2. Use a GPS receiver or a GPS application to collect GIS data for five of the following features: a park, a field, a school, a trail, a fire hydrant, and a tree. For each feature, record it three separate times. Input these data to a GIS, and compute the distances between them. Discuss how each object could have been represented differently, and comment on the variability in distance computations.
3. In the section on remote sensing, a study of sea ice in Yelverton Bay is described; it used both active and passive remote sensing. What would be the disadvantages of replicating this study using only Landsat data?
4. How would you acquire geographic information to study habitat loss of Grizzly Bears in Western Canada. Describe the decisions you would need to make, the types of information required, and how you could acquire this information.
5. How might networks of location sensors on mobile phones, cars, and products shift the way we use GIS. What are some new applications of GIS that could result from massive adoption of location sensors? What are likely to be some of the pitfalls or challenges?

Chapter 3

Coordinate Systems and Frames of Reference

3.1 Introduction

Locations on the surface of the earth are referenced to a three-dimensional, lumpy, mass of rock, water, sand, soil, and ice. In order to accurately locate positions on this surface, coordinate systems that attempt to bring mathematical order to this complexity have been defined. This chapter describes approaches for mapping positions on the surface of the earth in three dimensions and in two dimensions, and for translating coordinates within and between systems. All data used in GIS are referenced to the earth, and a solid foundational understanding of the underlying geoid, datum, and map projection is required in order to manipulate geospatial data between systems and to realize the inherent uncertainty and sources of error in many sources of geographic data.

3.2 Lumpy Ellipsoidal Earth, Flat Maps, and Projections

GIS are by definition concerned with storing, representing, and analyzing positions on the earth's surface. Determining locations on the earth is therefore an important part of GIS. Usually, we seek to describe the location of an object in space in two ways. We can refer to location in an absolute sense; for example, the coordinates in a coordinate system referenced to the earth (e.g., 47.101° N,−119.995° W or 11 North, 272740.3 E, 5220741.1 N). Alternatively, we can describe the location of an object relative to other objects (e.g., about 2 hours east of Seattle). Human language and reasoning about locations typically use the latter method, while information systems tend to be designed solely for the former. For example, asking a stranger for directions to the nearest gas station is more likely to produce a series of streets and intersections rather than a list of latitude and longitude coordinates. Because computers are generally good at storing and manipulating numbers, GIS are very good at handling coordinates in absolute space, whereas technologies for processing relative locational information are much less developed. We use this disjuncture—between the ways computers and humans naturally express and reason about spatial information—as a starting point for our discussion of map projections and coordinate systems, which are focused wholly on absolute space but are often used as the basis for human reasoning about space (or place) when using computers.

While the problem of locating positions on the surface of the earth is often simplified to positions on a sphere (or oblate-spheroid), in reality the shape of the earth is far from geometrically pure. Even defining what we mean by the shape of the earth is not entirely clear: Do we include every mountain and valley, the top of the forest canopy or the forest floor, the roofs of skyscrapers, the ocean surface or the underlying seabed? Generally, the shape of the earth is approximated by a theoretical body that is the shape defined by the earth's gravitational field. The earth's gravitational field varies with changes in surface topography, variations in materials underlying the earth's surface, and the earth's rotation and flattening at the poles. As such, the shape defined by the earth's gravitational field can only be estimated, approximated, and modelled from gravity measurements. The name given to this shape is the *geoid*, often described as global mean sea level, but more accurately it is the equipotential surface defined by gravity, or the surface that would arise if the earth were covered completely by water. The direction of gravity is everywhere perpendicular to the geoid, making mean sea level a good approximation.

Variations in gravitational fields are called gravity anomalies. With the gravity of earth approximated as g (9.8 m/s^2), anomalies are typically variations in magnitude around 1/10,000 of g. Accounting for known anomalies involves incorporating physically derived parameters into a mathematical model, usually using spherical harmonics (i.e., spherical version of a Fourier series described in Chapter 2) in what is called a geopotential model. The simplest effects to model are those resulting from the rotation of the earth about its axis, which makes the value of g a simple function of latitude

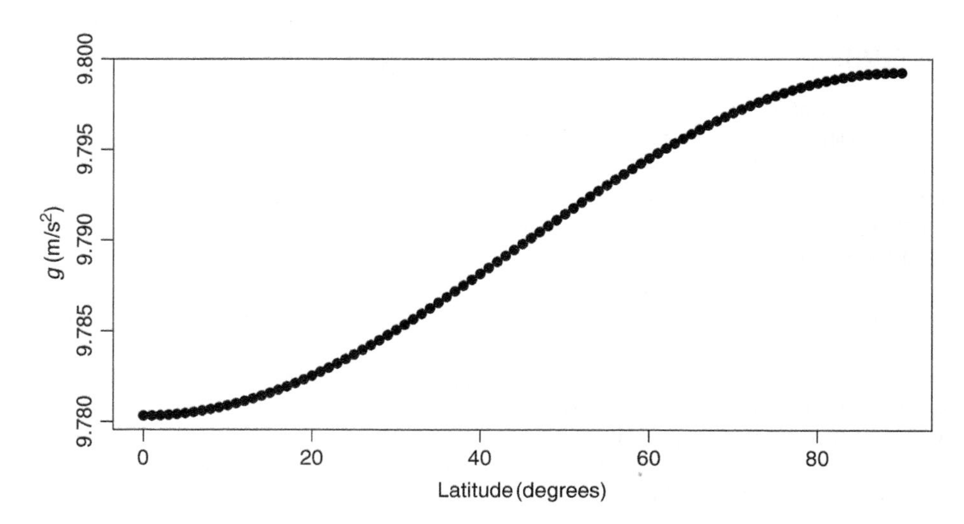

Figure 3.1 Gravitational variation with latitude.

(see Figure 3.1) and constants derived from the flattening ratio (i.e., equatorial radius and polar radius),

$$g = 9.7803267714 \cdot \left(\frac{1 + 0.00193185138639 \cdot \sin^2 \theta}{\sqrt{1 - 0.00669437999013 \cdot \sin^2 \theta}} \right) ,$$

where θ is the latitude for the position being estimated, and g is in m/s^2.

A full model of the geoid requires an understanding of the gravitational potential of every location on earth. The geoid is an equipotential surface such that it is perpendicular to the direction of the force of gravity. One way of obtaining this is a global gravity model, such as the Earth Gravitational Model 1996 (EGM96). The EGM96 is a spherical harmonic expansion model with 130,317 coefficients that incorporate a global network of gravity measurements and satellite-based measurements. The geoid is used as the reference model for definition of a global vertical datum. A *datum* is a physically derived point, line, or surface that serves as the reference for a coordinate system (either horizontally or vertically). All geographic positions are measured in reference to a datum. However, there are three possible surfaces that are used to determine position of a point on the surface of the earth: the geoid (equipotential surface of the earth's gravity field), a reference ellipsoid (horizontal datum), and the topographical surface (including terrain and seafloor). Figure 3.2 describes the relation between each of these surfaces. The difference in height between the geoid and reference ellipsoid is called the *geoidal height* (N_{geo}), or *geoidal separation*, and can be used to derive the *orthometric height* (H_{orth}).

Vertical datums can be local as well as global. For example, in Canada the Canadian Geodetic Vertical Datum 1928 (CGVD28), which was developed from traditional survey techniques (i.e., levelling), required an extensive network of over 90,000 survey benchmarks and 160,000 km of levelling lines. Since physical access is required to perform the levelling, the network is mainly distributed along roads and railways only, leaving major areas of the country unsurveyed (see Figure 3.3).

A geoid-based (CGG2010) vertical datum, CGVD13, is the new vertical datum for Canada. It takes advantage of GPS-derived ellipsoid heights and a geoid model for North America to determine orthometric heights (i.e., height above sea level). Differences between CGVD28 and CGVD13 are between −78 cm and +66 cm. The geoid in CGVD13 is the equipotential surface defined by 62,636,856.0 m^2/s^2, which is MSL

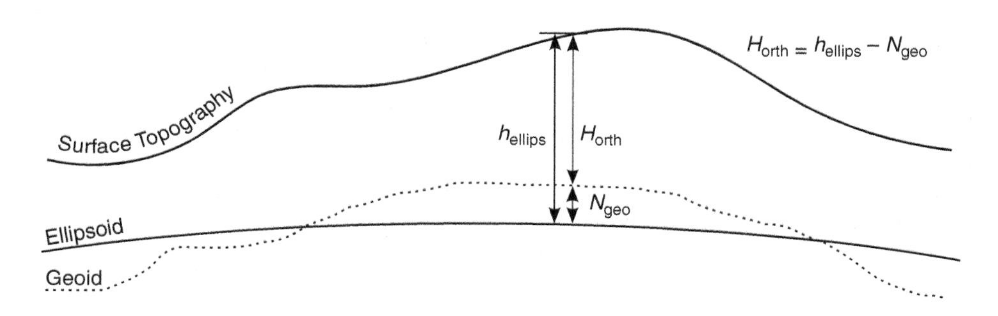

Figure 3.2 Geoidal height cross section model.

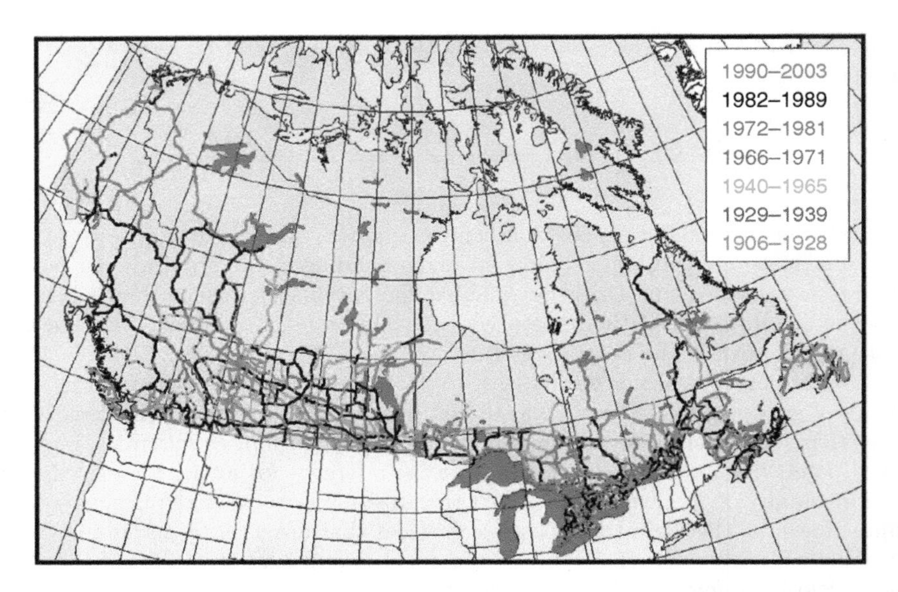

Figure 3.3 Survey levelling network used for validation of vertical datum (CGVD28).

Source: Natural Resources Canada. Use of Canadian Geodetic Survey products and data is subject to the Open Government Licence–Canada.

defined for coastal North America, whereas MSL in CGVD28 was derived from measurements made at a small number of tide gauges. As MSL is an average agreed upon by the US and Canada for all of North America, the coastline around Vancouver is technically slightly above sea level, while around Halifax it is slightly below sea level.

While the geoid serves as a reference for defining a vertical datum, a reference ellipsoid is required to provide a horizontal datum for positions on the earth's surface. An ellipsoid is to an ellipse what a sphere is to a circle. The reference ellipsoid is an oblate, flatter-at-the-poles ellipsoid of rotation, the rotation being about the axis connecting the north and south poles. The basic parameters of the earth ellipsoid are its *semimajor* (a) and *semiminor* (b) axes, typically measured in metres. Their ratio defines the *flattening* f (sometimes expressed as $1/f$):

$$f = \frac{(a - b)}{a} \quad .$$

Reference ellipsoids define a shape that is much simpler than a model of the geoid, so we often need to determine the separation of the geoid and the ellipsoid. This can be done locally and globally. Figure 3.4 represents a global map of the geoidal height with respect to the World Geodetic System 1984 (WGS84) ellipsoid. As can be seen, considerable variation exists between the two models of the earth. The WGS84 ellipsoid is the most widely used global horizontal datum and is used as the standard datum for positions recorded using GPS.

While the WGS84 ellipsoid is global (and used as the reference ellipsoid in GPS), other ellipsoids are defined to be accurate for specific regions of the world, termed

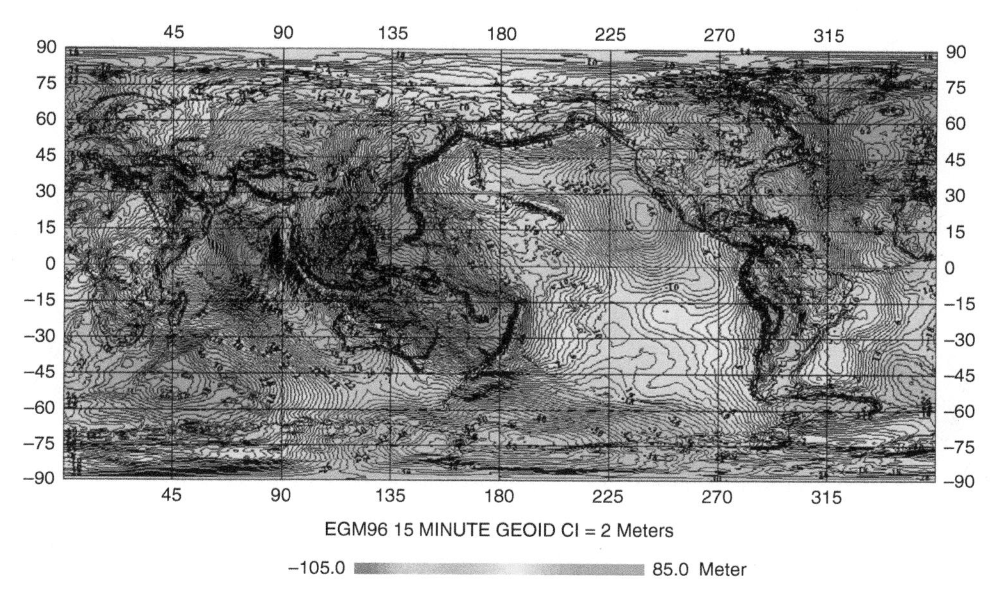

Figure 3.4 Geoidal Height above WGS84 ellipsoid.

Source: NASA/Earth Geopotential Model 1996.

local geodetic datums. The North American Datum 1982 (NAD82) is a commonly used horizontal datum for North America, whereas the European Terrestrial Reference System 1989 (ETRS89) is a commonly used datum for Europe. The reference ellipsoid used in NAD83 is GRS80 defined by ellipsoidal parameters $a = 6{,}378{,}137$ m, $b = 6{,}356{,}752.314$ m, while the WGS84 has modified b slightly since its original derivation. For many GIS applications, coordinates defined by NAD83 and WGS84 can be considered equivalent because the reference ellipsoidal parameters are almost identical. At the time these systems were introduced, they were practically identical in terms of origin, orientation, and scale of the reference frames. However, periodic updates to WGS84 and NAD83 to account for movement of tectonic plates mean that the difference between these two systems is not constant. Some estimates put the maximum difference between the two systems between 1.2 m and 1.5 m.

Positions on the earth are defined with respect to vertical and horizontal datums that model the shape of the earth. As such, datums define coordinates in geographic space: latitude ϕ, longitude λ, and height h above sea level (see Figure 3.5). *Latitude* is defined as the angle formed by a normal line (a line perpendicular to a tangent surface) through a point on the surface of the ellipsoid and the equatorial plane of the ellipsoid. *Longitude* is the angle between a meridian passing through a point on the surface and a plane defined by the reference meridian (e.g., Greenwich, England). *Height* in this case is height above the ellipsoid (not the geoid). When ellipsoidal height is combined with geoidal height (i.e., at the location in question, the separation between the geoid and the chosen reference ellipsoid is known), the orthometric height can be determined (recall Figure 3.2).

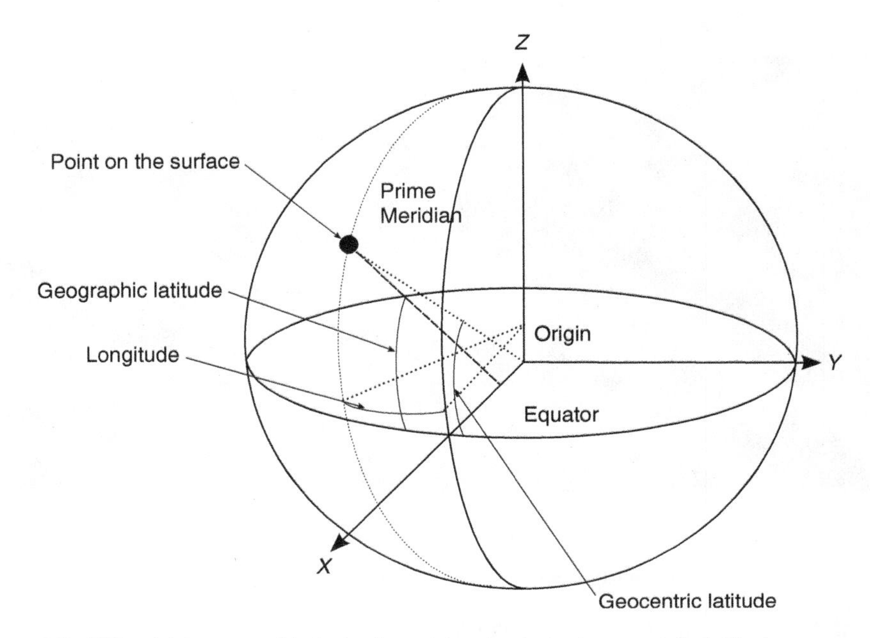

Figure 3.5 Ellipsoidal geographic latitude and longitude and geocentric latitude.

Ellipsoidal geographic latitude and longitude coordinates are not the only coordinate systems used to define geographic positions. Geocentric coordinate systems are defined with three axes (X, Y, Z) with an origin at the centre of the earth. Geocentric coordinates are typically in metres. Geocentric latitude differs from geographic latitude in that the angle is formed by a line through the centre of the earth, rather than a normal line (see Figure 3.5).

Conversion between ellipsoidal and geocentric coordinates is possible given ellipsoidal parameters

$$X = (v(\phi) + h) \cdot \cos\phi \cdot \cos\lambda$$
$$Y = (v(\phi) + h) \cdot \cos\phi \cdot \sin\lambda$$
$$Z = \left[\left(1 - e^2\right) \cdot v(\phi) + h\right] \cdot \sin\phi \quad ,$$

where h is ellipsoidal height, $v(\phi)$ is radius of curvature in the prime vertical (sometimes denoted R_n, and is perpendicular to the ellipsoid surface at a given latitude ϕ), and e^2 is eccentricity squared, defined as $(a^2 - b^2)/a^2$ (v is the radius that defines latitude in Figure 3.5, varying between 6,378,139 m at the equator and 6,399,592 m at the north pole). The radius of curvature at geodetic latitude ϕ, $v(\phi)$, can be calculated as

$$v(\phi) = \frac{a}{\sqrt{(1 - e^2 \cdot \sin^2\phi)}} \quad .$$

For more information on the derivation of these results, please see Appendix VI.

As an example calculation, consider the following problem. Given coordinates for the CN Tower and the WGS84 reference ellipsoid would give parameters $\phi = 43°38'$ (or approximately 43.63 in decimal degrees), $\lambda = 79°23'$, $a = 6{,}378{,}137$, $e^2 = 0.00669437999013$, which gives

$$v(43.63) = \frac{6{,}378{,}137}{\sqrt{(1 - 0.00669437999013 \cdot \sin^2 43.63)}} = 6{,}388{,}326 \quad ,$$

for the radius of curvature at $\phi = 43°38'$ and the coordinates

$$X = 851{,}865.3$$
$$Y = 454{,}4590$$
$$Z = 437{,}8751 \quad ,$$

for the geocentric domain coordinates. We could then apply offsets in each axis for a geocentric datum transformation (described below).

While the conversion from geodetic to geocentric coordinates is straightforward, the inverse conversion is more difficult and requires numerical methods (see Ligas and Banasik [76] for more details). Geocentric coordinate systems require knowledge of the shape and mass of the earth, and the location of its centre. Before such knowledge was available, *topocentric coordinate systems* were more common, where the origin of the system was located on the surface of the earth (e.g., Mead's Ranch, Kansas, for NAD27 based on the Clarke ellipsoid, 1866).

3.3 Datums

The best approximation of the shape of the earth is provided by a physically derived model of the geoid such as EGM96, and the most common mathematical representation of the earth is an oblate ellipsoid (defined by parameters a and b). An ellipsoidal coordinate system needs to be fixed and oriented to the earth, and this is the role that datums play in defining a coordinate reference system. A datum can be thought of as a set of parameter values that define the origin for a coordinate system. Thus, datums are based on physical geographic space and measurements (e.g., geodetic control surveys), while ellipsoids are mathematical constructs used for convenience in locating positions on earth.

Many countries have defined datums with parameters optimized for mapping their territories. Many of these coordinate systems are topocentric, with origins defined by prominent geographic features. Datums defined with parameters that optimize one region of the world are called local datums, whereas datums defined to best represent the world as a whole are called global datums (see Figure 3.6).

Worked Example: Local datum in Sri Lanka, Kandawala

Examining the location of the country of Sri Lanka, in Figure 3.4 (located off the southeast tip of India), it is clear that the WGS84 ellipsoid is a poor approximation to the geoid at this location. Local datums have been adopted for Sri Lanka's national

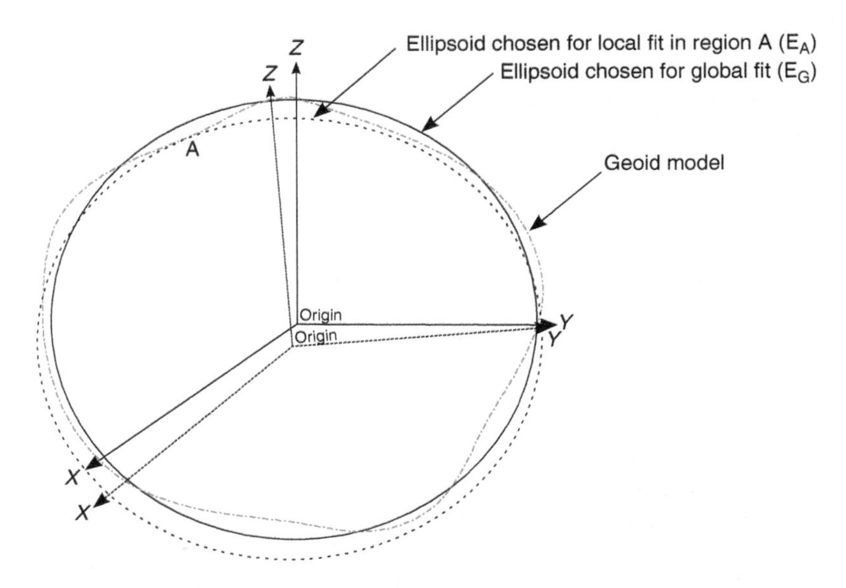

Figure 3.6 Local and global datums.

mapping. The local datum used in Sri Lanka is called Kandawala; it uses a reference ellipsoid called Everest 1830 (1937 Adjustment), with parameters $a = 6{,}377{,}276.345$, $1/f = 300.8017$.

The Kandawala datum was derived from a geodetic survey conducted between 1858 and 1906 at 110 control stations. The datum is used to define a national map grid that uses the Transverse Mercator projection and an origin based on the highest point in Sri Lanka, Mount Piduruthalagal. The coordinates of the Kandawala-based national mapping grid for Sri Lanka are outlined in Table 3.1. The baseline origin used in the original triangulation of Sri Lanka was at Kandawala, with coordinates fixed at latitude 7° 14′ 06.838″ N, longitude 79° 52′ 36.670″ E, which gives the Kandawala datum [2]. In the late 1990s, a new datum was developed, the Sri Lanka Datum 1999 (SLD99),

Table 3.1 Local datums and spatial reference systems used in the country of Sri Lanka: Kandawala and SLD99.

Parameter	Kandawala	SLD99
Longitude of origin (central meridian)	80° 46′ 18.16000″ E	80° 46′ 18.16710″ E
Latitude of origin	07° 00′ 01.72900″ N	07° 00′ 01.69750″ N
Central scale factor	0.9999238418	0.9999238418
False northing	200,000.000 m	500,000.000 m
False easting	200,000.000 m	500,000.000 m
Semimajor axis length	6,377,276.345 m	6,377,276.345 m
Reciprocal flattening	300.8017	300.8017

which integrated the existing Kandawala geodetic network with a new GPS-based geodetic control network. The SLD99 datum is also based on the Everest 1830 ellipsoid. Due to upgrading of the geodetic control with SLD99, the coordinates of its origin were modified. In the national mapping projection (Transverse Mercator projection—described later this chapter) used in conjunction with SLD99, the coordinates of the false origin (Mt. Piduruthalagal) are adjusted accordingly, to 50 km to the south and west. Understanding the local datums in Sri Lanka (and other parts of the world) is necessary for integrating local GIS data with data referenced to a global datum such as WGS84. Transforming data from one datum to another requires a procedure called a *datum transformation.*

3.4 Datum Transformations

Datum transformations involve translating coordinates between reference systems that are based on different datums. Coordinate conversions or reprojections refer to changing coordinates from one reference system to another when both systems share a common datum. Because datums are physically based (i.e., based on measurements made on the ground), they have inherent uncertainty, which can propagate during datum transformations. Coordinate reprojections on the other hand (with a common datum), are mathematically defined and therefore do not cause uncertainty or loss of accuracy in the output relative to the input data.

Transformation parameters are empirically derived values that describe change in each of the dimensions of the coordinate system. Note that the accuracy of the geodetic control network supporting the definition of each of the datums involved in the transformation is important, as distortions inherent in one datum can be transformed through to another datum during transformation. The transformation parameters are determined by modelling coordinates from a sample of survey points in each datum, generally using least-squares estimation techniques.

3.4.1 Geocentric Datum Transformations

Transformations between datums that are both geocentric adjust coordinates for differences in the origin of each datum, and can be realized through simply determining the changes in each of the dimensions. A simple geocentric transformation model (a translation) has parameters for change in X (ΔX), change in Y (ΔY), and change in Z (ΔZ) between the source and target datums. Because geocentric datums by definition have a common origin (i.e., the centre of the earth), one datum is denoted as source, with origin $X = 0$, $Y = 0$, $Z = 0$, and the origin of the target datums is referenced as a some distance away in each of the three respective dimensions. The parameters ΔX, ΔY, ΔZ are sometimes referred to as the translation vector but always refer to the values added to the source coordinates to translate the coordinates to the target datum. If going in the other direction, the signs have to be reversed.

$$\begin{bmatrix} X \\ Y \\ Z \end{bmatrix}_{\text{target}} = \begin{bmatrix} X \\ Y \\ Z \end{bmatrix}_{\text{source}} + \begin{bmatrix} \Delta X \\ \Delta Y \\ \Delta Z \end{bmatrix}_{S \text{ to } T}$$

The values for these parameters describe the change along the axis in each datum's origin (see Figure 3.7).

Examples of some geocentric transformation parameters are outlined in Table 3.2. Relying solely on geocentric datum shift, however, simplifies the differences between two datums to those related only to the position of the origin, ignoring other sources of distortion such as the orientation of the axes. As such, geocentric transformations incur errors typically in the range of 5–10 metres, which may be suitable for recreational grade GPS data, but not for surveying and mapping applications.

A more sophisticated transformation method uses additional parameters that account for differences in the rotations ($\mathbf{R}$) of each of the principal axes as well as differences in the scale (μ) of distance units, resulting in a 7-parameter transformation model (three translations, three rotations, one scale). This is sometimes called the *Helmert transformation*. In the geocentric domain, the transformation is now

$$\begin{bmatrix} X \\ Y \\ Z \end{bmatrix}_{\text{target}} = \mu \cdot \mathbf{R} \cdot \begin{bmatrix} X \\ Y \\ Z \end{bmatrix}_{\text{source}} + \begin{bmatrix} \Delta X \\ \Delta Y \\ \Delta Z \end{bmatrix}_{S \text{ to } T},$$

where rotation $\mathbf{R}$ and scale μ adjustments are incorporated. When datum transformation parameters are published, typically all seven are reported to enable more-accurate transformations. Note that rotation parameters can be reported in angular units or radians, and the rotation matrix $\mathbf{R}$ is expressed as

$$\mathbf{R} = \begin{bmatrix} 1 & -\alpha_Z & \alpha_Y \\ \alpha_Z & 1 & -\alpha_X \\ -\alpha_Y & \alpha_X & 1 \end{bmatrix},$$

whereas μ is a scalar value. For example, for a 7-parameter transformation from WGS84 to Krassovski 1940, the parameters are translations of $\Delta X = -24$, $\Delta Y = 123$, $\Delta Z = 94$, scale parameter of $\mu = -1.1$, and rotations $\alpha_X = -0.02, \alpha_Y = 0.26, \alpha_Z = 0.13$, with rotations expressed in arc seconds.

3.4.2 Ellipsoidal Datum Transformations

The systems of transformations above are for geocentric coordinates and can be used in an intermediate step between coordinate conversions from and to ellipsoidal geographic coordinates. That is, a common workflow for transforming between ellipsoidal geographic coordinates in different datums is first a conversion to geocentric, then transformation, followed by transformation back to ellipsoidal. But often, direct transformations using ellipsoidal coordinates are desired. Ellipsoidal datum transformations transform coordinates between reference ellipsoids by directly computing changes in ϕ, λ, and ellipsoidal height h.

The Molodensky approach requires parameters for changes in X, Y, Z as before, but it also requires changes in ellipsoidal parameters between source and target ellipsoids Δa, Δf. These parameters are used to derive the change in ellipsoidal coordinates,

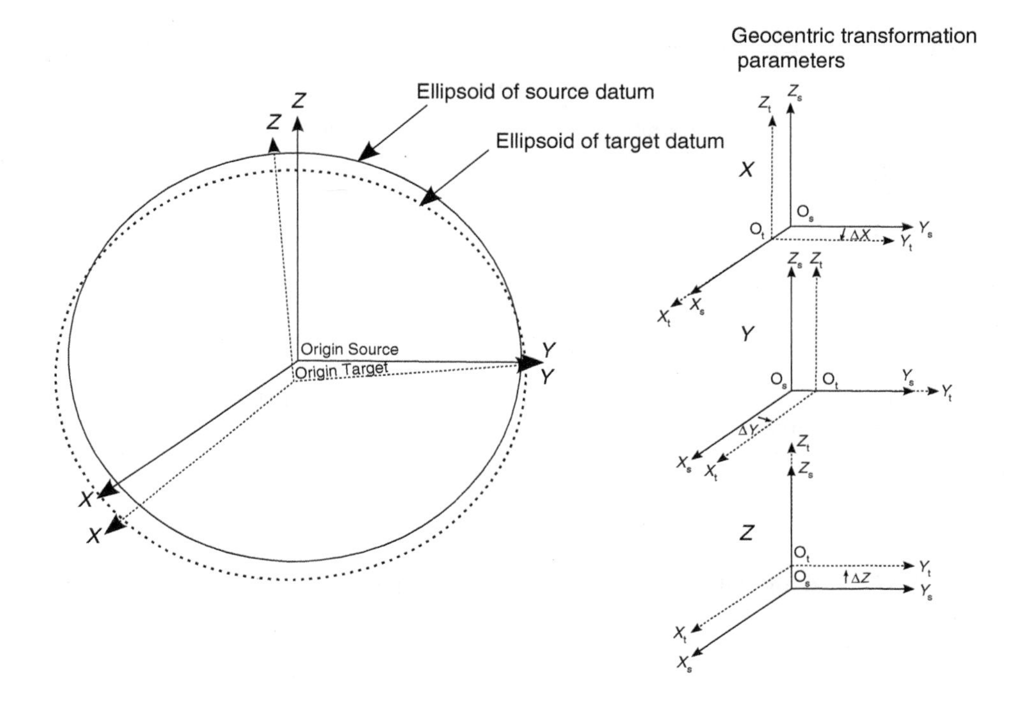

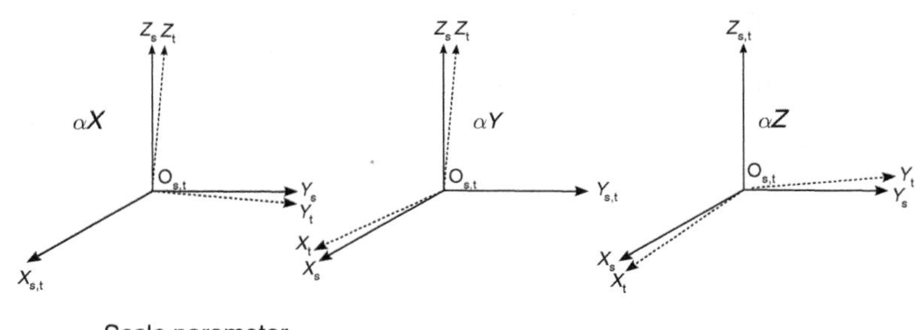

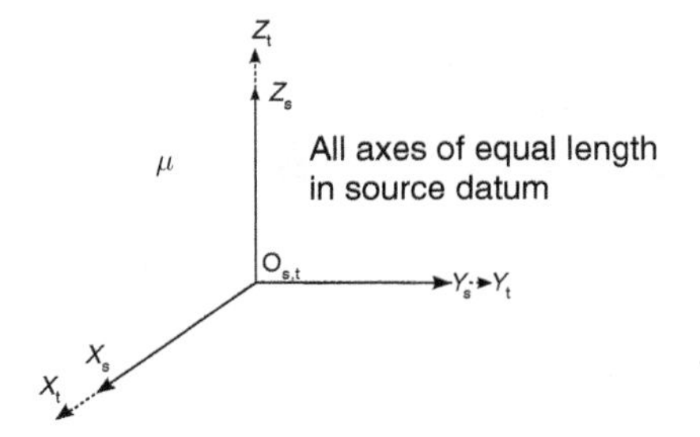

Figure 3.7 Datum transformation parameters for geocentric translation, rotation, and scale.

Table 3.2 Local datums and geocentric transformation parameters to WGS84.

Datum (local)	ΔX	ΔY	ΔZ
Kandawala	−97	787	86
Pulkovo 1942	28	−130	−95
Guam 1963	−100	−248	259
North American 1927	−7	162	188

and they work well over relatively small areas. One of the major sources of error of transformations is distortions in the underlying geodetic control surveys. Over larger areas, grid-based methods have become widely used for datum transformation on the scales of continents.

3.4.3 Grid-based Transformations

Where distortions vary greatly, transformation parameters can be derived locally by overlaying a grid and interpolating $\Delta\phi$, $\Delta\lambda$, and Δh parameters using an interpolation technique. The localized transformation parameters take advantage of variable resolution and quality of the underlying geodetic control to determine the best parameter estimates at each location. This greatly improves accuracy when doing datum transformations over large areas. For example, grid-based transformation is used in Canada for transforming from NAD27 to NAD83—published in a grid-based file format called NTv2. Table 3.3 outlines some examples of transformations in six locations in British Columbia, Canada (see Figure 3.8), using a grid-transformation method. The difference in the shifts in the X direction are on the order of 15–20 m and 5 m in the Y direction. Thus, identical coordinates in each datum describe locations over 20 m apart. The variance in these estimates is directly related to the amount of error that would be achieved if only single estimates of ΔX, ΔY, ΔZ were used, as in the transformations described earlier. The precision would be much lower, as there is extensive spatial variation over the province of BC, large parts of which are mountainous. The spatial distribution of

Table 3.3 Datum shifts from NAD27 to NAD82 for locations in British Columbia, Canada.

Point ID	Longitude λ	Latitude ϕ	ΔX in metres (error)	ΔY in metres (error)	$\Delta\lambda$	$\Delta\phi$
Gabriola Island	−123.864	49.190	−98.626 (0.089)	−20.245 (0.149)	−4.870897"	−0.655324"
Vancouver	−123.146	49.261	−96.193 (0.004)	−19.111 (0.002)	−4.757523"	−0.618615"
Tulameen	−120.759	49.546	−90.037 (0.083)	−15.934 (0.082)	−4.478924"	−0.515739"
Barriere	−120.129	51.180	−87.390 (0.222)	−12.492 (0.925)	−4.499136"	−0.404227"
Fort Nelson	−122.699	58.807	−86.698 (0.097)	−14.892 (0.196)	−5.400223"	−0.481286"
Liard River	−126.105	59.433	−88.925 (0.481)	−22.226 (0.651)	−5.640848"	−0.718230"

the error (in parentheses) ranges from 2 mm in Vancouver (a major city) to 92.5 cm in Barriere (a rural area). The precision of the transformations varies with both the density and age of the underlying geodetic control network in each area. Urban areas such as the City of Vancouver tend to have much more accurate surveying than more remote areas. For example, when compared to the BC Geodetic Control Monuments data, within a 5 km radius of the Vancouver point there are 1855 survey markers, whereas within a 5 km radius of the Barriere monument there are only 25 survey markers.

In addition to the localized transformation methods, an estimate of the accuracy of the transformation is provided. An additional advantage of grid-based transformations over large areas is that they standardize transformations.

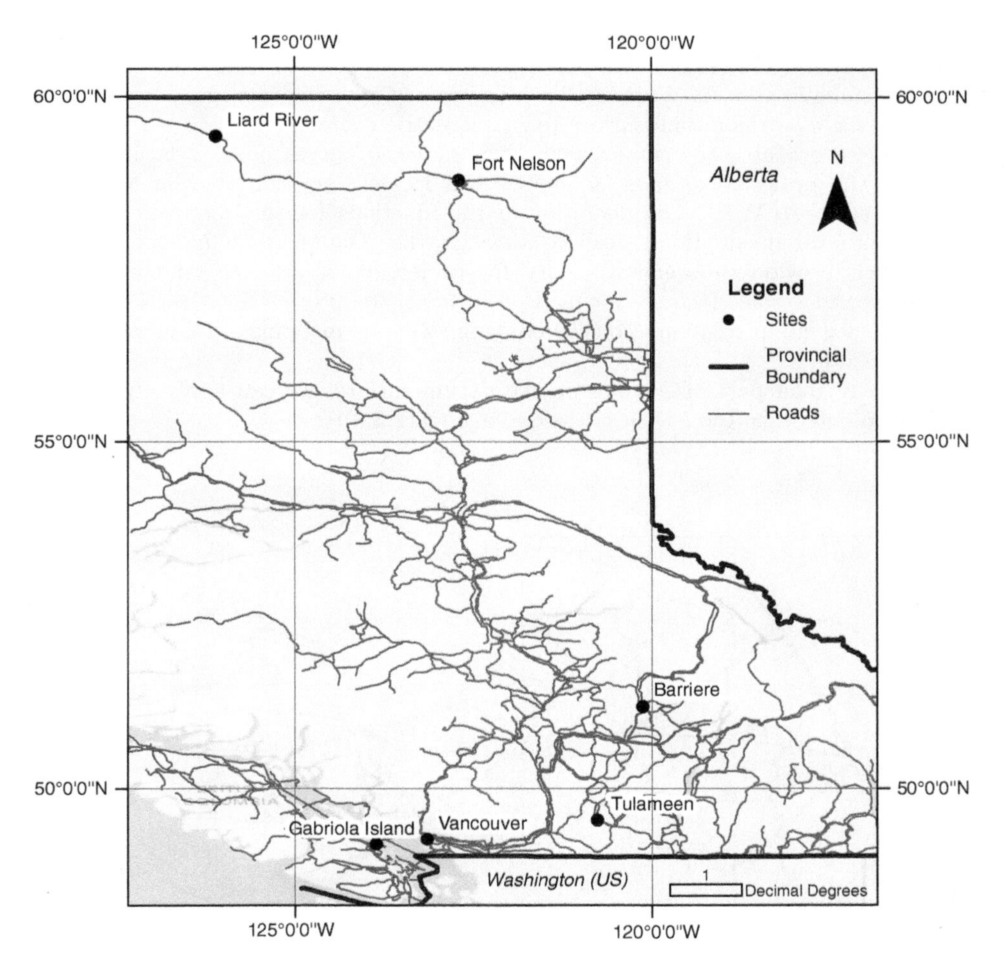

Figure 3.8 Datum transformation locations NAD27 to NAD83 using NTv2 grid-based transformation method.

3.5 Map Projections and Spatial Reference Systems

A map projection is a mapping from positions defined in three-dimensional space to two-dimensional space, typically to facilitate mapping on a plane and to simplify computations for spatial analysis of locational data. We represent this as a mapping of coordinates from a surface defined by a datum onto a projected surface. We denote coordinates on the datum surface as ϕ, λ, and the Cartesian coordinates on the projected surface as X, Y. Map projections are concerned with the set of functions

$$X = f_1(\phi, \lambda)$$
$$Y = f_2(\phi, \lambda) \quad .$$

In practice, we arrive at these functions indirectly. We note that there exists a reference system of meridians and parallels on the projected surface, with coordinates denoted generally as u, v. We can also construct a Cartesian coordinate system with coordinates x, y on the datum surface (e.g., geocentric coordinates).

Once we are able to map curvilinear parametric curves on the datum surface (ϕ, λ) to the projected surface (u, v), we can locate Cartesian coordinates on the projected surface (X, Y). The definition of the functions for this mapping are given by map projection equations, realized through the "Gaussian fundamental quantities," which provide the general theory for projection of any curved surface onto another curved surface (though for map projections, the projected surface is a plane). Note that for all projections considered here, we are mapping onto a plane, and therefore $Z = 0$.

Given a small part of a curve on the datum surface, we can apply Pythagoras' theorem to determine the length of the curve (Figure 3.9):

$$ds^2 = dx^2 + dy^2 \quad .$$

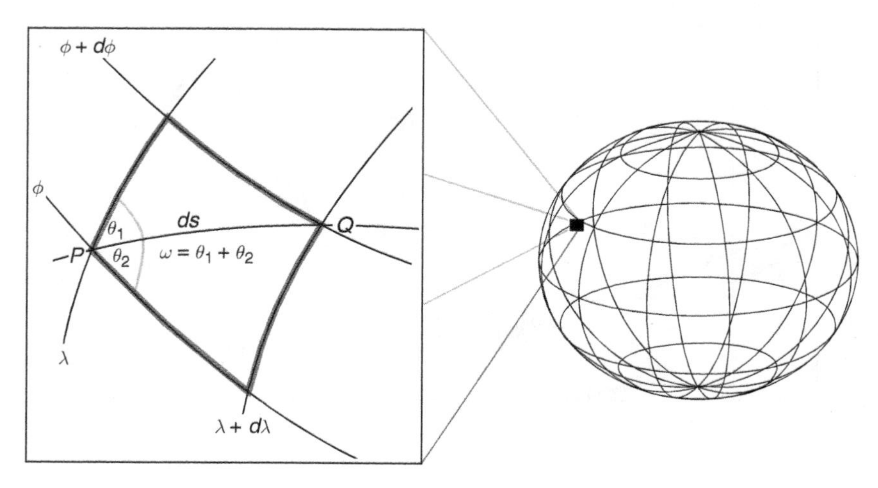

Figure 3.9 Differential rectangle example.

Note that the curvilinear coordinates are parametric curves that form an orthogonal mesh; the angle θ_1 is the angle formed by line ds and the meridian, while θ_2 is formed by ds and the parallel, which together sum to 90°. By differentiating the arbitrary functions f_1 and f_2 mapping coordinates X, Y to ϕ, λ as

$$dX = \frac{\delta X}{\delta \phi} \cdot d\phi + \frac{\delta X}{\delta \lambda} \cdot d\lambda$$

$$dY = \frac{\delta Y}{\delta \phi} \cdot d\phi + \frac{\delta Y}{\delta \lambda} \cdot d\lambda \quad ,$$

which can be substituted into the equation for the length of an elemental curve so that

$$\begin{aligned} ds^2 &= \left(\frac{\delta X}{\delta \phi} \cdot d\phi + \frac{\delta X}{\delta \lambda} \cdot d\lambda\right)^2 + \left(\frac{\delta Y}{\delta \phi} \cdot d\phi + \frac{\delta Y}{\delta \lambda} \cdot d\lambda\right)^2 \\ &= \left(\frac{\delta X}{\delta \phi} \cdot d\phi\right)^2 + 2 \cdot \frac{\delta X}{\delta \phi} \cdot d\phi \cdot \frac{\delta X}{\delta \lambda} \cdot d\lambda + \left(\frac{\delta X}{\delta \lambda} \cdot d\lambda\right)^2 + \\ &\quad \left(\frac{\delta Y}{\delta \phi} \cdot d\phi\right)^2 + 2 \cdot \frac{\delta Y}{\delta \phi} \cdot d\phi \cdot \frac{\delta Y}{\delta \lambda} \cdot d\lambda + \left(\frac{\delta Y}{\delta \lambda} \cdot d\lambda\right)^2 \\ &= \left[\left(\frac{\delta X}{\delta \phi}\right)^2 + \left(\frac{\delta Y}{\delta \phi}\right)^2\right] \cdot d\phi^2 + 2 \cdot \left[\left(\frac{\delta X}{\delta \phi} \cdot \frac{\delta X}{\delta \lambda}\right) + \left(\frac{\delta Y}{\delta \phi} \cdot \frac{\delta Y}{\delta \lambda}\right)\right] \cdot d\phi \cdot d\lambda + \\ &\quad \left[\left(\frac{\delta X}{\delta \lambda}\right)^2 + \left(\frac{\delta Y}{\delta \lambda}\right)^2\right] \cdot d\lambda^2 \quad , \end{aligned}$$

which if we parameterize the square brackets to simplify, gives an expression for ds^2 in terms of $d\phi$ and $d\lambda$.

This is simplified into a model for ds^2 with three coefficient terms, which are universally denoted in the study of map projections as follows, and referred to as Gaussian fundamental quantities (which exist on all surfaces with curvilinear coordinates):

$$E = \left[\left(\frac{\delta X}{\delta \phi}\right)^2 + \left(\frac{\delta Y}{\delta \phi}\right)^2\right]$$

$$F = \left[\frac{\delta X}{\delta \phi} \cdot \frac{\delta X}{\delta \lambda} + \frac{\delta Y}{\delta \phi} \cdot \frac{\delta Y}{\delta \lambda}\right]$$

$$G = \left[\left(\frac{\delta X}{\delta \lambda}\right)^2 + \left(\frac{\delta Y}{\delta \lambda}\right)^2\right] \quad ,$$

such that we now have definition of the length of the curve on the projected surface,

$$ds_{\text{proj}}^2 = E \cdot d\phi^2 + 2 \cdot F \cdot d\phi \cdot d\lambda + G \cdot d\lambda^2 \quad ,$$

giving us a way to to calculate ds^2 on the projected surface. Further, given knowledge of the datum surface (the earth), we can calculate ds_{datum} on the datum surface,

$$ds^2_{\text{datum}} = (R \cdot d\phi)^2 + (R \cdot \cos\phi \cdot d\lambda)^2 \quad , \quad (**)$$

which we can use with the above differential distances to determine the differences in lengths resulting from the projection process,

$$m^2 = \frac{ds^2_{\text{proj}}}{ds^2_{\text{datum}}} \quad .$$

This provides a very general way to examine and characterize map projections. Note that in the case of an ellipsoid, $E = \rho^2, F = 0, G = v^2 \cos^2 \phi$, where v is the radius of curvature in the prime vertical, and ρ is the radius of curvature in the meridian, defined as

$$\rho = \frac{a \cdot (1 - e^2)}{(1 - e^2 \cdot \sin^2 \phi)^{3/2}} \quad ,$$

and for a spherical earth, $E = R^2$, $F = 0$, $G = R^2 \cos^2 \phi$. On the ellipsoid datum, we use v and ρ as the radii instead of R, so (**) above becomes

$$ds^2_{\text{datum}} = (\rho \cdot d\phi)^2 + (v \cdot \cos\phi \cdot d\lambda)^2 \quad .$$

The above discussion provides just a flavour of the theory underlying the development of map projections but provides a window into the way we classify and understand map projections. A key concept in the above functions is the difference in measurements on the datum surface and the projected surface. The scale factor of a map projection is a numeric value—the ratio of the distance on the projected surface to the distance on the datum surface. Thus, when both distances are equivalent ($ds_{\text{proj}} = ds_{\text{datum}}$), the scale factor is 1 and there is no distortion. *There is always some scale distortion resulting from the projection*, and understanding the distribution of the scale distortion over the map is a way to decide which map projection is appropriate in a given context. For example, when

$$\frac{E_{\text{proj}}}{E_{\text{datum}}} = \frac{G_{\text{proj}}}{G_{\text{datum}}} \qquad \text{and} \qquad F_{\text{proj}} = 0 \quad ,$$

the scale factor m does not vary by direction, meaning the projection maintains the shape of mapped features (called a conformal projection). The scale factor m along the parallels is usually denoted k, while the scale factor along the meridians is denoted h.

A common way to group projections is according to the property of the map that is preserved upon projection. Map projection terminology can quickly become confusing, as there are many terms used interchangeably due to the many fields that use map projections (GIS, remote sensing, surveying, engineering, cartography, geodesy).

Here, we refer to a *map projection method* as a type of map projection, a *map projection* as a projection method plus a set of parameter values, and a *spatial reference system* as a map projection coupled with a datum.

3.5.1 Deformation Characteristics

Projections that are characterized as conformal preserve angles and in some cases the shape of features upon projection because the scale distortion is the same in all directions (Figure 3.10). This property means that angles measured on the datum surface are

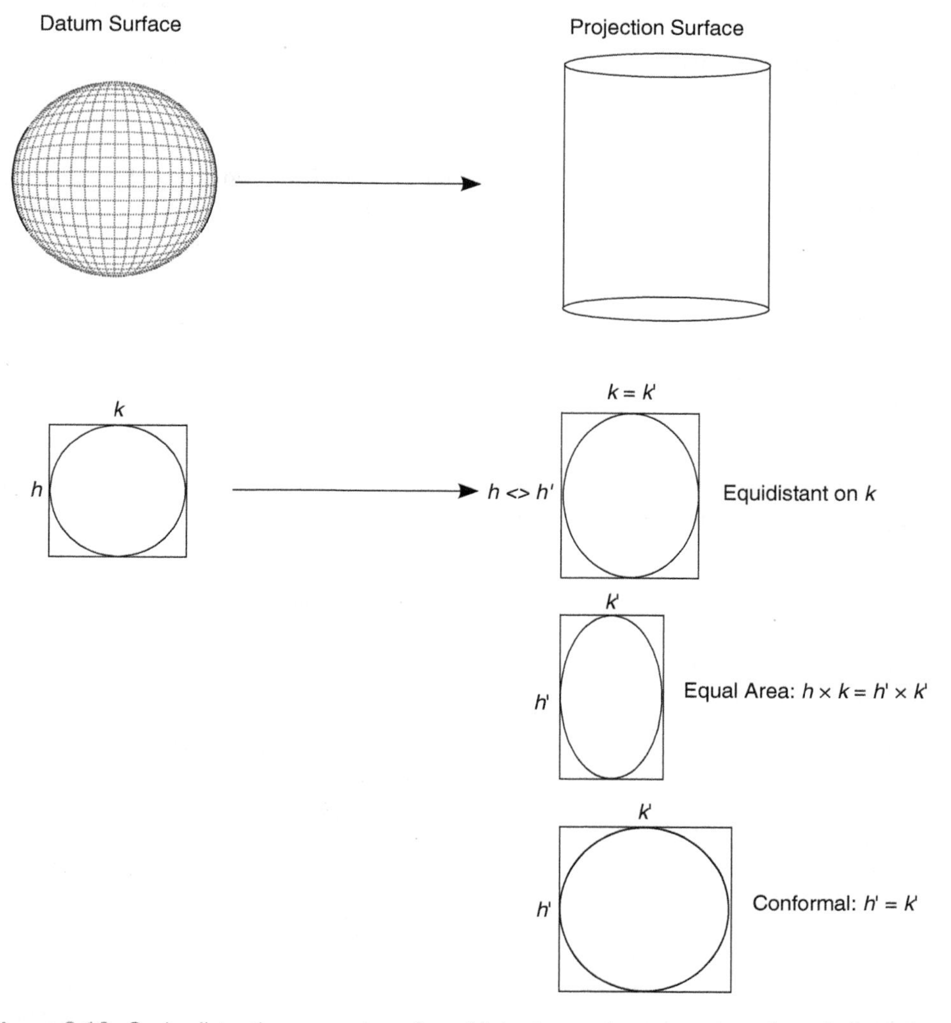

Figure 3.10 Scale distortion examples of equidistant, equal area, and conformal distortions upon projection.

maintained on the projected surface and thus have important uses in navigation maps. However, the size of areas can be severely distorted on these projections, especially for large areas. Scale is not constant across the map, and while distortion is isotropic (i.e., invariant under rotation), it is not stationary (i.e., invariant under translation). Therefore, the shapes of large areas like continents can be distorted on conformal map projections. Equal-area projections preserve area of mapped features. If we consider scale factors in the direction of the meridians and the direction of the parallels, where the product of these two scale factors is unity, we have preserved the overall area in mapped features in the projected surface (though the shape of mapped areas may be very distorted). Where distances in one direction are preserved, we have an equidistant projection, although distance can only be preserved from one (or sometimes two) points on the map to all other locations.

Deformation characteristics can be visually assessed by plotting a circle of unit size (on the earth) and plotting it onto the projected surface. On the projected surface, the unit circle will form an ellipse with semimajor and semiminor axes defined by the principal direction of maximum and minimum scale distortion. Thus the size, shape, and orientation of the ellipse relative to the unit circle provide an indicator of the amount and characteristics of distortion in a given map projection. Typically, this method is used only on global map projections, and in practice many circles are placed across the map and projected to visualize how scale distortion varies across the map. The method is due to French mathematician Nicolas Tissot and is called Tissot's Indicatrix (Figure 3.11).

3.5.2 Projected Surface Characteristics

A visual way of thinking about map projection methods is as a projected surface (i.e., the map sheet) being draped over a three-dimensional earth, the datum surface. The resulting surface is sometimes called a developable surface (see Figure 3.12). The point of contact between the datum surface and the projected surface is where distortions will be minimized. The intersection of a simple plane and the surface of a sphere will occur at a single point. This type of map projection method is called azimuthal. The point of contact between the plane and the sphere can vary depending on the area being mapped. For example, maps of polar areas may be azimuthal projections centred on the pole. An alternate developable surface for a map projection is formed by wrapping the projected surface around the datum surface to form a cylinder—a cylindrical projection. Finally, folding the mapped surface into a cone and placing this on the datum surface will form a conical projection. Each of these projection surfaces has differences in how the projection surface contacts the datum surface; that is, areas with minimal scale distortion.

The orientation of the projected surface with respect to the datum surface can vary as well. Three orientations are commonly employed. In a *normal projection*, the surface is parallel to the axis of the earth's rotation. If the orientation of the projected surface is perpendicular to the datum surface (i.e., earth's axis), this is called a *transverse projection*. Finally, some projections designed for specific regions of the world are oriented neither parallel nor perpendicular to the earth, and are termed *oblique projections*.

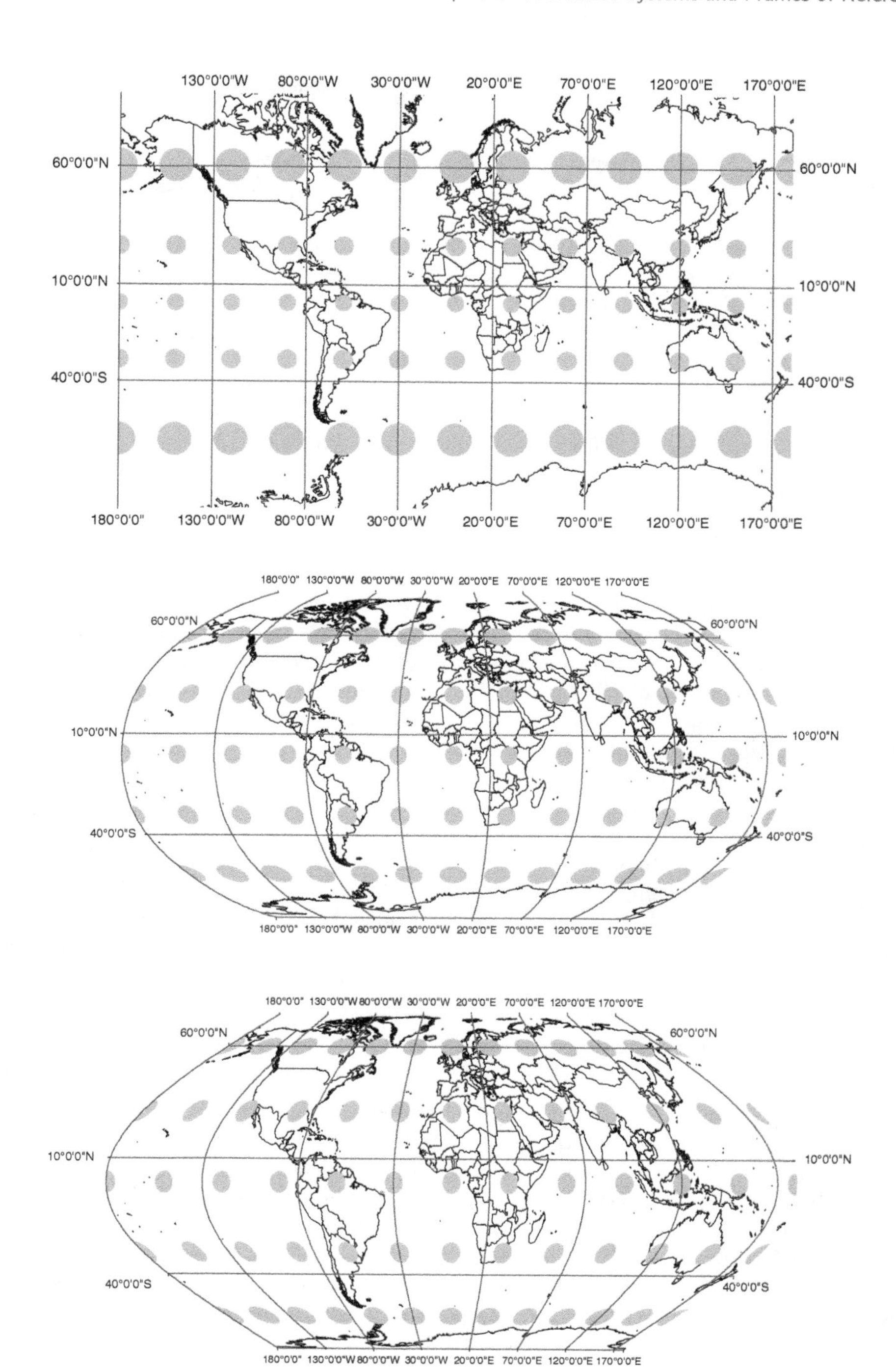

Figure 3.11 Tissot's indicatrix over world map projections.

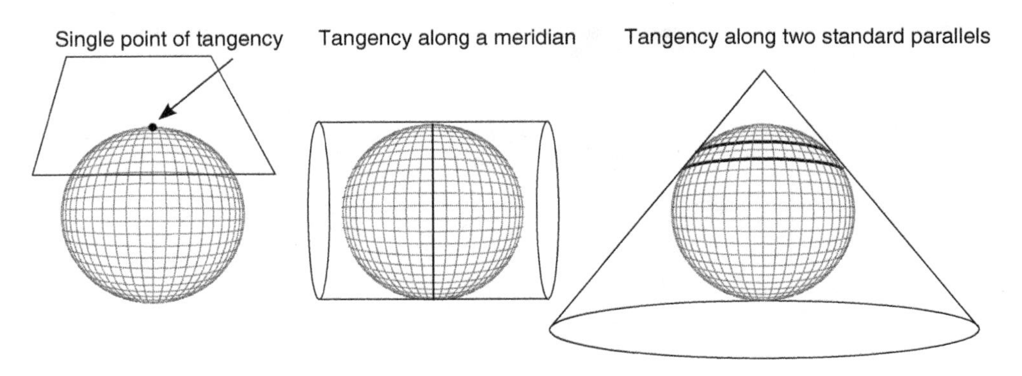

Figure 3.12 Developable surfaces used in common map projections.

3.5.3 Map Projection Methods

Azimuthal

Azimuthal map projection methods employ a planar developable surface for mapping. The origin of the projection is where the projected surface intersects the datum surface, usually at a single point (where the planar projected surface intersects the datum surface). If the tangent is along a circle the projection is said to be *secant*. As such, azimuthal projections are usually used for mapping polar regions. Meridians radiate outward from the plane origin, and parallels are concentric circles. Distortions increase greatly with distance from the origin (i.e., scale distortion increases directly with distance from the origin). The perspective of azimuthal map projections can vary as well, examples being the gnomic projection, where the perspective is the centre of the earth, and the stereographic, where the perspective is the point opposite the point of tangency (common for polar mapping).

A conformal azimuthal projection centred on polar areas is used as the standard mapping projection (instead of UTM) for areas with latitudes more extreme than 80°. Figures 3.13, 3.14, and 3.15 illustrate the Conformal Azimuthal (stereographic) projection and two other varieties of azimuthal projection.

Conic

Conic projections are formed by wrapping the developable surface into a cone and orienting it to a specific section of the earth. They are characterized by equally spaced meridians that form an orthogonal network with parallels that are concentric circles around the centre of the cone (Figures 3.16–3.18). Due to the circular nature of the conic projection, Cartesian coordinates are defined with respect to polar coordinates on the datum surface, with r the distance from the centre of the projection and θ the angle with respect to the central meridian. Thus, the relationship between Cartesian and polar coordinates is $X = r \cdot \sin\theta$ and $Y = r_0 - r \cdot \cos\theta$, where

$$r = \frac{R}{\sin\theta_0} \cdot \sqrt{1 + \sin^2\theta_0 - 2 \cdot \sin\theta \cdot \sin\theta_0} \quad ,$$

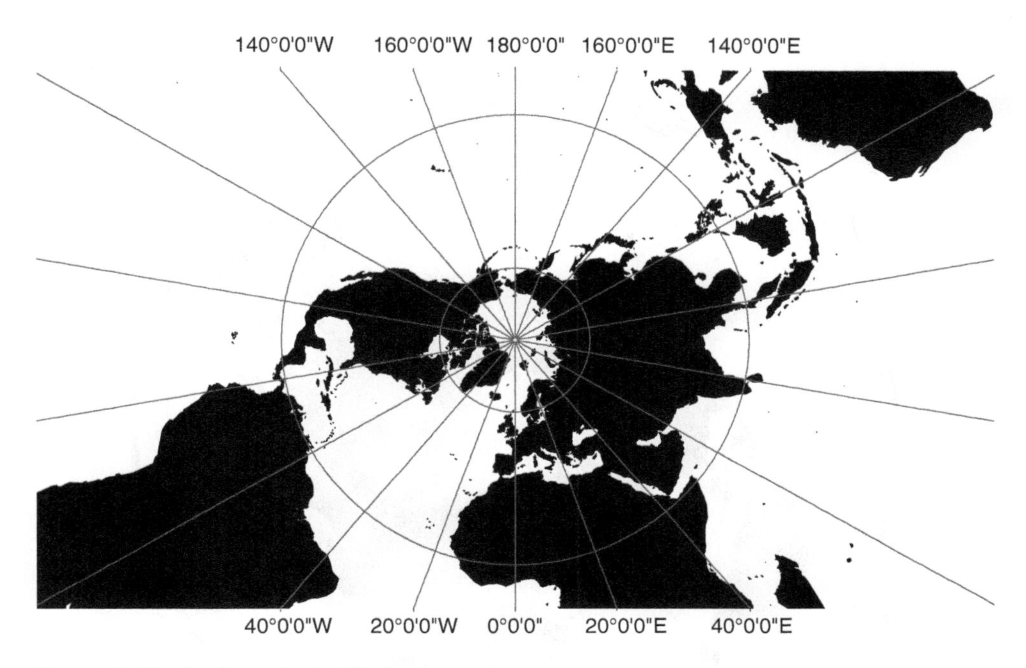

Figure 3.13 Conformal azimuthal polar projection.

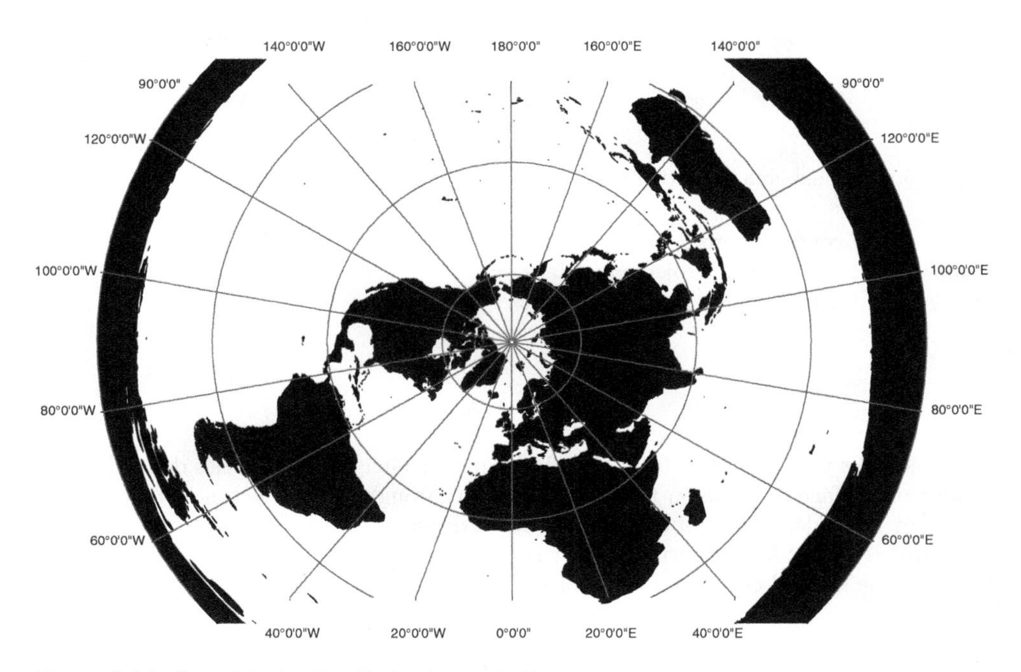

Figure 3.14 Equidistant azimuthal polar projection.

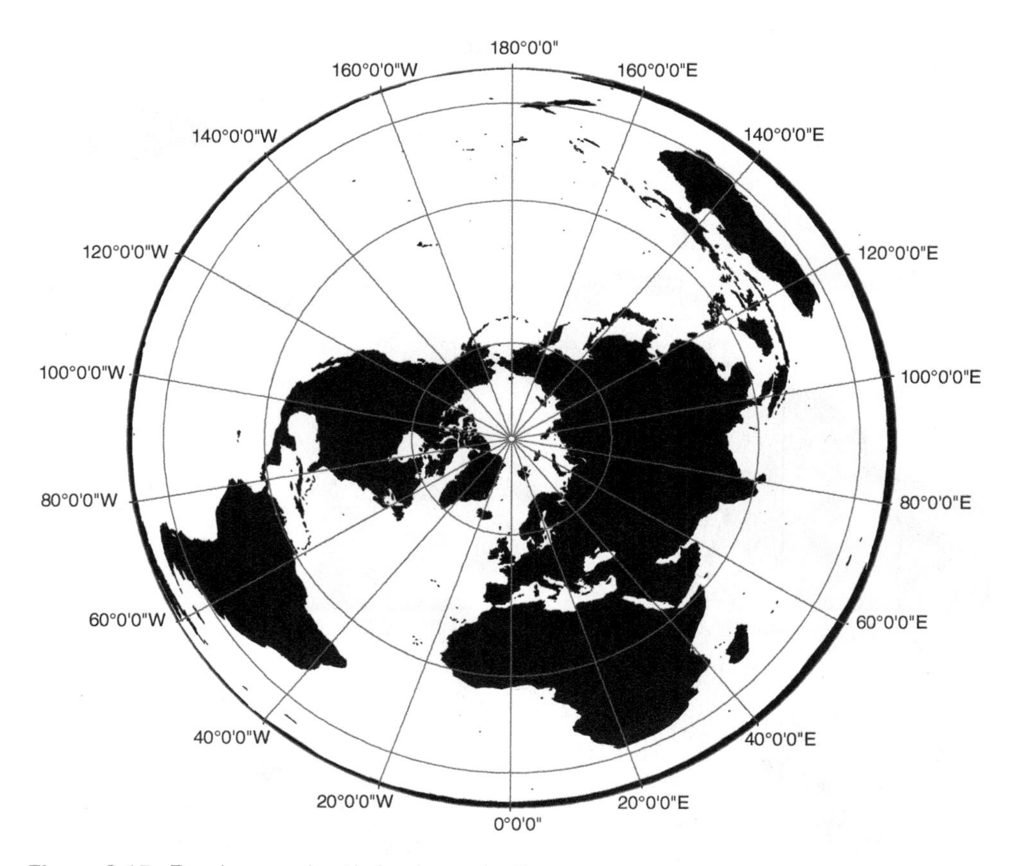

Figure 3.15 Equal-area azimuthal polar projection.

and $r_0 = R \cdot \cot \theta_0$, where θ_0 is the latitude of the origin. The Gaussian fundamental quantities for the conical projected surface are

$$E = \sin^2 \theta + \cos^2 \theta$$
$$F = r \cdot \sin \theta \cdot \cos \theta - r \cdot \sin \theta \cdot \cos \theta$$
$$G = r^2 \cdot \cos^2 \theta + r^2 \cdot \sin^2 \theta \quad ,$$

which equates to $E = 1$, $F = 0$, $G = r^2$. Relating the Gaussian quantities defined on the projection surface to those on the datum surface through r and θ is eased if we assume that they are independent of each other (i.e., r depends on ϕ only, and θ depends on λ only, which is true for all normal aspect conical projections). We can then determine r and θ as the general conformal conic projection equations:

$$r = C \cdot \left[\tan\left(\frac{\pi}{4} + \frac{\phi}{2}\right)\right]^{-n}$$
$$\theta = n \cdot (\lambda - \lambda_0) \quad ,$$

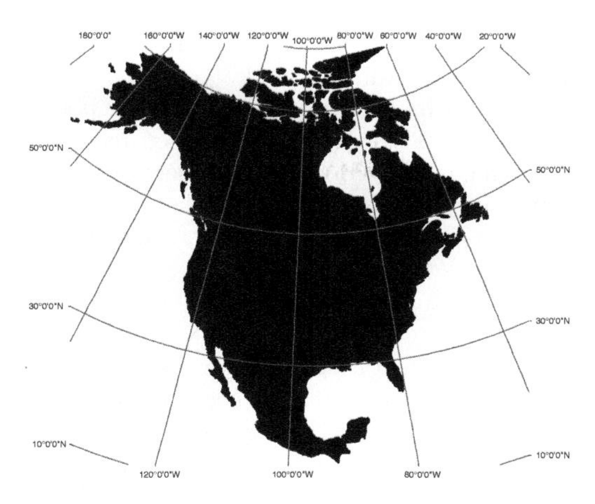

Figure 3.16 Albers equal-area conic projection of North America.

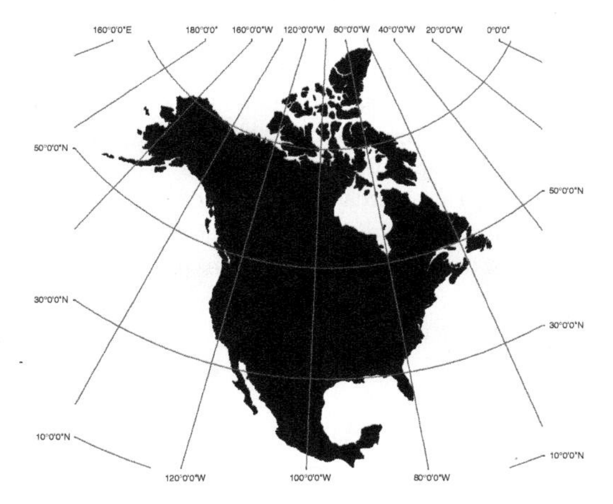

Figure 3.17 Lambert conformal conic projection of North America.

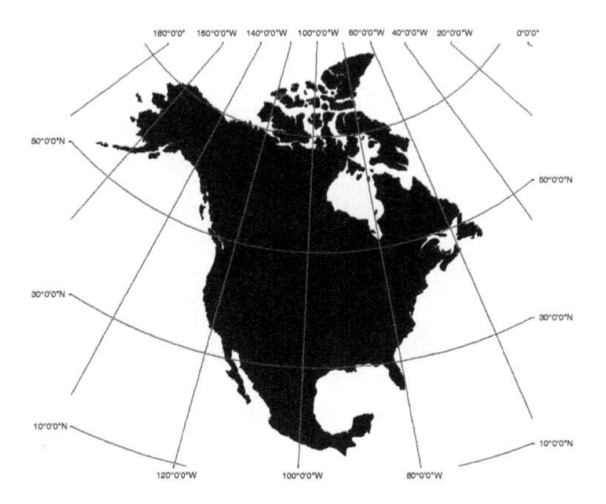

Figure 3.18 Conic equidistant projection of North America.

where n and C are geometric constants determined by the standard parallels used for the projection, and λ_0 is the longitude of the central meridian.

Conic projections are typically used for mapping mid-latitude areas near the poles, such as Canadian provinces and northern US states. They are parameterized by standard parallels, which are parallels along which scale is true. Most conic projections have two standard parallels (i.e., secant forms), and distortion increases with distance away from these latitudes. Multiple cones can be used in the construction of a map projection termed a *polyconic projection.*

Cylindrical

Cylindrical projection methods are characterized by an orthogonal network of meridians and parallels, where meridians are equally spaced straight lines and parallels are unequally spaced straight lines (Figures 3.19 and 3.20). The extent of cylindrical projections is rectangular, and Cartesian coordinates are defined with respect to ϕ and λ.

Mercator, Transverse Mercator, and Universal Transverse Mercator

The Mercator projection is a normal conformal cylindrical projection. A straight line drawn between two points represents a line of constant bearing (called a rhumb line or loxodrome), making this a popular projection for navigational charts. The projected surface is tangent to the datum surface only at the equator. Due to severe distortion in the size of areas with distance from the equator, the Mercator projection should not be used for mapping global phenomena. However, the Mercator projection is widely used for national mapping in equatorial regions, especially those with an east-west orientation.

The Mercator projection equations are typically written

$$x = R \cdot (\lambda - \lambda_0)$$

$$y = R \cdot \ln \left[\tan\left(\frac{\pi}{4} + \frac{\phi}{2}\right) \cdot \left(\frac{1 + e \cdot \sin\phi}{1 + e \cdot \sin\phi}\right)^{\frac{e}{2}} \right] \quad .$$

These equations derive from the fact that the scale factor is constrained to unity along the equator (recall Figure 3.10), such that we can write

$$k_0 = \frac{dX}{v_0 \cdot \cos\phi_0 d\lambda} = 1 \quad ,$$

and since $\cos\phi_0 = 1$ and $v_0 = R$ at the equator, we then have a relation between Cartesian and ellipsoidal coordinates in east-west:

$$dX = R \cdot d\lambda \quad .$$

Recalling that for an ellipsoid,

$$E_{\text{datum}} = \rho^2 \quad , \qquad G_{\text{datum}} = v^2 \cos^2\phi \quad ,$$

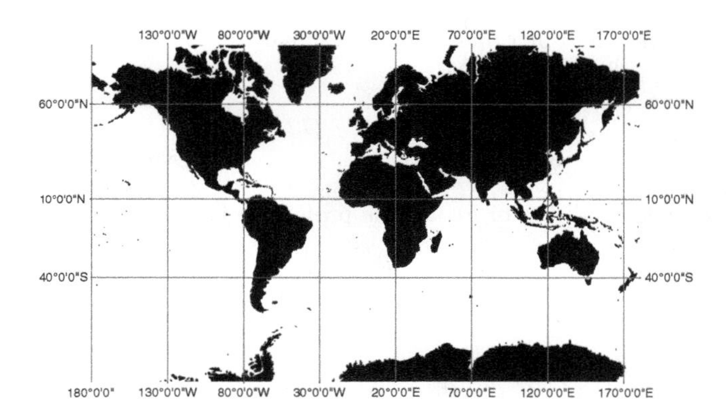

Figure 3.19 Transverse Mercator.

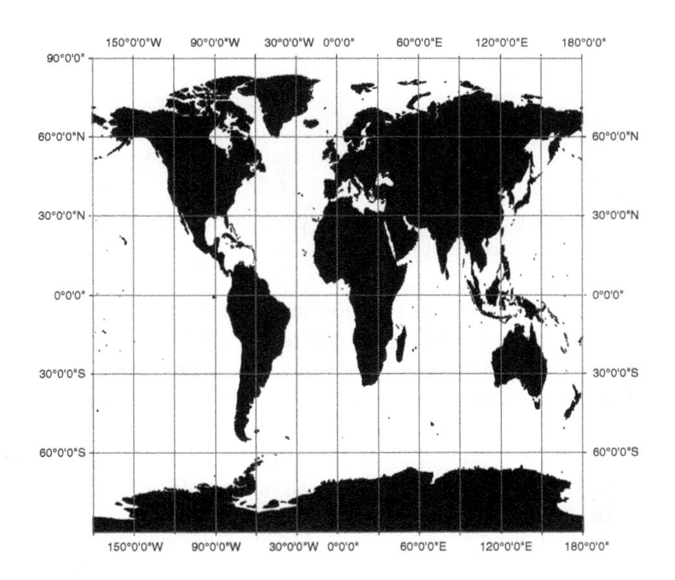

Figure 3.20 Equirectangular-cylindrical equidistant projection.

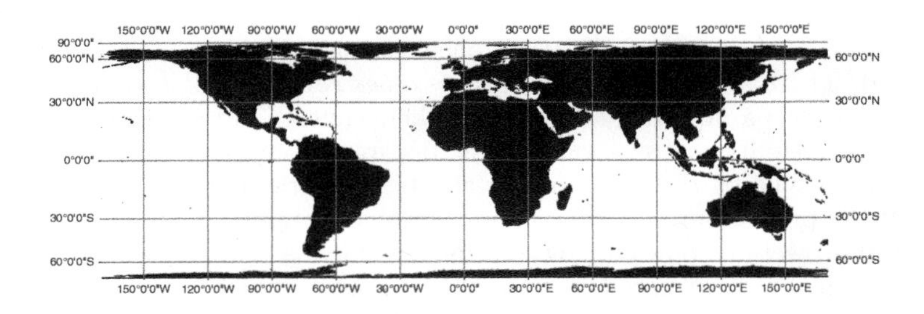

Figure 3.21 Cylindrical equal-area: Gall-Peters.

and since the parallels and meridians are orthogonal,

$$E_{\text{proj}} = \left(\frac{\delta Y}{\delta\phi}\right)^2 \quad , \qquad G_{\text{proj}} = \left(\frac{\delta X}{\delta\lambda}\right)^2 \quad ,$$

and due to the scale condition of $h = k$ (the projection is conformal),

$$\frac{\sqrt{E_{\text{proj}}}}{\sqrt{E_{\text{datum}}}} = \frac{\sqrt{G_{\text{proj}}}}{\sqrt{G_{\text{datum}}}} \quad ,$$

we can substitute for E and G to get

$$\frac{dY}{d\phi} = \frac{\rho}{\upsilon \cdot \cos\phi} \cdot \frac{dX}{d\lambda} \quad .$$

Substituting $R \cdot d\lambda$ for dX from above, we arrive at

$$dY = R \cdot \frac{\rho}{\upsilon \cdot \cos\phi} d\phi \quad ,$$

which through integration gives the Mercator projection equation for y [32]. The inversion equations for the Mercator projection are simple for X, but for Y require numerical approximations.

Another form of the Mercator projection is possible by rotating the projected surface with respect to the datum surface, which gives a *transverse* conformal cylindrical projection. The Transverse Mercator (TM) projection (also known as the Gauss-Kruger projection) therefore has a standard meridian rather than a parallel, called the Central Meridian. The TM projection is used widely for map projections around the world for equatorial and mid-latitude areas with north-south orientations. For example, as noted in Table 3.1, it is used as the basis for the national mapping grid in Sri Lanka, which uses a central meridian of 80° 46′ 18.16710″ E.

A major issue with the Mercator projections is that the distortions increase with distance from the standard parallel (Mercator) and central meridian (TM). As such, they are of limited utility for general purpose mapping over large areas. A special case of the TM projection addresses these issues by splitting the earth into a series of narrow map sheets called zones, each with its own central meridian. This projection, called the Universal Transverse Mercator (UTM), is one of the most widely used map projections in the world. There are a total of 60 narrow zones each 6° longitude in width. Each zone is divided into a northern and southern hemisphere, and the central meridian of each zone lies along its centre. UTM coordinates are denoted as eastings and northings and are accompanied by the zone number and hemispheric indicator (N or S).

Cylindrical projections are widely used for global mapping, many of which are pseudocylindrical projections that employ multiple cylinders at various latitudes, each with its own tangent standard parallel. Examples of pseudocylindrical projections include the sinusoidal, Mollweide, and Kavrayskiy's Fifth Projection.

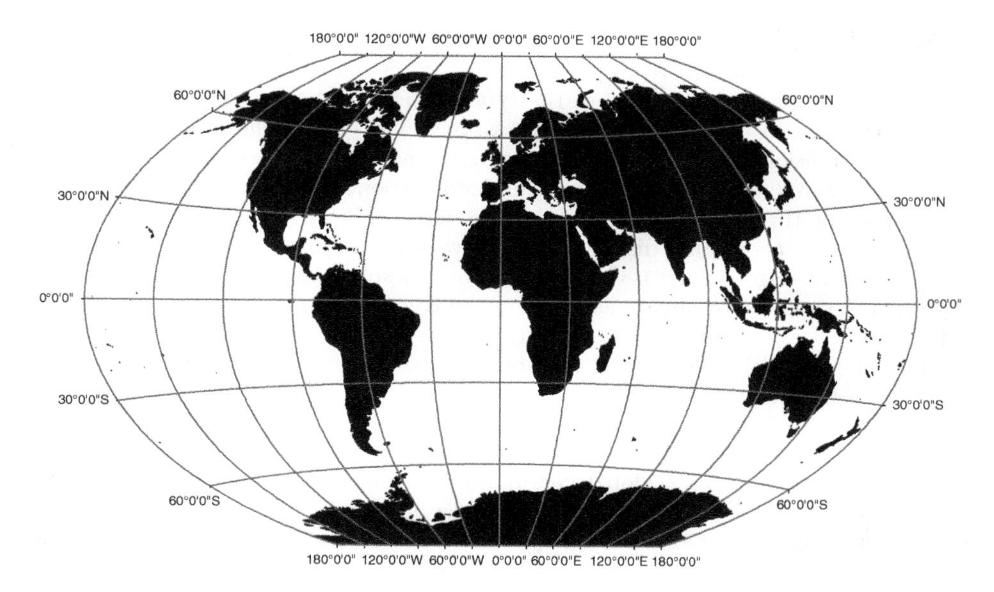

Figure 3.22 Winkel tripel-hybrid map projection attempting to minimize all three types of distortion.

Other projection methods

Not all projections preserve either distance, shape, or area. Many map projections are hybrids that attempt to optimize the distortion for particular purposes. For mapping world phenomena, hybrid projections are popular, as they strike a balance in distortion properties, particularly for shape and area. One widely used hybrid projection is the Winkel tripel projection, which combines an equirectangular projection and an azimuthal equidistant projection. The Winkel tripel projection is the official map projection used in National Geographic world maps.

A 3.6 Scale and Frames of Reference

An implicit assumption of the above discussion of map projections is that the viewer is working with geographic data at a fixed scale. This assumption is a carry-over from classical cartography and is counter to the reality of how we work with geographic data in GIS. One of the principal functions of exploring geographic phenomena in GIS is the ability to zoom in and out to varying levels of detail, often with multiple levels of generalization tied to fixed-scale thresholds. Using computer interfaces, we can have multiple frames of reference in which to view geographic data, and this can affect which projection should be used to represent the data. For example, a magnifier window is one way to implement variable-scale maps in GIS, where the user scans the map with the magnifier tool and the locations under the magnifier tool are represented in large scale with fine geographic detail, whereas locations on the periphery are represented in small scale with more generalized geographic detail.

What is the correct projection to use with such a map? How should the transition between scales be handled? What are the deformation characteristics of a variable resolution map? These are examples of the types of questions that new computer-based geographic representations offered by GIS brings, and of challenges for the development and extension of existing GIScience theory and methodology. Digital globes are another common way of viewing geographic information, where data can be visualized as projected or three-dimensional coordinates. Google Earth, for example, employs a projection for its coordinate display, and geographic coordinates for its computations (using spherical trigonometry). Google Maps uses a spherical variant of the Mercator projection that preserves angles, as its main use is for routing along street networks at large scales.

The increasingly diverse ways of capturing geographic information are changing the types of reference frames relevant for GIS data. A spatial frame of reference is the perspective and origin within which a set of geographic information measurements are embedded. The approaches described in this chapter for locating geographic information have been in reference to a global coordinate system, usually with its origin as the centre of mass of the earth. However, consider a sensor network of Wi-Fi sensors that detect object locations within a specified field of view. Each sensor has its own local coordinate system (x, y, z). Objects detected in a sensor's local coordinate system will have to be transformed to a global (geodetic) coordinate system in order to be mapped and integrated with other GIS data. The tools of control points and polynomial transformations can be used here as well to integrate data. However, if the location sensors are mobile (for example, on people), this becomes much more challenging. Real-time kinematic mapping methods can be used to differentially correct transformations to a global coordinate system.

Another example is geospatially tagged photos and videos. Whereas in traditional GIS, the spatial frame of reference is almost always top-down, we are now able to collect geographic information from a variety of perspectives. A geotagged photograph contains rich information about the landscape, built environment, natural features, and social and cultural processes, yet this information is typically excluded from a top-down perspective afforded by a map. Crandall and Snavely [29] describe techniques for combining geographic coordinates and image content to reconstruct 3D representations of landmarks that had been photographed and uploaded to the photo-sharing website Flickr. This process requires estimating the location of the viewer (through geographic coordinates) and matching structural features in pairs of images from that location. With enough photos, they demonstrate the 3D reconstruction of landmarks from 2D representations taken from different perspectives. Google Street View is another example of a non-traditional frame of reference for geographic information. If we consider these spatial reference systems for geographic information, interesting questions emerge pertaining to how to characterize spatial locations and relationships within these forms of geographic information. While these examples describe 2D representations that can be adjusted to provide alternate perspectives to geographic information, spatially referenced video adds the dimension of time variability, which further complicates things.

Spatial reference system (i.e., datums + projection + parameter values) information is required metadata for doing analysis with geographic information. The way metadata are stored and represented varies by data type and file format, and among software systems. However, standards have recently emerged that facilitate the automated interchange—including datum transformation and reprojection—of geographic data. One standard is the PROJ.4 strings, which describe spatial reference system definitions, transformations, and reprojections. PROJ.4 metadata are specified as text strings that accompany geographic data objects so as to identify the spatial reference system of their coordinates. Table 3.4 outlines commonly used PROJ.4 parameters. Online repositories of spatial reference system metadata such as spatialreference.org allow you to convert between formats.

Table 3.4 PROJ.4 specification of spatial reference system metadata.

Syntax Code	Description
+a	Semimajor radius of the ellipsoid axis
+alpha	Used with Oblique Mercator and possibly a few others
+axis	Axis orientation (new in 4.8.0)
+b	Semiminor radius of the ellipsoid axis
+datum	Datum name (see proj -ld)
+ellps	Ellipsoid name (see proj -le)
+k	Scaling factor (old name)
+k_0	Scaling factor (new name)
+lat_0	Latitude of origin
+lat_1	Latitude of first standard parallel
+lat_2	Latitude of second standard parallel
+lat_ts	Latitude of true scale
+lon_0	Central meridian
+lonc	Longitude used with Oblique Mercator and possibly a few others
+lon_wrap	Centre longitude to use for wrapping (see below)
+nadgrids	Filename of NTv2 grid file to use for datum transforms (see below)
+no_defs	Don't use the /usr/share/proj/proj_def.dat defaults file
+over	Allow longitude output outside −180 to 180 range, disables wrapping (see below)
+pm	Alternate prime meridian (typically a city name, see below)
+proj	Projection name (see proj -l)
+south	Denotes southern hemisphere UTM zone
+to_meter	Multiplier to convert map units to 1.0 m
+towgs84	3 or 7 term datum transform parameters (see below)
+units	meters, US survey feet, etc.
+vto_meter	vertical conversion to meters
+vunits	vertical units
+x_0	False easting
+y_0	False northing
+zone	UTM zone

Problems

1. Using the Grand River Watersheds dataset, reproject the data to a projection that uses a NAD27 datum. Create a map that shows the difference between the two datasets (before and after projection/transformation).
2. Use a GPS or an online map to identify the latitude and longitude coordinates for three locations around your home or school. Using formulas in the chapter, convert these coordinates to geocentric coordinates. Measure and report the Euclidean distances between each pair of coordinates.
3. Select five cities in Canada and look up their transformation parameters in the NTv2 grid transformation. Comment on why you think the values vary as they do.
4. How are parameters for datum transformations derived?
5. What are the advantages and disadvantages to increasing worldwide use of the global datum WGS84 in GIS data?

Chapter 4

Geographic Data Models

4.1 Introduction

In this chapter we review the various spatial data models that have been used for geographic data. We assume a basic familiarity with the ideas the reader might have garnered from an introductory level GIS course or hands-on experience with current software, so we come at the ideas from a slightly more formal view point than might be found in introductory texts. This should provide the reader not only enough of a theoretical foundation to be able to start to think critically about some issues at the base of present-day spatial data handling but also a way into the current literature. In the first two sections of the chapter we provide for each general class of spatial data model, a basic introduction, then focus on the geometry, attributes, topology, assumptions, and geometric processing in separate subsections. We briefly highlight the topic of transforming between representations in the next section. Next, we revisit the standard vector spatial data models, providing a historical overview of their development while standardizing their formal description. This discussion leads us to new ideas about future spatial data models based on the ideas of algebraic or combinatorial notions of space via oriented matroids.

4.2 Raster

4.2.1 Basic

A raster spatial data model is a regular tessellation of the plane consisting of cells that are homogeneous in size and orientation, usually a rectangular grid (see Figure 4.1). Each cell has an associated attribute value that need not be unique.

4.2.2 Geometry

The geometric, and in a georeferenced dataset geographic, origin of the raster cell is often taken as the top left corner of the cell (see Figure 4.2). The coordinates of the origin are given as a tuple of values: x, y, $[z]$, $[t]$. If the z value exists as an elevation, the raster can be considered a digital elevation model (DEM). The t value refers to a time dimension and is not widely implemented, although there are notable exceptions such

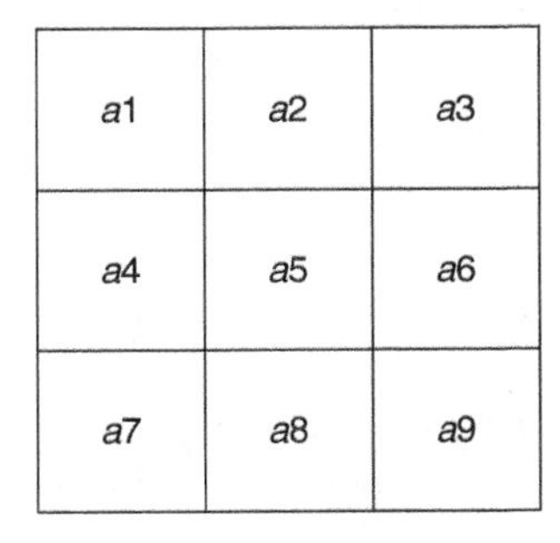

Figure 4.1 Raster geometry: raster grid.

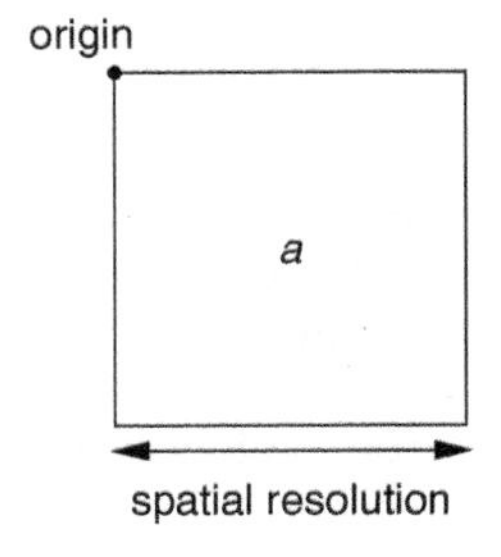

Figure 4.2 Raster geometry: raster cell.

as the netCDF data format, which uses time along with two spatial dimensions. The size of the cell is given as the spatial resolution of the raster.

Point data may be represented by a single raster cell. Line data may be represented by a contiguous chain of cells. Areal data may be represented by a cluster of contiguous cells. However, there is no explicit formal storage of these geometric primitives nor any relationships between them (cf. vector spatial data models described below).

4.2.3 Attributes

As mentioned above, each raster cell has an associated attribute value, a description of the content of the raster cell. The attribute is generally assumed to be homogeneous throughout the cell, although in certain contexts such as remote sensing, explicit recognition of heterogeneity (e.g., a mixed pixel) is considered. Several data types are possible for the attribute, although they must be the same data type across each raster dataset:

$$a_1, a_2, \ldots, a_n \in \begin{cases} binary \text{ or } , \\ number \text{ or } , \\ text \ . \end{cases} \tag{4.1}$$

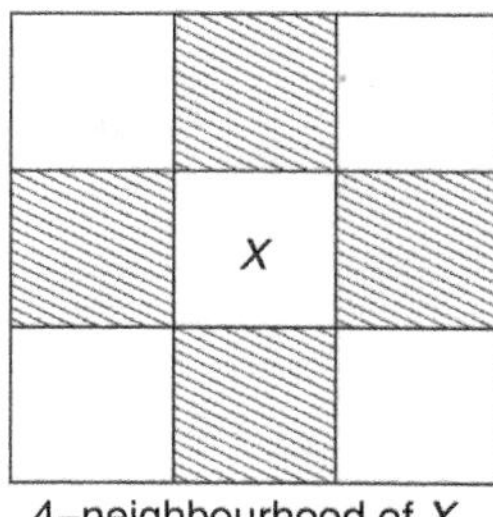

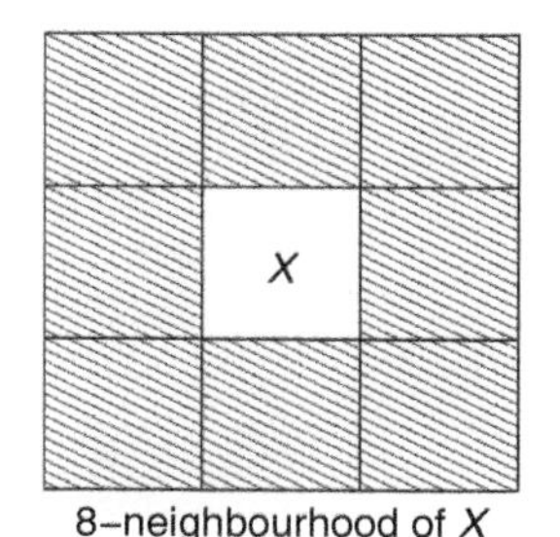

Figure 4.3 Raster neighbourhoods.

In a classified raster there are usually large blocks of raster cells with the same attribute data. This can be considered an inefficient way to store the spatial dataset. Various encodings of raster data at the file or data structure level seek to reduce this inefficiency. Run-length encoding file compression and quad-tree data structures (see Sections 2.2.7 and 5.6.3, respectively) are two such strategies.

4.2.4 Topology

The topological relationships possible on a raster are limited by the geometry of the tessellation. Most often the raster cells are square or rectangular, and the adjacencies considered are usually a 4-neighbourhood (rook's case) or an 8-neighbourhood (queen's case) (see Figure 4.3). Diagrammatically, the topology of a raster dataset may be represented as

$$\mathcal{R} \circlearrowleft \textit{neighbourhood} \quad .$$

4.2.5 Assumptions

The dataset is presumed to be exhaustive (i.e., a regular tessellation in the entire plane). This is why the raster spatial data model is often the representation of choice for continuous or field spatial data. Usually, though, only a subset of the data is captured, and thus dataset edge or boundary issues arise.

4.2.6 Geometric Processing

Map algebra

If we assume that all the input raster datasets are georeferenced to the same coordinate system, we may combine raster datasets to create new datasets. We may also perform operations on single rasters. There are four basic types of raster operations: local, focal, zonal, and global. In local (binary) operations the new values op(ai, bi) depend only on the values at each spatially corresponding cell in the input rasters (ai and bi), and this operation is repeated across all cells of the raster (see Figure 4.4). In focal (unary)

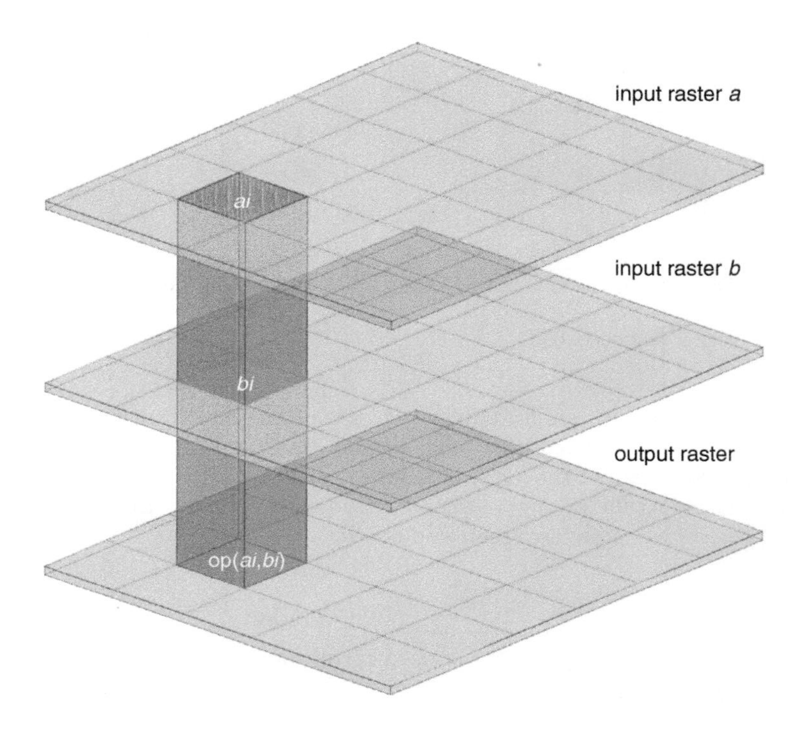

Figure 4.4 Raster operations: local binary.

operations the new values depend on the values in a neighbourhood of each raster cell. For zonal (binary) operations one of the input rasters defines a generalization of the concept of a regular neighbourhood to describe a set of zonal neighbourhoods that are used to aggregate (in some prescribed manner) the other input raster and so create the new raster. Global (unary) operations create new cell values based on some

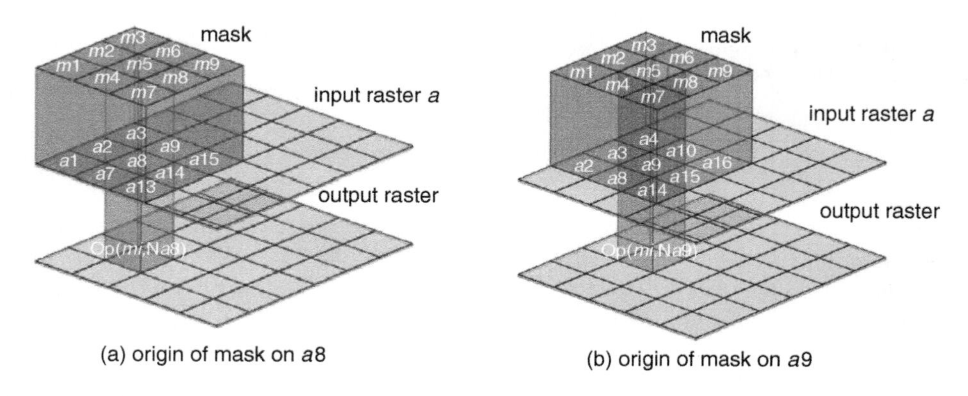

Figure 4.5 Raster operations: raster convolution.

combination of the values of all the cells of the input raster. See Chapter 11 for more details on specific operators.

Convolution—operator masks
In this case a new raster is created by moving a mask of values *mi* over the entire input raster and creating new values by combining the values in the neighbourhood defined by the mask N*ai* and the mask values *mi* (see Figure 4.5). The origin of the mask corresponds to the spatial position of raster cell value to be created, and this origin is usually at the centre of the mask; hence odd dimensions and square shapes prevail in mask designs.

4.3 Vector

4.3.1 Basic

A vector spatial dataset can be considered an irregular tessellation of a Euclidean plane. The partitions are assumed to be homogeneous. The partitioning can be made exhaustive if the concept of a world or universe polygon, PU, is introduced (see Figure 4.6). The usual basic geometric primitives are vertex (or point), edge (or line), and polygon (or area). Informally, datasets made strictly of vertices or edge and vertex data can be thought of as spatially exhaustive if the universe polygon is used to denote no vertex or no vertex and edge areas.

4.3.2 Geometry

In the most basic vector spatial data models, the vertices are georeferenced, one vertex represents a point feature, two vertices define a line feature, and n vertices define a polygon feature. As in the raster case above, x and y are coordinates in the plane, while z and t refer to height and time dimensions, respectively. Measure, m, is a generalized option for storing information to cover cases other than height or time; for example, cumulative length along a road.

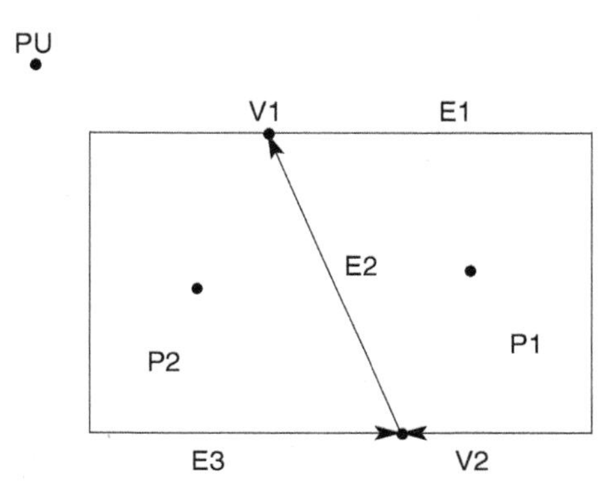

Figure 4.6 Vector geometry.

vertex:	0-dimensional	$x, y, [z], [t], [m]$
edge:	1-dimensional	$x_1, y_1, [z_1], [t_1], [m_1]$
		$x_2, y_2, [z_2], [t_2], [m_2]$
polygon:	2-dimensional	$x_1, y_1, [z_1], [t_1], [m_1]$
		$x_2, y_2, [z_2], [t_2], [m_2]$
		$\ldots$
		$x_n, y_n, [z_n], [t_n], [m_n]$

The following are some of the common additions found in more-advanced vector spatial data models. Shape in edges may be provided by intermediate vertices. Vertices at the end of shaped edges may be distinguished as nodes. Further, a vertex may be added as a pseudo-centroid of the polygonal partitions to act as an unambiguous attachment point for the partitions attributes. This helps in defining a hierarchical topological data scheme, as will be described below and in subsequent sections in this chapter.

4.3.3 Attributes

As with the raster spatial data models described above, characteristics of the partitions are stored as attributes of the geometric primitives. In this case however, there can be a hierarchy of such primitives, as illustrated below. Note that as attributes for the geometrical objects are typically stored in a separate database table it is possible, and in fact typical, to have a combination of data types implemented as columns in the table to characterize each partition in a vector spatial dataset. The topological relationships between the geometric primitives are discussed in the following section.

$$v_1, v_2, \ldots, v_k \in \{binary \cup number \cup text\}$$

$$e_1, e_2, \ldots, e_m \in \{binary \cup number \cup text\}$$

$$p_1, p_2, \ldots, p_n \in \{binary \cup number \cup text\}$$

4.3.4 Topology

In an early attempt to reduce storage requirements for vector spatial data, a nested hierarchy of geometric primitives was devised (see further elaboration of this progression below). A basic scheme is given in the diagram below. Each edge is prescribed by 2 vertices, and each polygon is prescribed by n edges.

$$\begin{array}{c}\text{vertex}\\ \Downarrow 2\\ \text{edge}\\ \Downarrow n\\ \text{polygon}\end{array}$$

As shown in the next diagram, this topologically aware representation meant that the edges $\mathcal{E}$ shared by adjacent polygon partitions $\mathcal{P}$ were stored only once, and each of the neighbouring polygons was reconstructed via its stored list of associated edges. An edge is the boundary of 2 polygons (one of which may be the universe polygon), and the orientation of the edge from its *begin* vertex $\mathcal{V}$ to its *end* vertex induces an orientation of *left* and *right* polygon.

$$\mathcal{P} \underset{right}{\overset{left}{\leftleftarrows}} \mathcal{E} \underset{end}{\overset{begin}{\rightrightarrows}} \mathcal{V} \tag{4.2}$$

4.3.5 Assumptions

The following assumptions generally hold for vector spatial datasets.

1. $\{Vertices\} \in \mathbb{R}^n$: the coordinates of the vertices are assumed to be embedded in a Euclidean plane.
2. (Usually) $\{Polygons\}$ are planar, meaning the polygons are simple (i.e., they have edges that intersect only at vertices).
3. (Historically) assume a world polygon, so that all edges have a left and right polygon topological relationship.

4.3.6 Geometric Processing

One of the key processing functions for vector spatial data is polygon overlay. Usually this consists of two input vector polygon datasets that are combined to create a new output vector polygon dataset. The output dataset carries attribute data derived from the input datasets. The main steps of polygon overlay are listed here, following Healey et al. [61, p. 267].

Polygon overlay

1. Transformation: input polygon datasets are both transformed to the same coordinate system.
2. Geometric intersection: a plane sweep algorithm is used to find intersections of the edges of the input data to create new edges.
3. Polygon linking: new edges are used to generate new polygon topology.
4. Output new polygon and edge records.
5. Database population: generate new attribute records for the split polygons.

Because of its central role in finding the intersections of vector spatial data, we spend a little time now briefly sketching the basic idea of a plane sweep algorithm.

Plane sweep description

All the edges that make up the polygons in each of the input polygon vector datasets are sorted, for example, by their minimum x-coordinate. Then a vertical line, the sweep line, is imagined to move across the spatial extent of the combined edge set. It creates event points at the endpoints of each of the edges and at intersection points of the

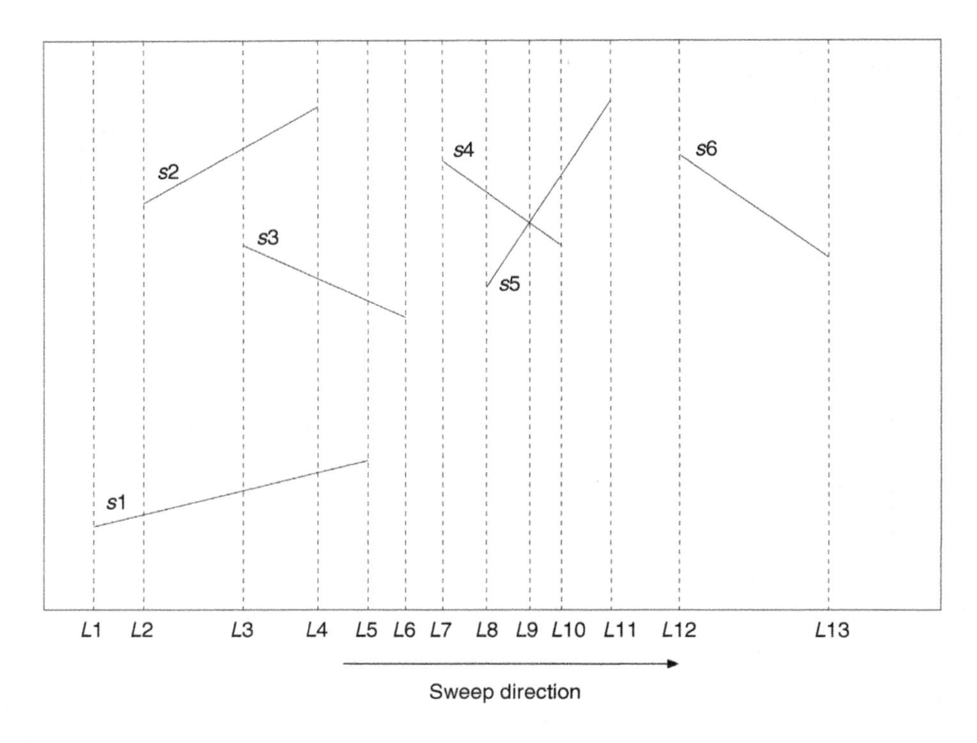

Figure 4.7 Event points in plane sweep algorithm.
Source: Adapted from Healey et al. [61].

edges. The search for intersections between edges is greatly eased in this approach, as only edges that mutually intersect the vertical sweep line need be tested for intersection. See Figure 4.7 for an example, in this case with a set of line segments. Here, $s1$ and $s2$ would be tested for intersection, as would $s1$ and $s3$, $s2$ and $s3$, and $s4$ and $s5$, but $s5$ and $s6$ and all the other possible pairwise combinations would not.

4.4 Dual Vector

A dual vector spatial dataset captures the adjacency (topological) information of a primal vector polygon partition. Historically this topological information was stored explicitly in non-spatial table data. However, this information could be usefully made spatially explicit using the information from these tables (see Figure 4.8). This section outlines the properties of such a spatial representation. More-recent systems create topological information as needed on the fly and thus do not by default store topological information. The dual edges, shown in Figure 4.8 as dashed lines, are usually schematic, but often polygon pseudocentroids are used to locate dual vertices. Usually the dual vertices inherit the primary polygon attributes. In terms of topology there is an isomorphism (one-to-one and onto mapping) between primary and dual edges and

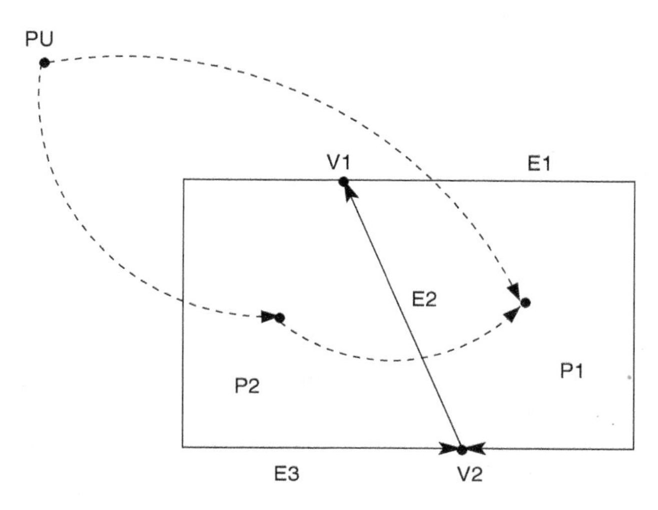

Figure 4.8 Dual vector geometry.

between primary polygons and dual vertices. It is assumed that the primary polygons are planar; that is, simple non-self-intersecting areas.

4.5 Other Models

Other models of spatial data exist but have not been adopted into mainstream GIS practice. These include multi-scalar models such as triangulations of spheres, Molenaar's Multi-Value Vector Maps [80], and category theory based Primal/Dual Multi-Value Vector Maps [98] (this last to be discussed briefly at the end of this chapter). However, the use of object-oriented approaches has gained favour in recent years in many aspects of computer science. Spatial data models adopting part or all of the object-oriented paradigm, along with some other pertinent issues, are addressed below.

4.6 Transforming Between Representations

It is often forgotten that there are options for representing spatial data. In this section we discuss in some detail moving between raster and vector representations. We end the section with some less-obvious transformations that nonetheless might prove useful.

4.6.1 Raster to Vector and Back

Raster datasets might be considered as rather arbitrary partitions of space that are then examined for patterns of raster cells to discover geographic objects. Conversely, a vector dataset might be considered as partitions derived from *a priori* identification of geographic or spatial objects. Figure 4.9 summarizes these ideas. It is important to note, however, that because computers are finite state machines, vector representations, practically, do not live in $\mathbb{R}^2$ but are approximated by finite resolution fields $\mathbb{Q}^2$.

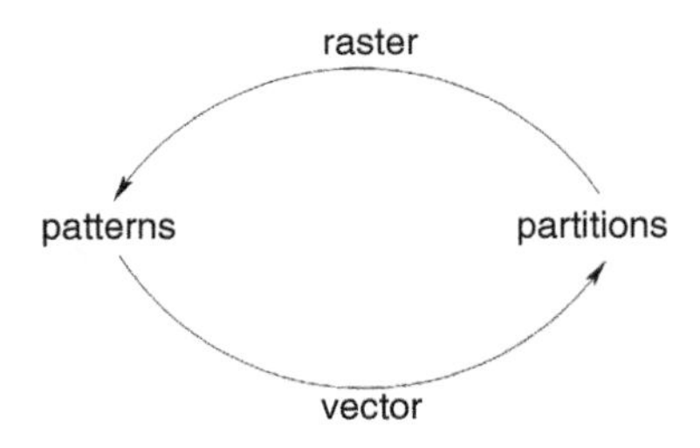

Figure 4.9 A conceptualization of raster and vector representations.

In later chapters of the text we discuss in more detail the problems inherent in both finite, discrete representations and floating point arithmetic approximations to Real numbers. The following subsections describe raster-to-vector and vector-to-raster transformations, respectively.

Raster to vector

To convert a raster image to a vector spatial dataset, perhaps a scan of a paper map (ignoring for a moment the issue of geo-referencing), the following steps can be carried out. Of course manual or semi-automatic digitizing is possible, although the hardware for this is becoming increasingly unavailable. Following this list of steps is an elaboration of some of the details of each step.

1. Smooth to remove random noise (speckle); use, e.g., a median low-pass filter.
2. Edge detect; use, e.g., a high-pass filter to create an edge map, then threshold to a binary image.
3. Thin lines to one pixel width; use, e.g., Zhang-Suen thinning algorithm.
4. Chain code pixels to create a collection of chains of pixels, each representing an arc.
5. Vector reduction to convert chains of pixels into sequences of vectors, which can be generalized, e.g., by using the Douglas-Poiker algorithm.

Median Filter

A low-pass median filter convolves a mask over each raster cell and finds the median value among the values covered by the mask. This value is used to fill the position of the raster at the origin of the mask. See Figure 4.10 below.

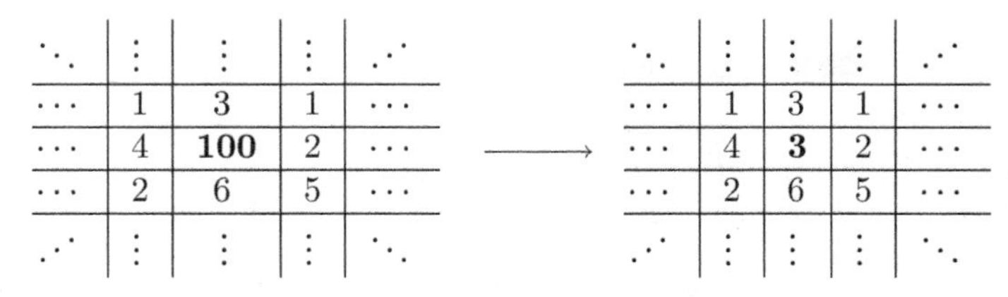

Figure 4.10 A local result of the application of a 3×3 median filter.

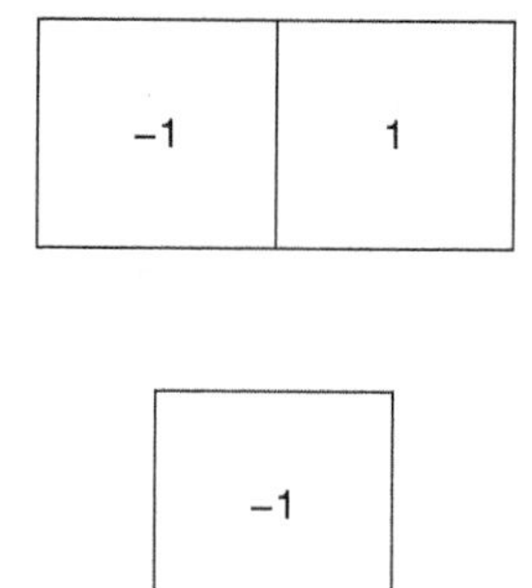

Figure 4.11 Edge detection filter components.

High-pass Filter

In a typical high-pass filter, one or more masks are convolved over the data and possibly combined, or the results otherwise processed, to create a new value at the location of the origin of the masks. An edge in a digital image can be considered as the boundary between two adjacent pixels with significantly different brightnesses (digital number values). In order to understand the form of the convolution masks used in these types of filters, this gradient can be thought of as a slope:

$$slope = \frac{rise}{run} = \frac{f(b) - f(a)}{b - a} \quad . \tag{4.3}$$

For digital images, $b - a = 1$. Thus,

$$slope = f(a + 1) - f(a) \quad . \tag{4.4}$$

Therefore, basic masks for edge detection filters are built from the following forms.

Many variations on this form are seen in popularly employed high-pass filters. Figure 4.12 illustrates using a Sobel 3×3 edge detection filter—showing the result on one raster cell.

Zhang-Suen Thinning Algorithm

The basic goal is to thin, but not so much as to change topological relationships between edges (i.e., disconnecting pieces). Note that the algorithm was developed based mainly on empirical tests and thus is best illustrated by an example (see Figure 4.15), which is given following a description of the method.

Zhang-Suen Erosion Algorithm for Raster Thinning

Input: $m \times n$ binary raster (0 = white, 1 = black)
Output: $m \times n$ thinned binary raster (black is thinned)

Procedure:

Step 1.0 For each cell p in the raster, do Step 1.1
 1.1 If $((2 \leq N(p) \leq 6)$ and $(T(p) = 1)$ and $(p_N \cdot p_S \cdot p_E = 0)$ and $(p_W \cdot p_E \cdot p_S = 0))$, then mark p
Step 2.0 For each cell p in the raster do Step 2.1
 2.1 If (there are any marked points)
 Then (set all marked points to 0)
 Else (halt the process)
Step 3.0 For each cell p in the raster do Step 3.1
 3.1 If $((2 \leq N(p) \leq 6)$ and $(T(p) = 1)$ and $(p_N \cdot p_S \cdot p_W = 0)$ and $(p_W \cdot p_E \cdot p_N = 0))$, then mark p

3×3 Sobel Filter

$$G_x = \begin{array}{|c|c|c|}\hline -1 & 0 & 1 \\ \hline -2 & \mathbf{0} & 2 \\ \hline -1 & 0 & 1 \\ \hline \end{array} : A \qquad G_y = \begin{array}{|c|c|c|}\hline -1 & -2 & -1 \\ \hline 0 & \mathbf{0} & 0 \\ \hline 1 & 2 & 1 \\ \hline \end{array} : A \qquad G = \sqrt{G_x^2 + G_y^2} \quad ,$$

where : indicates the Frobenius product of two equal dimension matrices (in this case the 3×3 mask and the raster cells that fall under the mask). The Frobenius product is the sum of the entries of the Hadamard product of two equal-dimension matrices, and the Hadamard product is simply the element-wise multiplication of the two matrices. The bold typeface indicates the origin of the masks.

Example Calculations

Note: the bold typeface indicates the raster element being processed.

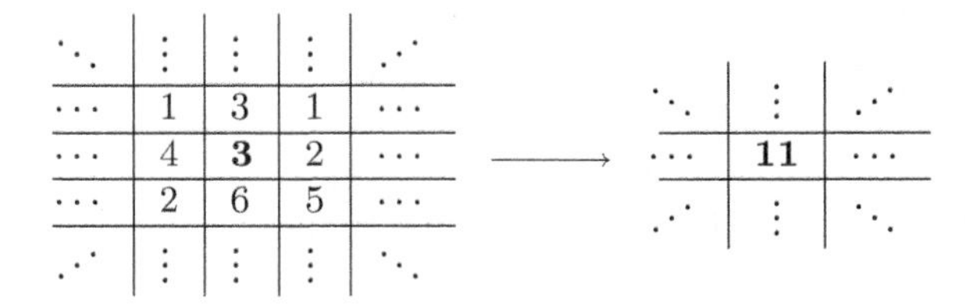

$$G_x = \begin{array}{ccc} -1 \cdot 1 & +0 \cdot 3 & +1 \cdot 1 \\ -2 \cdot 4 & +0 \cdot 3 & +2 \cdot 2 \\ -1 \cdot 2 & +0 \cdot 6 & +1 \cdot 5 \end{array} = -1 \qquad G_y = \begin{array}{ccc} -1 \cdot 1 & -2 \cdot 3 & -1 \cdot 1 \\ +0 \cdot 4 & +0 \cdot 3 & +0 \cdot 2 \\ +1 \cdot 2 & +2 \cdot 6 & +1 \cdot 5 \end{array} = 11$$

$$G = \sqrt{(-1)^2 + 11^2} \doteq 11$$

Figure 4.12 A local result of the application of a 3 × 3 Sobel filter.

Step 4.0 For each cell p in the raster do Step 4.1
4.1 If (there are any marked points)
Then (set all marked points to 0)
Else (halt the process)
Step 5.0 Go to Step 1.0

where, $N(p)$ is the number of black raster cells (1's) in the queen's neighbourhood of the raster cell p (see Figure 4.13), $T(p)$ is the number of 0 to 1 transitions as you travel clockwise in the queen's neighbourhood around p (see Figure 4.14), and p_N, p_S, p_E, p_W are the (0 or 1) raster cell values in the rook's neighbourhood raster cells (labelled as compass directions).

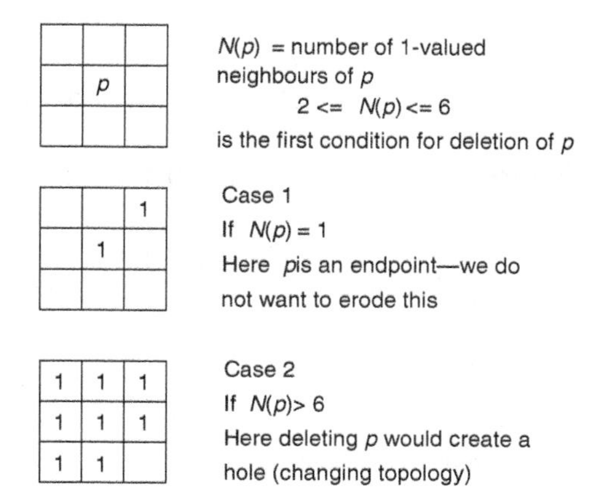

Figure 4.13 Zhang-Suen $N(p)$ operator.

	p	

$T(p)$ = number of 0 to 1 crossings in a sequential cycle of the neighbours of p

$T(p) = 1$

is the second condition for deletion of p

1		
1	1	1
1		1

An example,
Here $T(p) = 2$
and deleting p would create two pieces (changing topology)

Figure 4.14 Zhang-Suen $T(p)$ operator.

Chain Coding

Using a standard compass of discretized direction and distance (one raster cell distance, adjusted for diagonals), the one-raster-cell-width lines from the previous thinning step are coded into character strings. The strings also include at the beginning the (x, y) coordinates of the starting raster cell for each edge. Note that there are techniques to deal with coding arc intersections, but they are not covered here. Further note that chain coding allows for some rudimentary lossless data compression, as illustrated in the example given below.

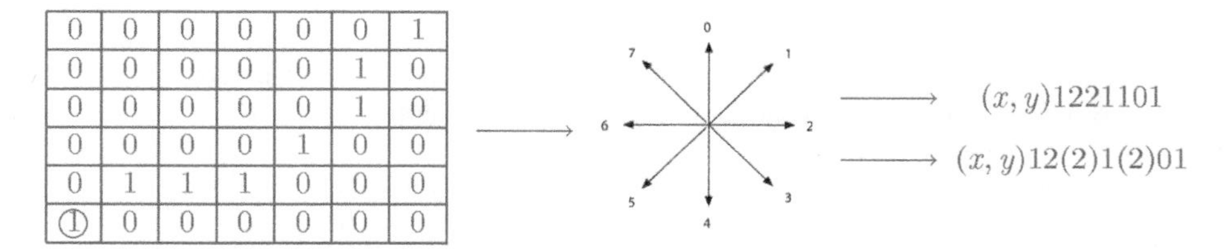

Vector Reduction

The chain codes generated in the previous step can be readily converted to a vector representation, which may then be generalized to smooth the edges (see Figure 4.16).

An example of Zhang-Suen erosion algorithm
Note: the subscripts on the 1's refer to the calculation examples given below.

0	0	0	0	0
0	1_1	1_2	0	0
0	0	1_3	1_4	0
0	0	1_5	1_6	0
0	0	0	0	0

$\longrightarrow$

0	0	0	0	0
0	0	1_2	0	0
0	0	1_3	0	0
0	0	0	0	0
0	0	0	0	0

Calculations:

p	$N(P)$	$T(p)$	mark or no mark
1	$N(1) = 2$	$T(1) = 1$	$p_N \cdot p_S \cdot p_E = 0$, $p_W \cdot p_E \cdot p_S = 0$, mark
2	$N(2) = 3$	$T(2) = 2$	no mark
3	$N(3) = 5$	$T(3) = 2$	no mark
4	$N(4) = 4$	$T(4) = 1$	$p_N \cdot p_S \cdot p_E = 0$, $p_W \cdot p_E \cdot p_S = 0$, mark
5	$N(5) = 3$	$T(5) = 1$	$p_N \cdot p_S \cdot p_E = 0$, $p_W \cdot p_E \cdot p_S = 0$, mark
6	$N(6) = 3$	$T(6) = 1$	$p_N \cdot p_S \cdot p_E = 0$, $p_W \cdot p_E \cdot p_S = 0$, mark

Figure 4.15 A local result of the application of the Zhang-Suen algorithm.

For already georeferenced digital spatial raster datasets, other techniques are employed, such as the polygon cycling algorithm [61, pp. 253–55], which assumes a raster with reasonably large homogeneous regions as input.

Vector to raster

Convex polygons in vector spatial data can be rasterized in a series of horizontal spans. Non-convex polygons can be decomposed into convex pieces, and those pieces rasterized. There are three classes of algorithms for vector-to-raster conversion: frame buffer, image strip, and scan line [61].

4.6.2 General Transformations

Other useful transformations of representations exist, and we now present two examples. The first is a Hough transformation from $\mathbb{R}^2$ to a slope versus length space, with an example of the medial axis of drumlins from two drumlin fields in Southern Ontario (see Figure 4.17). Note that the lengths and slope for this example are from page units of a scanned map, not a projected coordinate system. This transformation, illustrated in Figure 4.18, indicates the clustering in both the direction and length of the drumlins from the two fields. Field 1 displays generally steeper negative slopes than Field 3.

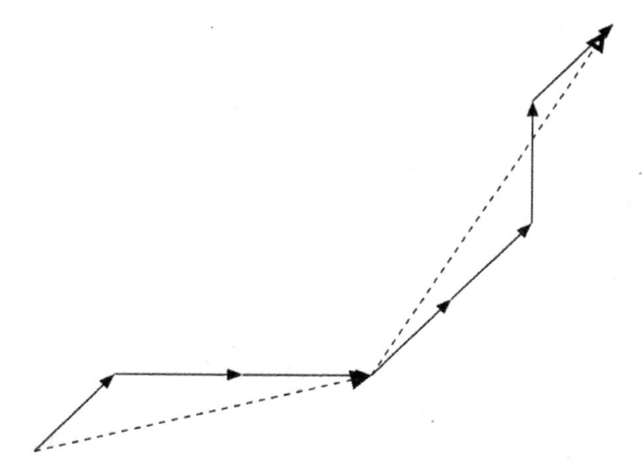

Figure 4.16 Vector representation and reduction.

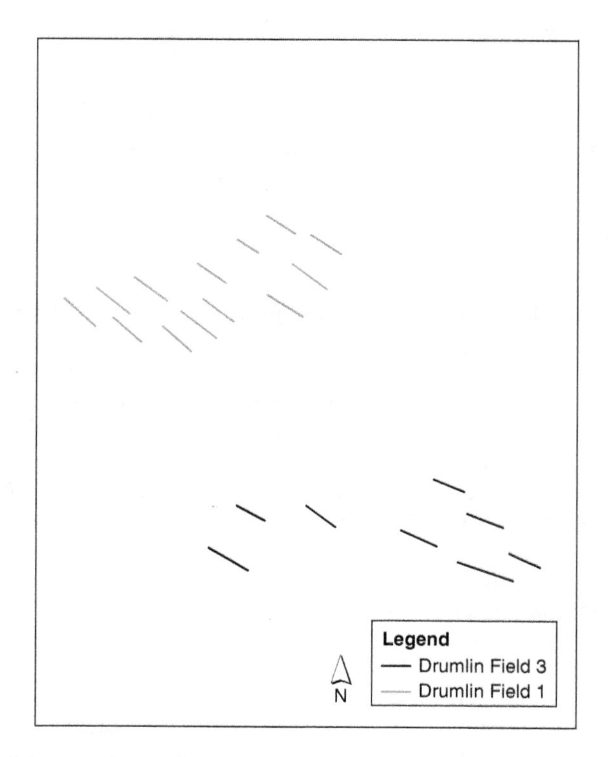

Figure 4.17 Medial axes of drumlins.

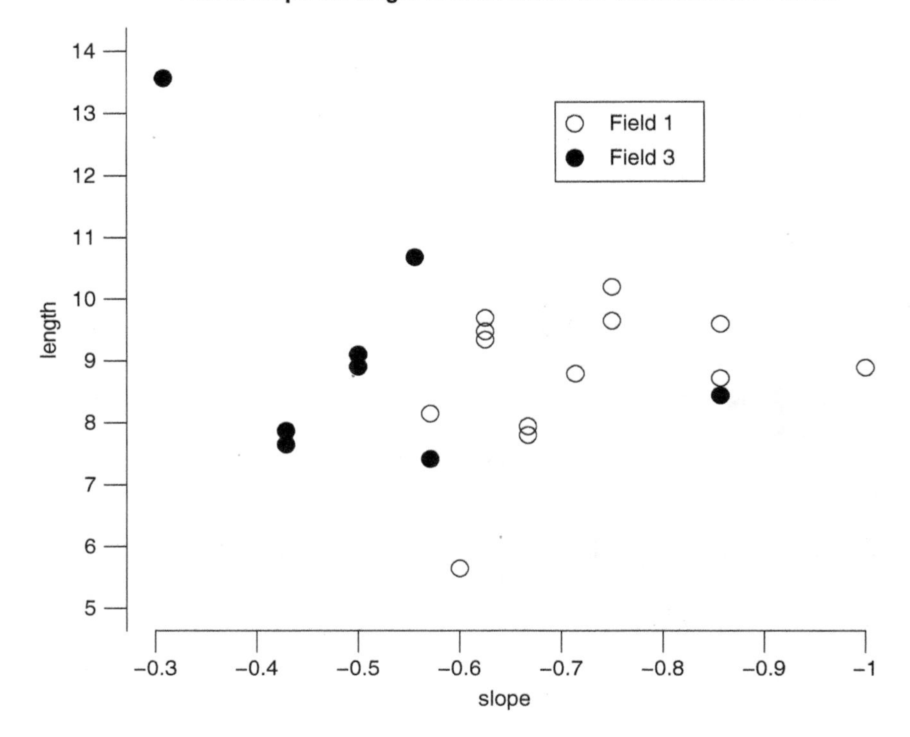

Figure 4.18 Transformed drumlin data.

In our next example we show how shape spaces can be used to change the representation of k-ads (a group of k points) in $\mathbb{R}^2$ to (k-2)-ads in the complex projective plane. This allows for shape metrics or an explicit distance to be defined between shapes. *Shape* is the property of a geometric figure that is invariant under translation, rotation, and scaling. Early efforts to use shape metrics to analyze polygonal boundaries in landscape spatial data utilized Kendall coordinates. This works well for shapes that have well-defined landmarks. In this approach, three points in $\mathbb{R}^2$ are transformed to one point in $\mathbb{C}^2$. A more detailed description of Kendall coordinates is presented in Chapter 10.

It has been demonstrated by one of the authors how to use statistical shape-analysis tools for the automatic parsing of morphological features from GPS tracking points of Grizzly bears provided by Foothills Model Forest Grizzly Bear Research Project in the northeastern slopes of Alberta, Canada. In this approach, three consecutive points of the trajectory in $\mathbb{R}^2$ are transformed to one point in the trajectory in $\mathbb{C}^2$. The position of this point in $\mathbb{C}^2$ can be used to interpret aspects of the local morphology of the original trajectory (see Figures 4.19, 4.20, and 4.21). Here you can see that starting at point 39 of a counterclockwise (CCW) turn, followed by a clockwise turn (CW) in $\mathbb{R}^2$, shows up as a transition from the lower to upper half planes in $\mathbb{C}^2$. This is automatically labelled by turn direction and the sequence of changing turn types; for example, CCW to CW can be automatically labelled as a zig-zag portion of the trajectory.

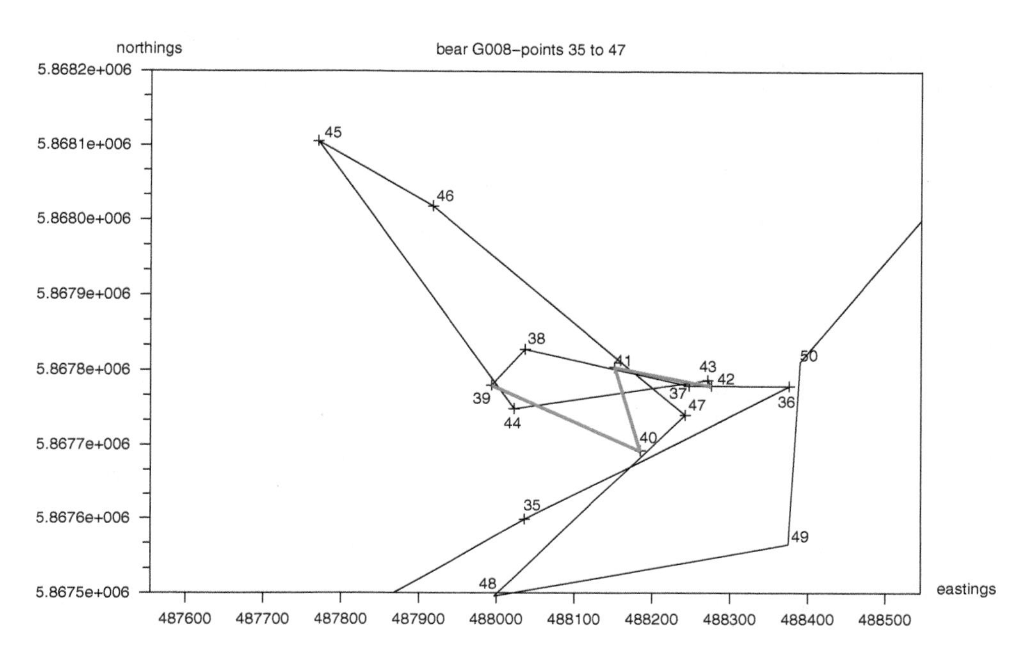

Figure 4.19 Original trajectory data in $\mathbb{R}^2$.

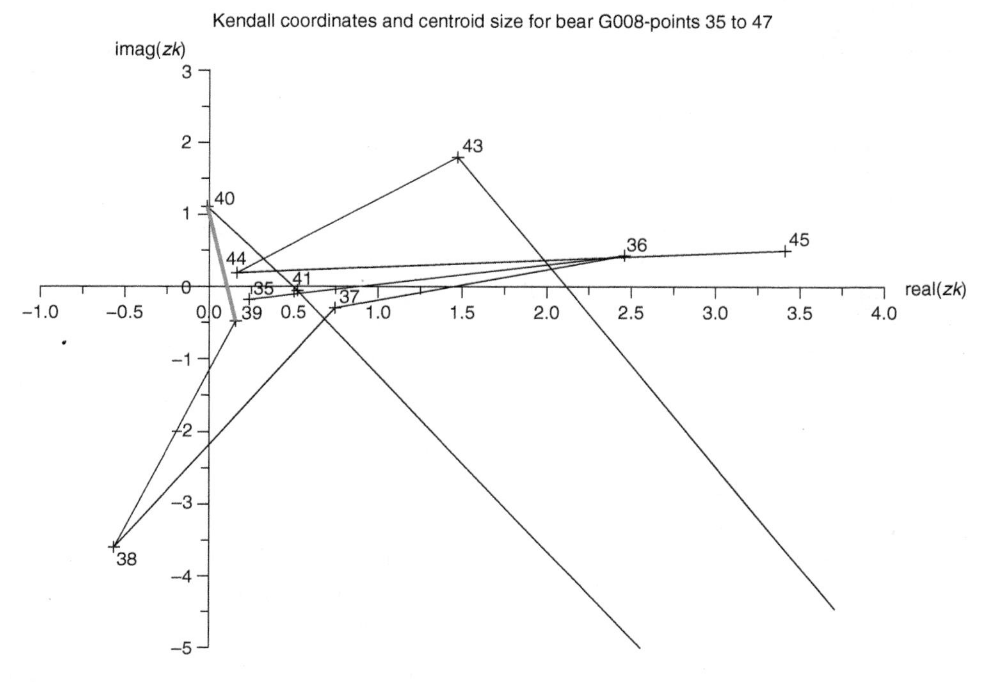

Figure 4.20 Transformed trajectory data in $\mathbb{C}^2$.

Identification number	Size	Turn type	Other information
35	M	CCW	**
36	M	CW	*F (flat triangle)
37	M	CCW	**
38	S	CCW	**
39	S	CCW	**
40	S	CW	R* (equilateral triangle)
41	S	CCW	**
42	M	CCW	**
43	M	CW	**
44	M	CW	**
45	M	CW	*F (flat triangle)

Figure 4.21 Automatically extracted morphological properties of trajectory.

4.7 Revisiting Standard Models: Topology, Embedding, and Oriented Matroids

This section discusses the evolution of spatial data models. The discussion here is decidedly theoretical, and its aim is to trace the progression of ideas about representing spatial relationships. Especially important is the interplay of geometric, geographic, and topological information. To help illustrate the conceptual differences, Gold's PAN graphs are introduced [45]. They illustrate the pointers or stored relations that exist between the geometric features for a particular spatial data structure. In the diagrams that follow, $\mathcal{P}$ denotes polygons, $\mathcal{E}$ denotes edges, and $\mathcal{V}$ denotes vertices.

Note that this is a slight paraphrasing of Gold's original terminology. Gold referred to edges as arcs and vertices as nodes, following the standard arc-node terminology used to describe an existing spatial data model. Furthermore, the term *diagram*, referring to a PAN graph, is used here to be consistent with the ideas and terminology from mathematical category theory. We will come back to this idea briefly in the later parts of this chapter.

4.7.1 Computer Aided Drawing/Drafting/Design (CAD)

CAD systems have been used for automated and digital cartography and general graphics tasks. The topological relationships maintained by CAD systems are restricted to relationships from vertex to geometric primitive (edge or polygon); for example, "*this* list of points comprises the vertices of *this* polygon," as represented in Diagram 4.5. CAD systems do not generally store higher-order relationships among geometric primitives; for example, "these three lines form the boundary of this triangle." Extensions are available that allow for feature coding and projected geographic (real world) coordinates, and in the case of feature codes these extensions are usually implemented as indices in the internal file structure of the shapes that are in turn linked to external database tables that contain the labelling information. In practice, however, in these systems attributes are usually coded in the graphical display properties; for example, "streams are blue and on level 23."

$$\mathcal{P} \longrightarrow \mathcal{V} \longleftarrow \mathcal{E} \tag{4.5}$$

CAD can bring data together for visual display—overlaying the boundaries of a woodland and a road alignment, for instance—but does not provide a direct means to analyze boundary conflicts. Custom code or add-ons are required to do this type of spatial analysis because of the limited built-in topology inherent in CAD-structured data; that is, no explicit adjacency information is stored with the data.

4.7.2 Geographic Information Systems (GIS)

GIS were introduced over 40 years ago [21, pp. 9–13]. Some early GIS were raster based, where data are stored in a regular rectangular tessellation—like pixels in a digital image. These approaches have been referred to as continuous-field models, in contrast to entity or discrete-view models such as those implemented in vector-based systems [16]. Many of the most popular recent systems are primarily vector based—meaning data are stored as geometric objects defined by vertices and the relationships between them. Moreover, advanced systems usually have both raster and vector functionality as well as other tools for working with triangulated irregular networks (TIN), digital elevation models (DEM), and linear networks.

In the following subsections, discussion concentrates on the discrete or vector models, in particular on representations of the subdivision of the plane into polygons. Information on the historical approaches relies on two sources: Worboys [123] and Laurini and Thompson [74]. Following Worboys [123, p. 201], it is indicated for each particular model whether the model captures topological relationships, the embedding of the geometry into a particular space (usually with a coordinate system), or both. These models may be implemented via tables or linked lists. Tables are commonly used in GIS implementations, and linked lists are commonly used in computer graphics applications. Tables are usually implemented as database tables in a relational database system. Linked lists are usually implemented using pointer data structures in the implementing programming language (e.g., pointer variables in C).

Spaghetti model

In the spaghetti model, polygons are defined as lists of vertices of their bounding edges. This model captures the embedding, since the vertices are usually mapped to a geographic coordinate system (denoted *coord* in Diagram 4.6), but has no explicit topological relationships. The edges and vertices are redundantly represented when polygons are adjacent, but such common vertices allow for the derivation of adjacency relationships between polygons [123, p. 193]. Diagram 4.6 and Figure 4.22 capture the main features of this model.

$$\mathcal{P} \longrightarrow \mathcal{V} \longrightarrow \mathit{coord} \tag{4.6}$$

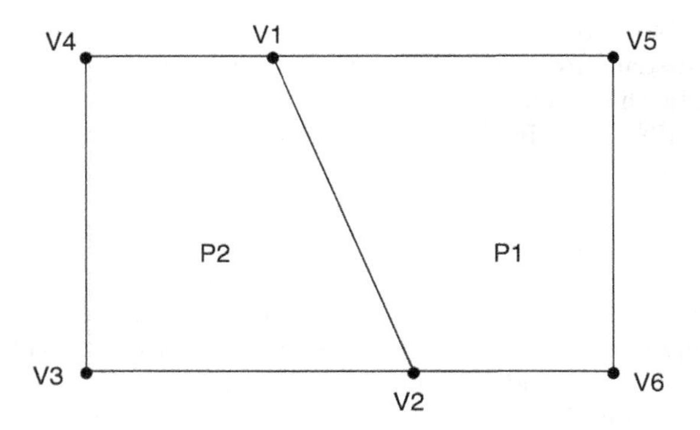

Figure 4.22 Geometry of spaghetti model.

For example, polygon P1 is defined by the set of vertices V1,V2,V6,V5. The SYMAP application, from Harvard in the late 1960s, implemented this type of model [74, p. 206].

Node-Arc-Area (NAA) model

In the NAA model, redundant edges (or arcs) and vertices (or nodes) are eliminated. This model captures edge and polygon (or area) topology but does not embed the geometry in a space. The model represented in Diagram 4.7, consists of a *planar graph* of *directed edges* (edges having a *begin vertex* and an *end vertex*). Although a formal definition of these terms appears in Appendix II, their meaning should be clear from Diagram 4.7 and Figure 4.23. The only term requiring some clarification at this point is "planar," which means none of the edges intersect, except at vertices. The orientation of the edges induces an orientation of the polygons relative to the edges that bound them. This is apparent from inspection of the simple example given by Figure 4.23. Note that there are no intermediate vertices to refine the shape of the edges in this model, so the

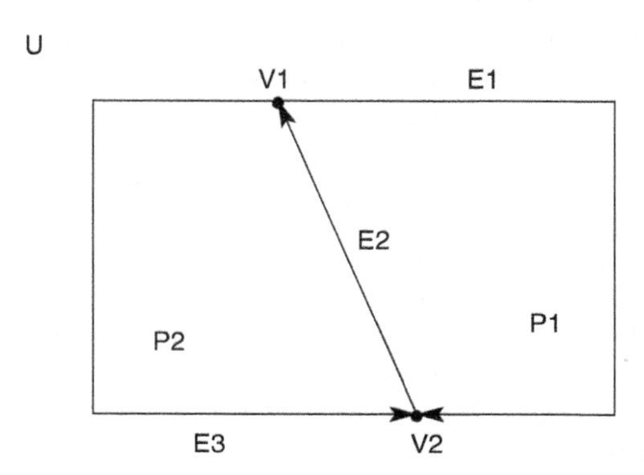

Figure 4.23 Geometry of NAA model.

diagram is a little misleading in this regard, but the form of the diagram is retained to aid in comparing the different models. Further, note that the inclusion of a label U for the area external to the given polygons. This identification keeps the model consistent by allowing all edges to have defined left and right polygons.

$$\mathcal{P} \underset{right}{\overset{left}{\longleftarrow}} \mathcal{E} \underset{end}{\overset{begin}{\longrightarrow}} \mathcal{V} \tag{4.7}$$

Extended Node-Arc-Area (ENAA) model

The ENAA model adds some additional relationships, beyond the NAA model, that define the sequence, actually a cycle, of edges that make up each polygon boundary and the sequence of vertices that define each edge and embedding of the vertices into a geographic coordinate system. (see Diagram 4.8 and Figure 4.24).

$$\mathcal{P} \underset{right}{\overset{\overset{sequence}{\longrightarrow}}{\overset{left}{\longleftarrow}}} \mathcal{E} \underset{end}{\overset{\overset{sequence}{\longrightarrow}}{\overset{begin}{\longrightarrow}}} \mathcal{V} \longrightarrow coord \tag{4.8}$$

This model captures polygon and edge topology and embedding. The ENAA model is at the heart of the Arc/Info polygon coverage data model.

Digital Line Graph (DLG) model

The DLG model was created and used by the United States Geological Survey agency for its topographic map sheet series. It models topology and embedding. Diagram 4.9 and Figure 4.25 illustrate the main features of the model. Diagram 4.9 introduces the notation $\mathcal{A}$, which represents non-spatial attribute data of the DLG model. Further, *reference_pnt* are arbitrarily placed special reference points associated with each polygon,

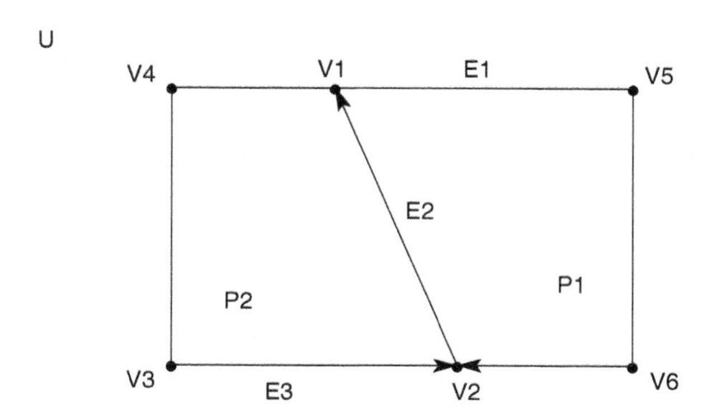

Figure 4.24 Geometry of ENAA model.

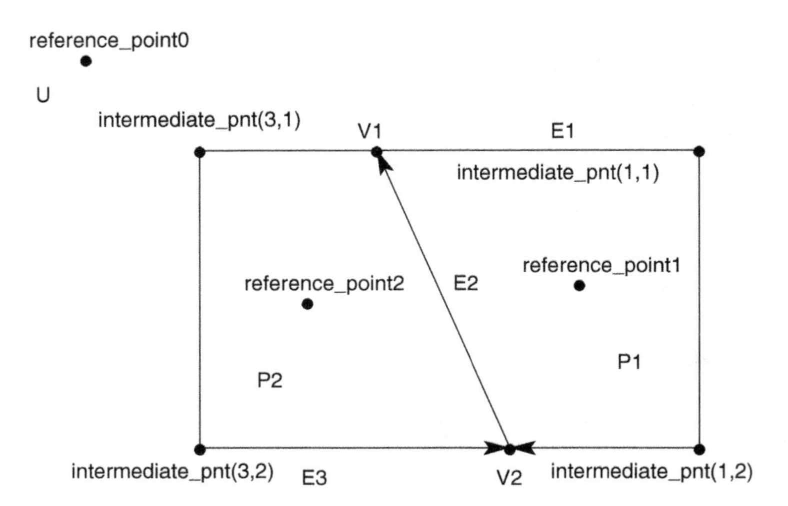

Figure 4.25 Geometry of DLG model.

and *intermediate_pnt* are intermediate vertices that give explicit shape to each edge object.

$$\begin{array}{ccccccc} coord & & coord & & & & \\ \uparrow{\scriptstyle reference\_pnt} & & \uparrow{\scriptstyle intermediate\_pnt} & & & & \\ \mathcal{P} & \overset{left}{\longleftarrow}\underset{right}{\longleftarrow} & \mathcal{E} & \overset{begin}{\longrightarrow}\underset{end}{\longrightarrow} & \mathcal{V} & \longrightarrow & coord \\ \downarrow & & \downarrow & & \downarrow & & \\ \mathcal{A} & & \mathcal{A} & & \mathcal{A} & & \end{array} \tag{4.9}$$

Doubly Connected Edge List (DCEL) model

The DCEL model [83] captures topology but not embedding. It extends the NAA model by adding next and previous edge relationships (see Diagram 4.10 and Figure 4.26). Note that this relationship is already implicit in the begin and end vertex relationships. Explicitly capturing these relationships allows for simpler algorithms in cases where the systematic traversal of edges is needed, such as in shortest-path problems.

$$\mathcal{P} \overset{left}{\longleftarrow}\underset{right}{\longleftarrow} \overset{\overset{next}{\curvearrowright}}{\underset{\underset{previous}{\curvearrowleft}}{\mathcal{E}}} \overset{begin}{\longrightarrow}\underset{end}{\longrightarrow} \mathcal{V} \tag{4.10}$$

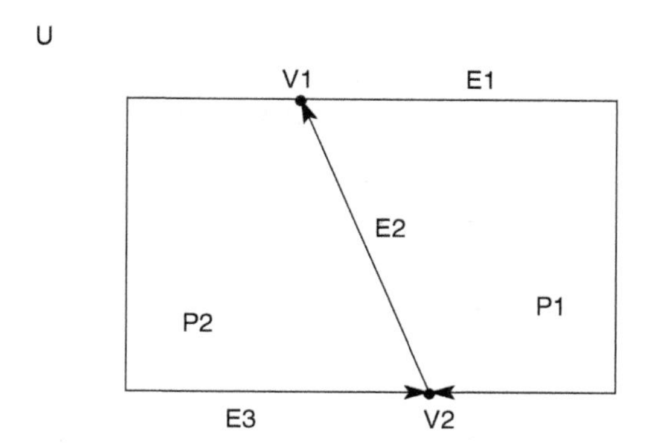

Figure 4.26 Geometry of DCEL model.

Twin-edge model

The twin-edge model [88, p. 146] is a variation on the DCEL, where each edge is split into two oppositely oriented half-edges. Each half-edge is in a list that uniquely bounds a polygon, and pointers are defined from each face to one of its incident half-edges and from each vertex to one of its incident half-edges. Diagram 4.11 and Figure 4.27 illustrate the idea. The defined relationships allow for traversal of the edges as in the DCEL model. In comparison to the DCEL, the twin-edge model allows for easier reconstruction of individual polygon boundaries due to the redundant edge lists. Note that the twin relationship identifies the corresponding half-edges.

$$\mathcal{P} \underset{\xleftarrow{\ bounds\ }}{\xrightarrow{one\_edge}} \frac{\mathcal{E}}{2} \circlearrowright_{twin,\ next,\ previous} \xleftarrow{one\_edge} \mathcal{V} \tag{4.11}$$

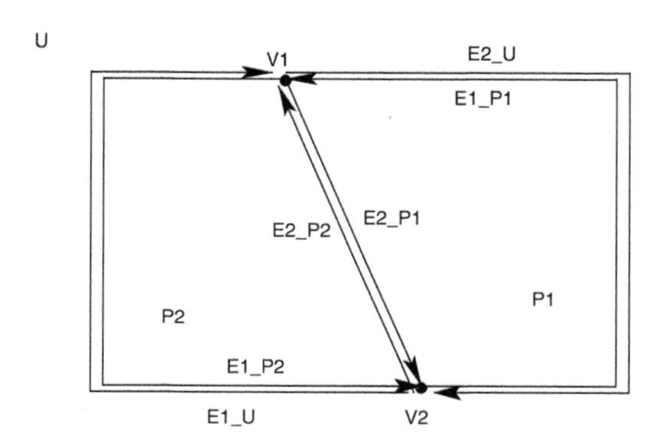

Figure 4.27 Geometry of twin-edge model.

Winged-edge model

The winged-edge model [9] is an extension of the DCEL model and captures topology but not embedding. Pointers are defined from each face to one of its incident edges and from each vertex to one of its incident edges (see Diagram 4.12 and Figure 4.28). Loops or isthmuses are not allowed in this basic model but are allowed in Weiler's extension [120] (not shown).

$$
\begin{array}{ccccc}
 & & \overset{next\_clockwise}{\curvearrowright} & & \\
 & \overset{one\_edge}{\longrightarrow} & \overset{next\_anti-clockwise}{\curvearrowright} & \overset{one\_edge}{\longleftarrow} & \\
\mathcal{P} & \overset{left}{\longleftarrow} & \mathcal{E} & \overset{begin}{\longrightarrow} & \mathcal{V} \\
 & \underset{right}{\longleftarrow} & \underset{previous\_clockwise}{\curvearrowleft} & \underset{end}{\longrightarrow} & \\
 & & \underset{previous\_anti-clockwise}{\curvearrowleft} & &
\end{array}
\tag{4.12}
$$

Quad-edge model

The quad-edge model [55] captures topology but not embedding. Each edge is associated with four circular lists, as shown in Diagram 4.13. The interesting feature of this model is that by considering the two vertex lists (*next_vertex* and *previous_vertex*) as polygon lists and the two polygon lists (*next_polygon* and *previous_polygon*) as vertex lists gives the topology of the dual graph to the planar graph that was originally modelled (see Figures 4.29 and 4.30).

$$
\begin{array}{ccccc}
 & & \overset{next\_vertex}{\curvearrowright} & & \\
 & & \overset{next\_polygon}{\curvearrowright} & & \\
\mathcal{P} & \overset{one\_edge}{\longrightarrow} & \mathcal{E} & \overset{one\_edge}{\longleftarrow} & \mathcal{V} \\
 & & \underset{previous\_vertex}{\curvearrowleft} & & \\
 & & \underset{previous\_polygon}{\curvearrowleft} & &
\end{array}
\tag{4.13}
$$

Dual Independent Map Encoding (DIME) model

DIME was created by the US Census Bureau in 1979 and is based on a schematic representation of a single-line street network and its enclosed street blocks. This model

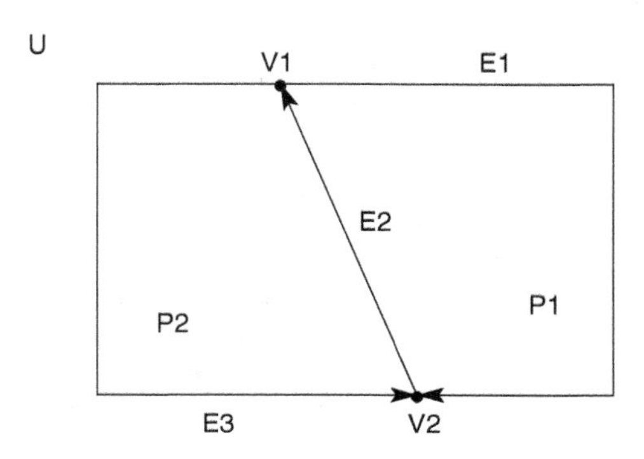

Figure 4.28 Geometry of winged-edge model.

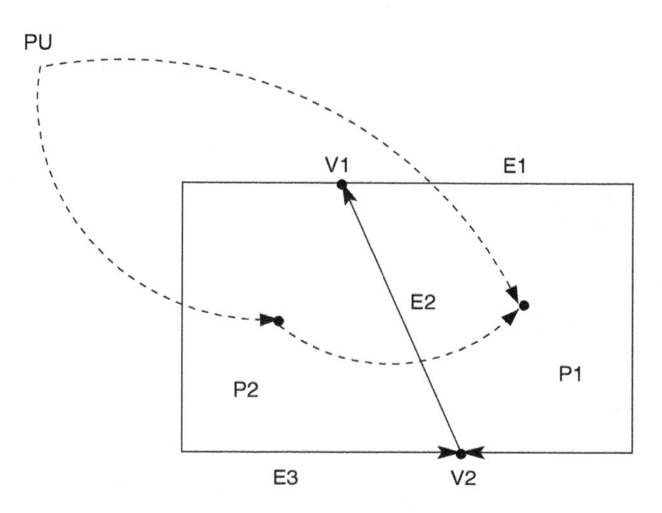

Figure 4.29 Geometry of quad-edge model.

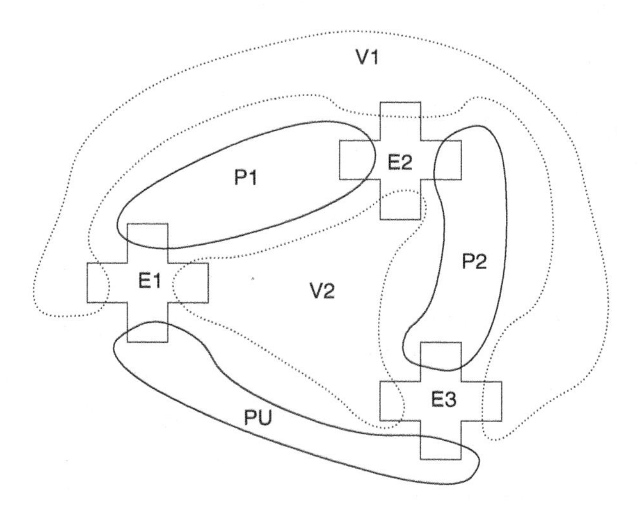

Figure 4.30 Concept of quad-edge model (after O'Rourke [88]).

captures topology and embedding of the edge intersection vertices. The address ranges of each street segment (*low_left_add., high_left_add., low_right_add., high_right_add.*) are captured as attributes. Diagram 4.14 and Figure 4.31 present the model for comparison.

$$\begin{array}{ccccc} \mathcal{P} & \underset{right}{\overset{left}{\leftleftarrows}} & \mathcal{E} & \underset{end}{\overset{begin}{\rightrightarrows}} & \mathcal{V} \longrightarrow coord \\ & & \begin{smallmatrix} low\ left\ add. \downarrow \ \downarrow high\ left\ add. & low\ right\ add. \downarrow \ \downarrow high\ right\ add. \end{smallmatrix} & & \\ & & \mathcal{A} & & \end{array} \tag{4.14}$$

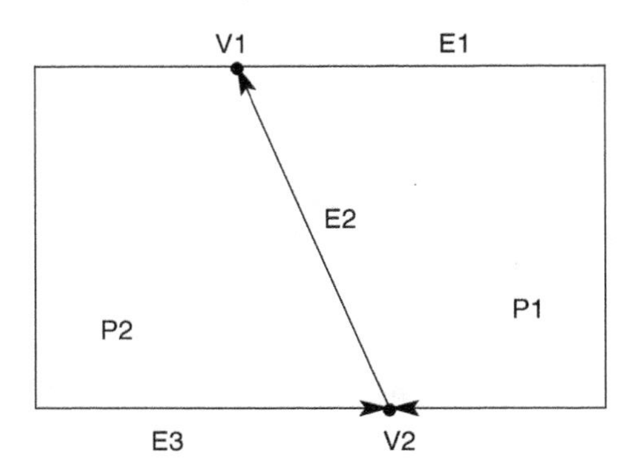

Figure 4.31 Geometry of DIME.

POLYVRT model

The POLYVRT model was used in Harvard University's ODYSSEY system. It includes intermediate vertices between edge endpoints and edge and polygon topology but no embedding. See Diagram 4.15 and Figure 4.32.

$$\mathcal{P} \underset{right}{\overset{left}{\longleftarrow\!\!\longleftarrow}} \mathcal{E} \underset{end}{\overset{begin}{\longrightarrow\!\!\longrightarrow}} \mathcal{V} \tag{4.15}$$

Topologically Integrated Geographic Encoding and Referencing (TIGER) model

The TIGER model was evolved from DIME by the United States Census Bureau (1986–1990). It captures topological relationships between polygons and edges and embedding. The geometric part of the model is based on *simplicial* structured sets. There are many technical details about such mathematical objects that are interesting for spatial data

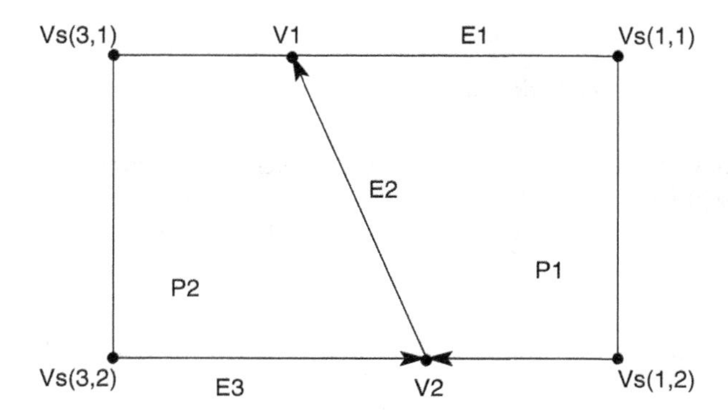

Figure 4.32 Geometry of POLYVRT.

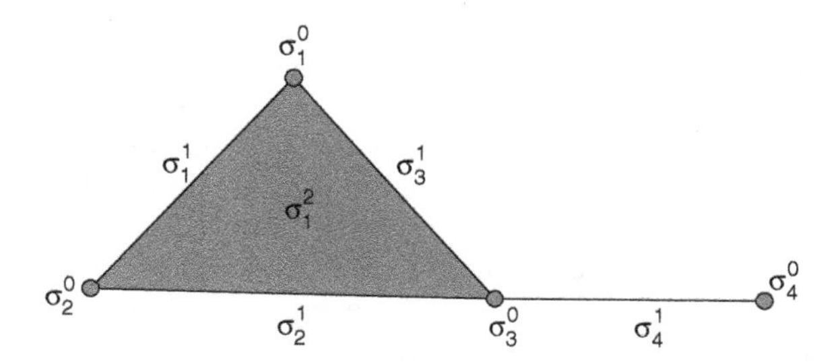

Figure 4.33 An example of a simplicial complex.

handling, but for now consider that vertices are modelled by 0-cells, edges are modelled by 1-cells and polygons are modelled by 2-cells. Figure 4.33 sketches the main idea that such structures have a nested hierarchy. In the diagram, σ labels the simplexes in an example simplicial complex, with superscripts indicating the dimension of the simplex. Lower-dimensional simplexes form the boundaries for higher-dimensional simplexes. Turning back to the TIGER model in the simplified Diagram 4.16 (see [74, pp. 470–71] for the complete model), the terminology established at the beginning of this section is used.

$$\begin{array}{ccccccc} & & & & coord & & \\ & & & & \uparrow & & \\ geographic\_cover & \overset{\longrightarrow}{\longleftarrow} & \mathcal{P} & \overset{\xrightarrow{intermediate\_points}}{\longleftarrow} & \mathcal{E} & \overset{\longrightarrow}{\longleftarrow} & \mathcal{V} \longrightarrow coord \\ & & \downarrow & & \downarrow \cdots \downarrow & & \\ & & \mathcal{A} & & \mathcal{A} & & \end{array} \tag{4.16}$$

Conference on Data Systems Languages (CODASYL) model

The CODASYL model uses a network database structure, which is not in common use today although it was popular 25 to 30 years ago. The CODASYL model is related to a pre-categoric approach to spatial data models based on directed graphs, which will be briefly discussed near the end of this section [98]. Note that in Diagram 4.17 only the polyline part of the model is shown.

$$\begin{array}{ccc} & \longleftarrow & \\ & \xrightarrow{begin} & \overset{next}{\curvearrowright} \\ \mathcal{E} & \longleftarrow & \mathcal{V} \\ & \longleftarrow & \underset{previous}{\curvearrowleft} \\ & \underset{end}{\longrightarrow} & \end{array} \tag{4.17}$$

Arc/Info Coverage model

The following model captures most of the key features of the spatial data model for polygon map layers referred to as coverages. The geometric primitives, shown in Diagram 4.18, include vertex (point or node), edge (arc), polygon (partition), and region (denoted $\mathcal{R}$). Edges are made by connecting vertices, polygons by connecting edges, and regions are groups of possibly overlapping polygons. This model is commonly referred to as an arc-node data structure—explicitly stated as "maps as graphs."

Networks are represented in this model by a set of connected directed edges. Note that the attributes of the polygons are linked to the centroid vertex associated with each polygon. Additionally, the sequence relationship between polygons and edges is a list of the edges that bound the polygon, while the sequence relationship between the edges and vertices is a list of intermediate vertices between the *begin* and *end* vertices that demarcate the shape of the edge. Further, a sequence of polygons defines a region. The core of this model seems to be a combination of features from the DLG and ENAA models (see Diagram 4.18).

$$\begin{array}{ccccccccc} & & & \multicolumn{3}{c}{\xrightarrow{\quad\quad\quad centroid \quad\quad\quad}} & & & \\ & & & \xrightarrow{sequence} & & \xrightarrow{sequence} & & & \\ \mathcal{R} & \xrightarrow{sequence} & & \xleftarrow{left} & \mathcal{E} & \xrightarrow{begin} & \mathcal{V} & \longrightarrow coord & \\ \downarrow \cdots \downarrow & & \mathcal{P} & \xleftarrow[right]{} & \downarrow \cdots \downarrow & \xrightarrow[end]{} & \downarrow \cdots \downarrow & & \\ \mathcal{A} & & & & \mathcal{A} & & \mathcal{A} & & \end{array} \tag{4.18}$$

Shapefiles

Shapefiles are a popular file format and spatial data model introduced by ESRI but with an open specification so anyone can write to or read a shapefile format by following the published standards. Shapefiles are organized as a collection of one type of spatial feature. The available feature types are point, polyline, polygon, multipoint, all the previous types with a height z specified, all the previous types with a general measure m specified, and multipatch. We will focus on the polygon feature type here. Diagram 4.19 and Figure 4.34 capture the main features of this model. We have introduced a new notation here, $\mathcal{LR}$ = LinearRing, which is a list of points that specify a set of line segments between the points that are closed to form a ring by requiring the last point to equal the first, and we let $\mathcal{V}$ = Point (in previous models we refer to Vertex).

$$\begin{array}{ccc} \mathcal{P} & \xrightarrow{Parts} & \mathcal{LR} \longrightarrow \mathcal{V} \longrightarrow coord \\ \downarrow & \downarrow & \\ numParts & NumPoints & \end{array} \tag{4.19}$$

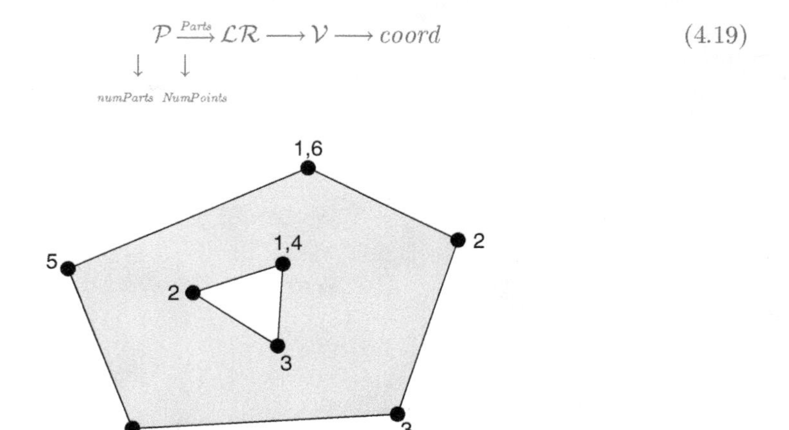

Figure 4.34 Geometry of shapefile polygon model.

Topological information is encoded by the standard that the outside boundary of a polygon has its linear ring component points ordered clockwise, whereas any holes in the polygon have their linear ring component points ordered counterclockwise.

KML

The Keyhole Markup Language (KML) is a scripting language for Google Earth. Diagram 4.20 and Figure 4.35 capture the main features of the polygon part of this model. All polygon boundary points are linear rings with component points ordered counterclockwise. Interior hole linear rings are identified by a label in the markup language.

$$\mathcal{P} \overset{outerBoundaryIs}{\underset{innerBoundaryIs}{\rightrightarrows}} \mathcal{LR} \longrightarrow coord \tag{4.20}$$

Object-relational hybrids

Current commercial desktop GIS software packages are designed with some degree of object orientation (OO) by utilizing component-based software architectures. On some earlier packages (Arcview, for example), OO was restricted to interfaces and scripting languages. However, systems have been described in the literature that utilize object-oriented data modelling more completely (see Pantazis [90], Tang et al. [110], and Bouillé [14]), and at least one fully object-oriented commercial GIS system, Smallworld 3 (see [105]), uses this approach. However, the latter software is now almost exclusively developed for network datasets and analysis (i.e., one-dimensional topology).

GIS data models structure spatial data either thematically (a map or layer paradigm) or as objects. In thematically oriented GIS, topological relationships are generated within the themes but not between themes, are restricted to Euclidean metrics, and do not allow explicit representation of time [39]. Object-oriented GIS use class definitions for geometric (representational) primitives and interfaces but still retain thematically organized datasets and consequently store only within-theme topology. Some current GIS software includes a multi-feature planar topology model.

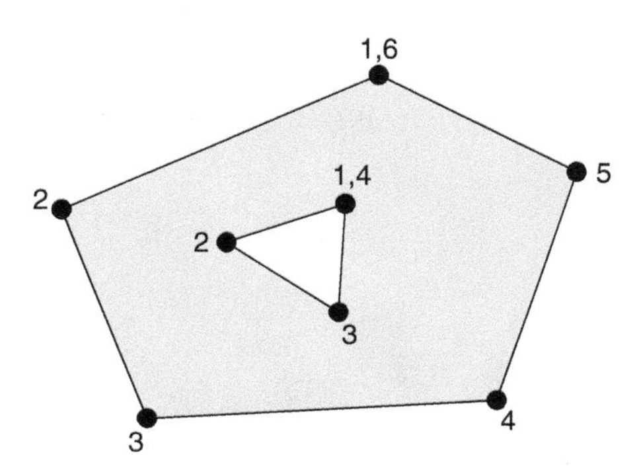

Figure 4.35 Geometry of KML polygon model.

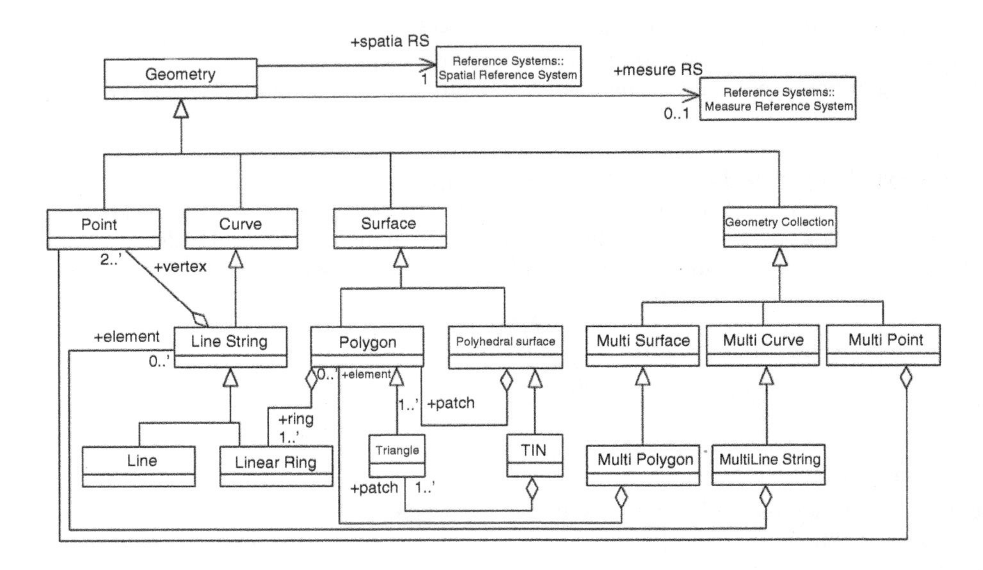

Figure 4.36 OGC Geometry Class Hierarchy.

Source: Open GIS Consortium Inc.

Most major commercial GIS implementations have implemented OO within an existing commercial relational database architecture. An open source implementation of this object-relational design is provided by PostGIS with the PostgreSQL open source relational database. More recently, the very lightweight and portable open source GIS Spatialite spatial extension to SQLite has become available. These follow the Open GIS Consortium (OGC) implementation standards (see [86]). Figure 4.36 shows the overall geometry class hierarchy from the OGC spatial data model. Note that this diagram is given from the source in Unified Modelling Language (UML). In UML, the rectangles denote the objects, the triangle arrowheads refer to object generalization, the diamond arrowheads refer to object composition, and $x..y$ enumerates the possible relationships between objects. For rough comparison, we now present the polygon portion of the OGC model, using the conventions and format we used for the other spatial data models (see Diagram 4.21). Here the following new symbols are introduced: $\mathcal{PS}$ = PolyhedralSurface, $\mathcal{LS}$ = LineString, and $\mathcal{C}$ = Curve.

$$
\begin{array}{ccccccccc}
\mathcal{PS} & \underset{BoundingPolygon}{\overset{PatchN(k)}{\rightrightarrows}} & \mathcal{P} & \underset{interiorRingN(j)}{\overset{exteriorRing}{\rightrightarrows}} & \mathcal{LR} & \longleftrightarrow & \mathcal{LS} \longleftrightarrow & \mathcal{C} & \underset{endpnt}{\overset{PointN(i)}{\underset{startpnt}{\longrightarrow}}} \; \mathcal{V} \\
\downarrow \quad \downarrow & & \downarrow & & & & \downarrow & \downarrow \quad \downarrow \quad \downarrow & \downarrow\downarrow\downarrow\downarrow \\
isClosed \quad numPatches & & numInteriorRing & & & & numPoints & length \quad isClosed \quad isRing & x \; y \; z \; m
\end{array}
\tag{4.21}
$$

Note that there is linear interpolation between the points (vertices) in the LineString objects and that if two polygons share a common boundary in a

PolyhedralSurface object, their respective LinearRings must be oriented in opposite directions to enforce a consistent planar orientation across the PolyhedralSurface. Further note that the LinearRing ($\mathcal{LR}$), LineString ($\mathcal{LS}$), and Curve ($\mathcal{C}$) objects combined could be considered comparable to the edge ($\mathcal{E}$) features discussed above. Similarly, the combination of PolyhedralSurface ($\mathcal{PS}$) and Polygon ($\mathcal{P}$) are comparable to the polygon features discussed above in other spatial data models. The OGC model offers a quite different approach to topological relationships among polygonal features from most of the older historical developments, but you can clearly see the influence of the shapefile and KML standards in this model. Though the OGC model is based on rigorous mathematical topological ideas, it discards some of the useful, or perhaps just overly familiar, explicit topological relationships (the orientation induction between oriented edges and polygons, for example).

Despite these advances, to address adequately the challenge of modern spatial analysis we need to consider more-general notions of topology and boundaries [23], beyond standard metrics and including time [64]. In this context, Pantazis [90] intends to extend his object-oriented GIS research to consider indeterminate boundaries, time, and moving objects. Spatial models that specifically address indeterminate boundaries have yet to be widely implemented.

Summary of major GIS spatial data models and other possibilites

The inventory of selected spatial data models reviewed in the preceding subsections gives a picture of the range of approaches that have been developed. Most of these models have been implemented either as linked lists (more common in image processing and computational geometry applications) or as tables (more common in GIS applications and primarily now within relational database management systems).

Spatial data models can be formulated from the perspective of a pre-category; that is, using the directed graph diagrams of simple categories to create a meta-spatial data model [98]. In practice, this turns out to be the union of simple classes of directed graphs (see Diagram 4.22).

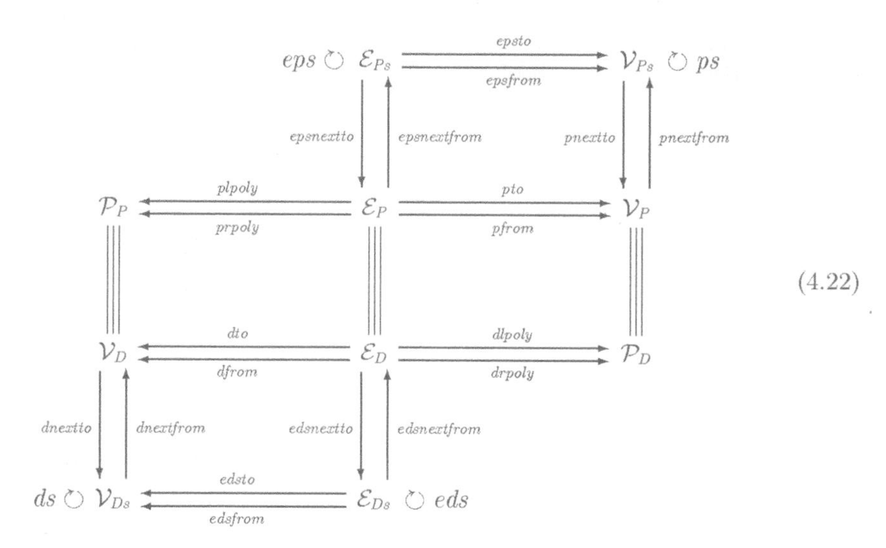

(4.22)

In fact these are, in a few cases, the same directed graphs that are implicit in the NAA model discussed above. Note in the diagram that the triple lines represent isomorphisms (one-to-one and onto maps) between the objects, the subscript P refers to primal objects, and the subscript D refers to dual objects. The model may also use the category of simplicial complexes. Simplicial set structures are used in the TIGER model, but in that case they are used for the spatial partition of space. However, the simplicial set structures could be used for both attribute modelling and edge modelling, based on attributes. In that case the distinction between attribute and spatial data is blurred, as is appropriate for the Liebnizian view of the partition of space [5].

A quite different approach to spatial data modelling is provided by cellular automata and this topic is discussed in the next section.

4.7.3 Cellular Automata

Cellular Automata (CA) are models of dynamic processes, driven and constrained by local interactions. In a highly speculative monograph, CAs have been proposed as a model for all physical processes in the universe [122]. Diagram 4.23 and Figure 4.37 indicate the neighbourhood topology of adjacent partitions (e.g., grid cells) in a CA for comparison with the previously discussed CAD and GIS data models and topological structures. In fact, in CA and raster spatial data, some processing is easier due to the regular structure of the tessellation (i.e., the regularity of various neighbourhood structures for each cell). It is this feature that is often used in image processing in the various convolution operators such as image smoothing or edge detection [30, Ch. 3]. A good introduction to CA may be found in Weisbuch's *Complex Systems Dynamics* [121].

$$\mathcal{P} \overset{neighbourhood}{\circlearrowleft} \qquad (4.23)$$

Some work has been done in creating decision support systems that integrate the methodologies of CA, raster GIS, and system dynamic models—Wu's work on modelling land conversion [124] and Theobald and Gross's EML project for modelling landscape

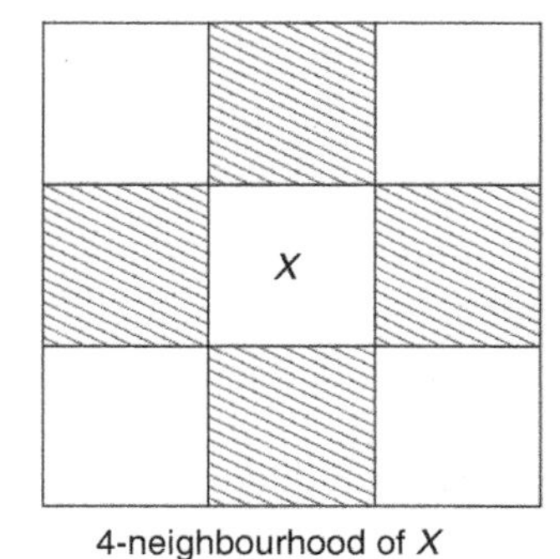

4-neighbourhood of X

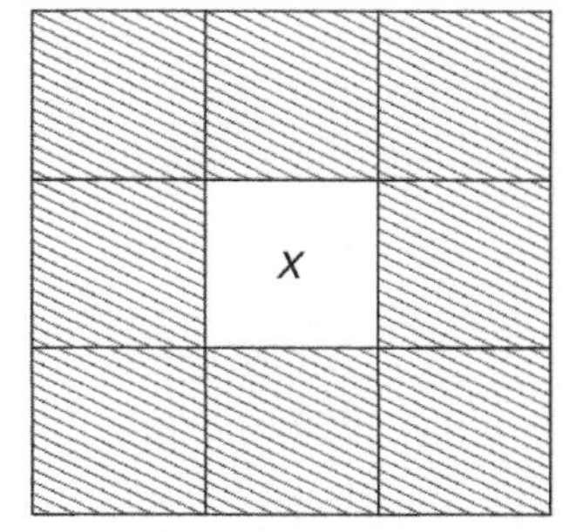

8-neighbourhood of X

Figure 4.37 CA neighbourhoods.

change [112], for example. Related work uses Diffusion Limited Aggregation as a dynamic model of urban growth [8]. However, since CA-based models are tied to grid-like data structures, they lack the degree of generality needed to deal with indeterminate boundaries or qualitative reasoning. Overcoming this to a certain extent are agent-based models where the agents are free to move about the grid and have an even richer set of behaviours and states available to them (see [44]). Boots and others have examined the use of recursive Voronoi diagrams as a generic model of dynamic spatial processes [13] [12]. This approach is also not bound to a regular grid and works within vector or discrete, as well as raster, spatial models.

4.7.4 Oriented Matroids

Discrete representation

In computational geometry, and thus GIS and remote sensing, we are not working in continuous Real space $\mathbb{R}^n$; we are stuck with discrete representations with generally finite precision numbers as our coordinates. We will talk more about the latter issue in Chapter 8, but for now we can consider the implication that we cannot assume that topological results that hold in a Real space hold for the discrete space. Examine Figure 4.38. Now, if we snap the intersection from p (actual) to q (closest location we have a point for), is the dotted line to the left or right of the intersection of lines ab and cd?

Topological paradoxes

The result of this discretization of representation is many-fold, but the following two specific cases should give the reader some flavour of the issues involved.

Jordon's Curve Theorem

An extremely straightforward but surprisingly important result from analysis is Jordon's curve theorem, which states roughly that every simple closed planar curve partitions the plane into one area inside the curve and one area outside the curve (see Figure 4.39).

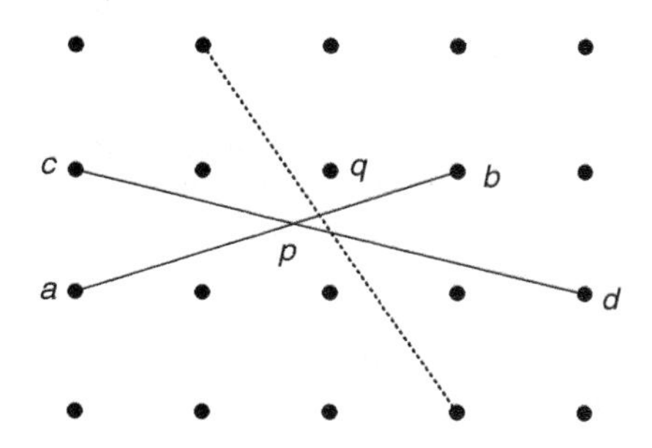

Figure 4.38 Intersection problem.

Source: Reprinted from Stell and Webster [106] with permission of Elsevier.

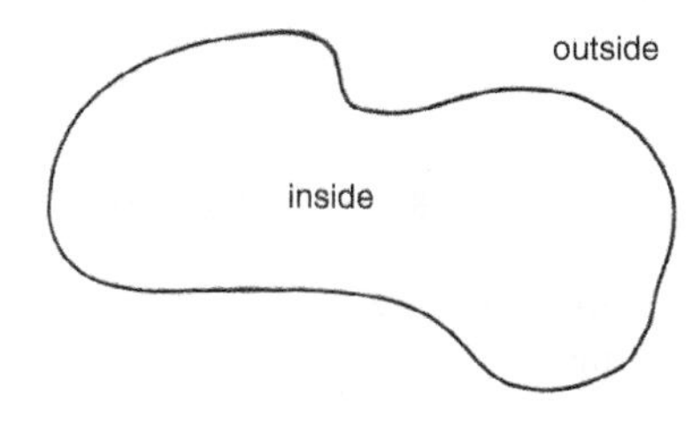

Figure 4.39 Jordan's theorem.

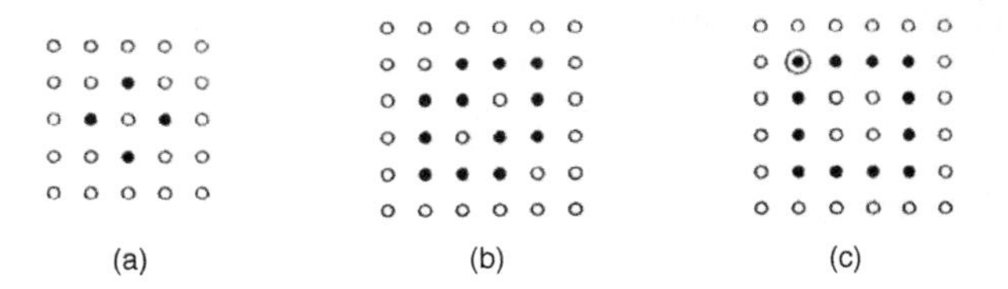

Figure 4.40 Connectivity paradox.
Source: From Schneider [101].

Unfortunately, this theorem does not hold in discrete digital spaces, as Figure 4.40 illustrates.

Looking at Figure 4.40 (a), the example on the left, if we consider a rook's neighbourhood to define connectivity, then the digital curve (black points) is not connected (closed), yet it defines one area inside the curve (a single white point) and one area outside the curve. Alternatively, if we consider a queen's neighbourhood to define connectivity, then the black points describe a closed curve, yet the white points remain all connected. Thus the closed curve does not partition the space into two disjoint areas. Looking at Figure 4.40 (b), the example in the middle, and considering a rook's neighbourhood to define connectivity, the black points make up one simple closed curve, yet they partition the white points into three areas, two inside the the curve, and one outside. In the final example, Figure 4.40 (c), if we consider a rook's neighbourhood, the white points remain in two partitions even if the circled corner black point is removed, effectively opening the curve.

Helly's Theorem

As a final example of these issues, consider Helly's theorem. This is a statement about intersections of convex sets in a continuous space. The theorem is as follows:

> Let C be a finite family of convex sets in $\mathbb{R}^n$ such that, for $k \leqq n+1$, any k members of C have a nonempty intersection. Then the intersection of all members of C is nonempty.

In Figure 4.41 left, rotating the set of grey cells by 90 degrees three times to create four digital convex sets violates Helly's theorem. Figure 4.41 right does not violate Helly's theorem, as the central point is in all four convex sets. Solutions such as the structure on the right have been proposed by Schneider and others [101]. However, another promising avenue for addressing these foundational problems is offered by an oriented matroid perspective. This approach is sketched out below, based on research presented by Webster [119] and by Stell and Webster [106].

Addressing the issues

An oriented matroid approach to spatial representation takes seriously the knowledge that machine computation is not done in $\mathbb{R}^n$ but with a subset of machine representable

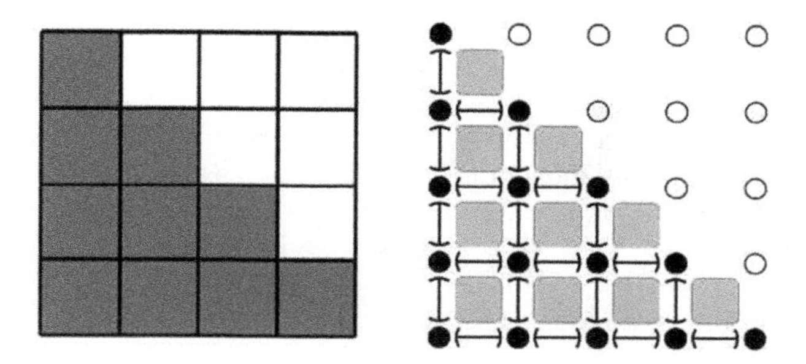

Figure 4.41 Helly's theorem.

Source: Reprinted from Webster [119] with permission of Elsevier.

coordinates. This should lead to more-robust computational geometry than is currently available [106]. Further, this approach directly addresses the problems of the embedding space in practice being discrete and not continuous. This leads to the possibility of algebraic (combinatorial) models of relative positions, to quote Stell and Webster [106]:

> ... capturing geometric information in a purely finite combinatorial structure that makes no mention of $\mathbb{R}^n$ or $\mathbb{Q}^n$, even though the intended models are finite subsets of these spaces.

To give a flavour of how such a scheme works, we work through an example based on Knuth's CC-systems definitions [70]. Note that in this case, based on Stell and Webster [106], we talk in terms of clockwise ordering rather than the counterclockwise ordering common in the computational geometry literature, including Knuth (hence the "CC").

Following Stell and Webster [106], a CC-system is a finite set E together with a ternary relation $B(x, y, z)$ such that the following five properties hold:

CC1. $B(x, y, z) \Rightarrow B(y, z, x)$
CC2. $B(x, y, z) \Rightarrow \neg B(x, z, y)$
CC3. $B(x, y, z) \vee B(y, x, z)$
CC4. $B(z, x, y) \wedge B(w, z, y) \wedge B(w, x, z) \Rightarrow B(w, x, y)$
CC5. $B(z, y, v) \wedge B(z, y, w) \wedge B(z, y, x) \wedge B(z, v, w) \Rightarrow B(z, v, x)$

This system has the following geometry. A straight line is any pair of distinct points $\{a, b\}$, and its corresponding open half spaces are respectively the sets $\{x \mid B(a, b, x)\}$ and $\{x \mid B(a, x, b)\}$. The convex hull of a set X is a set of points such that x is in the convex hull if there is no open half space that contains x and does not intersect the set X. Thus we can model some key geometric relationships based on the combinatorial CC-structure without reference to an underlying embedding space.

To make this more concrete in a more traditional GIS setting, consider any finite subset E of the plane such that no three points are collinear. We let $B(x, y, z)$ hold if

x, y, z are distinct points and if, when travelling clockwise around their unique circumcircle, upon hitting x you next hit y before z. This is the clockwise ordering, and this also satisfies CC1-5 above, and thus is a CC-system.

To see this in practice we present an example given in Stell and Webster. Let there be six points in the plane, as shown in Figure 4.42.

In the plane with floating point homogeneous coordinates we can test $B(x, y, z)$ by checking the sign of the $det(x, y, z)$. If $det(x, y, z) > 0$, where

$$det(x, y, z) = \begin{vmatrix} x_1 & x_2 & 1 \\ y_1 & y_2 & 1 \\ z_1 & z_2 & 1 \end{vmatrix} = x_1(y_2 - z_2) - y_1(x_2 - z_2) + z_1(x_2 - y_2) \quad , \tag{4.24}$$

then that triple of points has a clockwise ordering. Thus in the example one of the open half spaces corresponding to the line $\{a, b\}$ is $\{e\}$ since $\{e \mid B(a, b, e)\}$. Similarly,

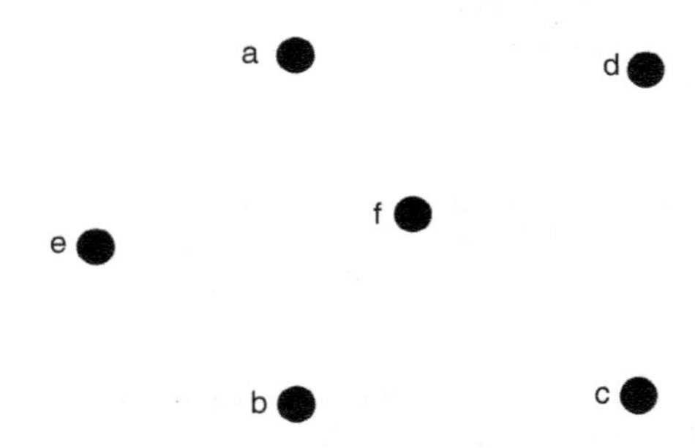

Figure 4.42 Clockwise ordering example, points.

Source: Adapted from Stell and Webster [106] with permission of Elsevier.

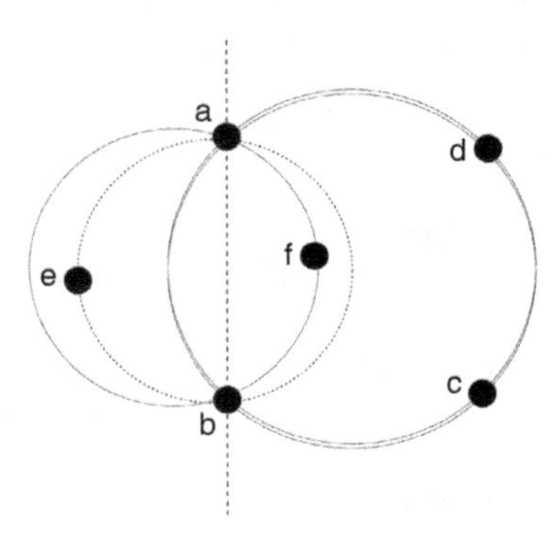

Figure 4.43 Clockwise ordering example, derived geometry.

Source: Reprinted from Stell and Webster [106] with permission of Elsevier.

the other corresponding open half space is $\{c, d, f\}$ since, for example, $\{f \mid B(a, f, b)\}$. Further, the convex hull of $\{a, b, c, d, f\}$ is $\{a, b, c, d\}$ (see Figure 4.43).

Knuth [70] notes that his CC-system is equivalent to a certain type of oriented matroid. Oriented matroids may be defined in many equivalent ways and are in fact a significant abstraction of many important mathematical concepts, including linear dependence. It is beyond the scope of this text to delve any deeper into oriented matroids. Suffice it to say that they may be formally defined in an axiomatization, as was the CC-system given above, and there is a rich theory of results using these structures.

Note also that there are categorical structures associated with oriented matroids. So given the earlier discussion in this chapter, we can envision dealing with both geometry and topology (two of the key ideas in spatial data) in terms of categorical representations.

Problems

1. Use the City of Victoria Landsat dataset to calculate the edge image by convolution using the Sobel 3×3 edge detection filter.
2. Find or create your own binary image and thin it using Zhang-Suen thinning.
3. Sometimes we implement an asymmetric convolution mask for our raster operations. For what reasons, related to the physical features or processes we are modelling or analyzing, might we do this?
4. Create or use an existing street network for a small town. Consider two representations, a single line street network and a polygon representation of the streets. Create a comprehensive description of the analytic methodologies available with each representation. Which one is better?
5. Describe the geometry, geography (embedding), and topology of the GIS software you most commonly use. Can you identify similarities to any of the historical spatial data models presented in this chapter? Describe some advantages and disadvantages to these representations.

Chapter 5

Geographic Representation and Data Modelling

5.1 Introduction

There are two dominant views of what a GIS is. There is the view of GIS as a computerized extension of a paper map, what might be called the *map layer view*. This perspective considers geographic information as inherently "special," and maps as a core medium for the communication of geographic information through GIS. This might be the view of GIS adopted by cartographers and mapping professionals. Alternatively, there is the view of GIS as an information management system designed for geographic data, or what we will call the *database view*. In this perspective, maps are simply an information product that can be produced by the system with data from a geographic database, not all that different from other information products such as graphs, tables, or summary statistics. This might be the perspective of a traditional database administrator tasked with managing a GIS. A tension exists between these two perspectives. While one view posits that GIS are no different than any other information management system, simply providing traditional information management functionality for geographically referenced spatial data, the other perspective considers maps and mapping as the core reason for GIS. While these two perspectives represent extremes, they impact training and education, design choices in software packages, and wider perceptions of what GIS is and should be capable of. In our following discussion of database management systems, we will try to highlight both of these views and how the ideas and concepts associated with each view have evolved as the technology for geographic information storage has developed.

5.2 Relational Databases and GIS

Through the 1960s and 1970s, databases and database management systems (DBMS) became the dominant mode of data storage, access, and manipulation. A variety of architectures emerged during this time, including network, hierarchical, and relational models. Many of the early DBMS were developed for specific applications, and progress was incremental owing to a lack of a general theoretical framework. For example, in hierarchical databases, queries had to be designed into the development of a DBMS itself

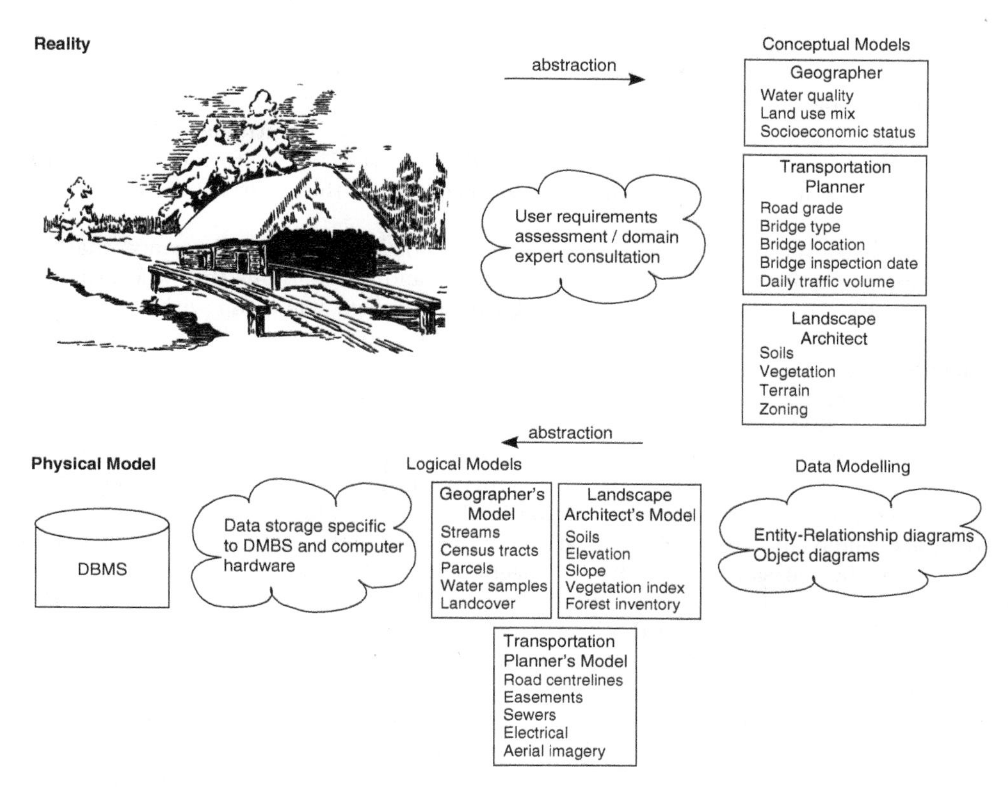

Figure 5.1 Steps in the data modelling process, at increasing levels of abstraction from the real world.

(all possible queries had to be known before the system was made operational). Such limitations severely limited widespread adoption of DBMS as well as their adaptation to storing geographic information. DBMS origins were rooted in file-based processing systems, whereby transactional operations were performed sequentially on a series of files at regularly scheduled intervals. Such processes might be used to update tax data with recent collections, or to update an agricultural inventory system with recently obtained harvesting data.

The design of databases can be thought of as a series of abstractions that simplify the complexity of the real world into computer representations that support the information needs of a set of users. Figure 5.1 outlines database design stages for some possible types of information users (geographer, transportation planner, and landscape architect). The first level of abstraction is a conceptual model that defines the components of the system being modelled. Note that this requires knowledge of the processes being modelled, at the outset of database design. The information needs at a conceptual model level will almost always differ depending on the intended application or user needs. For example, a transportation planner may be interested in defining a database—one that can handle dynamic, real-time traffic volume data obtained from traffic sensors and interpolate continuous traffic volume estimates at unsampled locations—as a key information need, whereas a geographer examining landuse pattern and economic development may only need to represent static vector geometries of road centrelines. Information needs can be highly specific or generalized, and they have to be mapped to a logical structure of observations and measurements that can be stored in the database.

The next level of abstraction is logical data modelling, whereby the conceptual information needs are mapped onto tables, columns, and relationships that will form the database. Physical database (i.e., physical schema) design has to do with the allocation of memory for data storage, query optimization, data type definitions, indexing schemes, and hardware and software configuration. Physical database design tasks are usually done by software developers, GIScience researchers, and other information scientists.

Records in a DBMS are a collection of individual pieces of data (sometimes called a row or a tuple). Considering the origins of DBMS in document storage and batch processing, records might conceptually be bits of data—notes of text, numbers, photographs, audio and video clips—that are grouped together into a single collection and stored in the same place. Each collection might describe a common object or phenomenon, (e.g., houses, forest stands, street lights, Netflix users etc.), usually one instance of what would be called in statistics the *unit of analysis*. The way that records in a collection are stored can affect how that information can be accessed and used. For non-geographic information, records might be sorted based on an attribute of interest or, in a geographic context, ordered on a geometric property such as cardinal direction or area. Collections can also be grouped together by the objects or phenomena they describe, into tables. Thus, a database is analogous to a set of collections about a common topic or concept. A key information need in most real-world applications is the ability to access records based on their relationship to other records. The development of DBMS and, specifically, *pointers* between records enabled relationships to be encoded into the database that reflected reality. In hierarchical DBMS, relationships are restricted to one-to-many (i.e., parent–child), whereas in network DBMS, relationships can also be many-to-many (see Figure 5.2).

There are obvious problems with this sort of structuring, as noted by Codd [24]. If the ordering of the records changes, applications designed to access those records will fail (i.e., there is an ordering dependence). Similarly, if an index is used to access records in a table, similar dependence will result (i.e., there is an indexing dependence). Finally, when access to records depends on hard-coded access paths (hierarchies or networks), programs designed to use these records depend on these access paths not changing (i.e., there is an access path dependence). All of these dependencies create DBMS that do no support true *data independence*: the separation of application programs from internal data representations. What is required is random access to data, independent of applications built on top of the database.

Principles of DBMS design aim to support application-independent data storage, retrieval, and management. Database design is achieved through the specification of a schema language called the data definition language (DDL), first described by the Data Base Task Group associated with the COBOL programming language. A schema language gives users the ability to design the logical structure of the database without

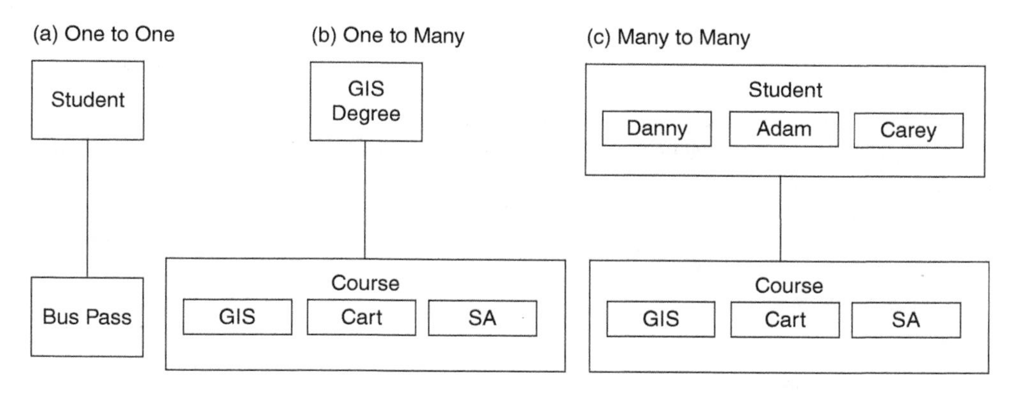

Figure 5.2 Cardinality of relationships that might need to be represented within a DBMS. (a) Each student is assigned one and only one bus pass, (b) each GIS degree has multiple prerequisite courses, and (c) each course can be taken by many students, while each student can take many courses.

reference to the underlying physical data storage or access paths. The purpose of DDL was to specify the records, attributes (i.e., data types), and constraints on relationships in a database. A subschema of DDL enabled redefinition of some of the properties of data structures defined in the schema through DDL upon creation. Another language, called the data manipulation language (DML), was specified to enable access to data and relationships between records without manipulating the address pointers themselves. What the DDL and DML facilitated was separation between a *physical* and a *logical* view of the database. This separation was later extended into a multiple-schema approach encompassing three views of the database. It is worth highlighting that these schema languages kept the relational schemas distinct from the actual tables and data. Tsichritizis and Klug [114] described the internal, external, and conceptual views of the database relating to physical storage, a user-view of the data and/or relationships embedded within the database, and a conceptual view of the phenomena being represented by the database. These ideas have persisted through the development of DMBS and continue to be important today. DDL and DML were later integrated into constructs within the Structured Query Language (SQL), the most widely used query language for databases today.

Relational DBMS are based on a *relational view* of data. A relation R is defined as a set of records, composed of sets D. The number of sets in R are said to describe the degree of R; for example, unary, binary, ternary, and n-ary being common degrees of R. The model is derived from set theory. If we consider a k-tuple of data containing a sequence of elements $D_1 = a, b, c$ as a set, a collection of sets $D_1 \ldots D_n$ can be termed a relation R if it is a set of k-tuples each of which has its first element from D_1, its second element from D_2 and so forth. A relation is a subset of the Cartesian products of sets. For a Cartesian product of two sets D_1, D_2, the the following relationship holds:

$$|D_1 \times D_2| = |D_1| \times |D_2| \quad ,$$

where $|D|$ denotes the cardinality (or number of rows) in D. As such, the Cartesian product of two sets can grow very large. Relational databases express relations as subsets of the Cartesian product of two or more sets.

Five operations were introduced in the relational algebra for the relational model. The union $R_1 \bigcup R_2$ of two k-ary relations R_1 and R_2 is such that tuples in R_1 or R_2 are retained in $R_1 \bigcup R_2$. The difference $R_1 - R_2$ of two k-ary relations is such that tuples in $R_1 - R_2$ are in R_1 but not in R_2. As described above, the Cartesian product of k-ary R_1 and j-ary R_2 is the set of all possible ordered pairs where the first element is from R_1 and the second element is from R_2. For example, given $R_1 = a, b, c$ and $R_2 = 1, 2, 3$, the Cartesian product is

$$R_1 \times R_2 = \begin{bmatrix} (a,1) & (a,2) & (a,3) \\ (b,1) & (b,2) & (b,3) \\ (c,1) & (c,2) & (c,3) \end{bmatrix} .$$

The projection operation is a unary operation that removes elements that are not supplied as arguments. Projection is denoted π such that $\pi_{j,k}R_1$ returns the jth and kth sets of R_1. Note that ordering of j and k is arbitrary. The final operation in relational algebra is selection, denoted $\sigma\Theta R_1$, where Θ is a conditional statement that can be applied to each tuple of R_1. The conditions are built from comparison operators and Boolean logic operators, and depend on the type of information stored in R_1.

5.2.1 Emergence of RDBMS in GIS

In common parlance, we can think of R as a table, composed of records, defined by attributes (or fields, columns etc.). The formal relational algebra described above gives the following properties to R:

- Each row represents a unique n-tuple of R.
- The ordering of rows in R is not important.
- The ordering of columns in R is important.
- Columns are named with their domain label.

This organization of data allows insertion of additional records or deletion of existing records from each n-ary R. A major emphasis of the original relational concept was for users of the system not to interact with R directly (in which column ordering is important), but to work with "relationships," or ad hoc relations. Thus, columns must be uniquely identifiable across the DBMS. In modern relational database management systems (RDBMS), this is accomplished through accessing column names by prefixing with their relation (i.e., table) name.

The concept of keys is fundamental to the relational model. A *primary key* (PK) is a set of one or more columns in a table that uniquely identify each row in the table. Typically, on creation of the table, a column is assigned as the PK and populated with an auto-incrementing long integer. Consider a table describing land parcels in the Nanaimo municipal GIS dataset. In the *parcel* table structure in Figure 5.3, the field

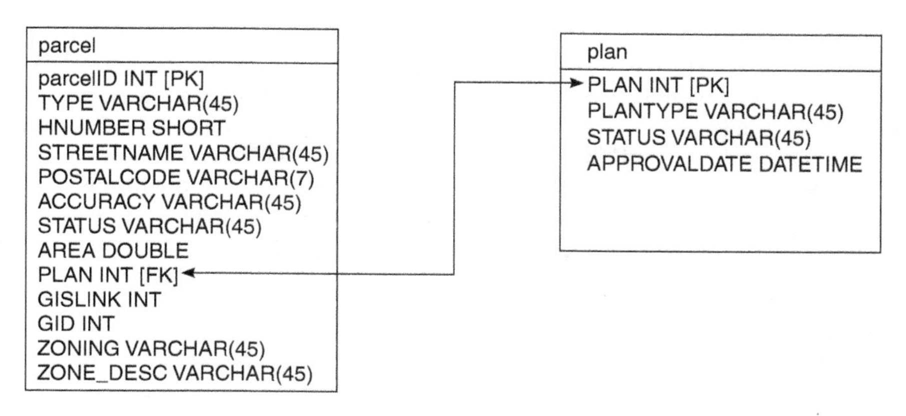

Figure 5.3 Graphical representation of database tables used in a municipal DBMS.

Table 5.1 A sample of a projection of data stored in the parcel table (not all columns shown).

parcelID	HNUMBER	STREETNAME	PLAN	ZONING1	ZONING1_DESC
289497	520	CHESTNUT ST	9448	R1	SINGLE DWELLING RES.
289498	521	DRAKE ST	9448	R1	SINGLE DWELLING RES.
289499	502	CHESTNUT ST	9448	R1	SINGLE DWELLING RES.
289500	511	DRAKE ST	9448	R1	SINGLE DWELLING RES.
289501	500	CHESTNUT ST	9448	R1	SINGLE DWELLING RES.
289502	555	DRAKE ST	9448	R1	SINGLE DWELLING RES.
289503	437	DRAKE ST	10791	R1	SINGLE DWELLING RES.
289504	568	DRAKE ST	6188	R1	SINGLE DWELLING RES.
289505	545	BRECHIN RD	6188	R1	SINGLE DWELLING RES.
289506	558	DRAKE ST	5606	R2	DUPLEX RES.

parcelID is a unique identifier functioning as the PK (although other candidate PKs also exist). When a PK is referenced in another table, it becomes a foreign key (FK) in that table. When multiple candidate primary keys exist, one is arbitrarily selected and denoted the PK. In the sample data in Table 5.1, the *parcelID* column is the PK.

The relation *parcel* contains columns such as *STREETNAME* among others that contain geographic information. While the locational descriptors may be unique to the parcels table, the columns *TYPE* and *PLAN* relate to the parcel type (e.g., park, parcel, strata) and the plan number that the parcel falls within. As such, *TYPE* and *PLAN* are *foreign keys* (i.e., they are PKs in another table). To access the *PLAN* column from the *parcel* table, we prefix with the table name so that *parcel.PLAN* is distinguishable from the table in which *PLAN* is a primary key (i.e., *plan.PLAN*). Figure 5.3 depicts a common way of showing a table definition, with table name at the top, followed by a listing of column names and their data types (database specific).

Classic relational theory describes relations with columns (called domains in Codd [24]) that can be defined by non-atomic elements, such as other relations. In this way, complex hierarchical relations can be embedded within such a system. Given relation R defining the columns in Figure 5.3 in the *parcel* table, it might be desirable to store the history of tax adjustments made to each parcel as a new column $TAXADJHIST$. In relational theory, this could be accomplished through a non-atomic domain that stores all of the tax adjustments made for each parcel in the *parcel* relation. However, the number of adjustments would vary by parcel (older parcels having more than new ones), and some may have none at all. This makes it impossible to store $TAXADJHIST$ within R if R is structured as a two-dimensional tabular array of rows and columns.

A naive solution to this would be to store duplicate records for each parcel for each distinct value of $TAXADJHIST$. However, this would be a very inefficient approach to data storage that incurs many issues. For example, if geometry were stored in one of the columns, spatial data would be duplicated multiple times. The better solution to this problem is to break up the data into multiple tables such that only atomic elements are stored in each relation (which can then be stored as a two-dimensional table), and tables are linked through key columns. The process of reorganizing the structure of data in a database to minimize redundant storage (often by breaking up tables into multiple tables) is called *normalization*. But before defining the logical view of tables, keys, and columns, we first have to have a firm understanding of *what* it is we are modelling—through a process called data modelling. Currently, the most widely used data modelling methodology is entity-relationship modelling, which maps our conceptual information needs to a logical design of database tables. Modelling methodologies designed for object orientation, such as the Unified Modelling Language (UML) and variants thereof, are also increasingly being adopted for object-oriented and object-relational databases.

5.2.2 Entity-Relationship Modelling

Entity-relationship (ER) modelling is a graphical method of designing a structure of tables in a database that meets the information needs of the database's users. For GIS databases (i.e., databases that contain geographic information), ER modelling is a critical step in spatial representation and integration of topological relationships and constraints. An entity in an ER model is a *thing* about which information is being stored, such as employee, road, forest stand, or student. A *relationship* in an ER model is a connection between two entities, such as works-at, connects, manages, or enrols-in. *Attributes* describe properties of entities, such as employee name, road length, forest stand area, or a student's registration status. Commonly, entities are nouns described by attributes, and relationships are verbs. With GIS databases, relationships can be spatial, and it is important to distinguish spatial and aspatial relationships. Specifically, topology can play an important role enforcing spatial relationships among entities, such as connects, adjoins, and is-within, which ensures that database representations match information needs associated with our view of the geographic world.

ER modelling is a graphical modelling method (see Figure 5.4). Entities are represented as rectangles, attributes as ellipses, and relationships as diamonds. Relationships are somewhat more complicated than entities, as they have a number of important

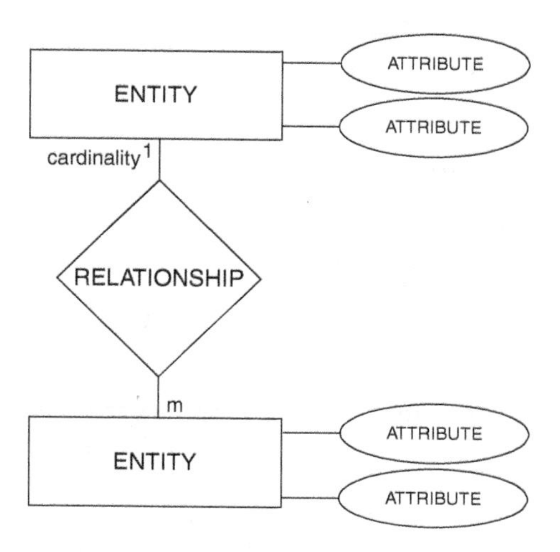

Figure 5.4 Basic components of an entity-relationship model.

properties. Relationship cardinality is the numerical relationship between instances of each of the entities involved in the relationship (see Figures 5.2, 5.4). For example, in the classic example of a student enrolling in a class, we determine the cardinality by considering for each student, How many courses can they be enrolled in? Typically, we define this simply as none, one, or many. We then consider the inverse—how many students are enrolled in each class—which takes values none, one, or many. Because in most schools one student enrols in many classes, while a class is made up of many students, the cardinality is many-to-many. Figure 5.5 outlines an ER diagram for the enrols relationship between entities student and class. Relationship cardinality is denoted by 1, 2, and m placed next to the entity rectangle; every relationship should have its cardinality clearly denoted.

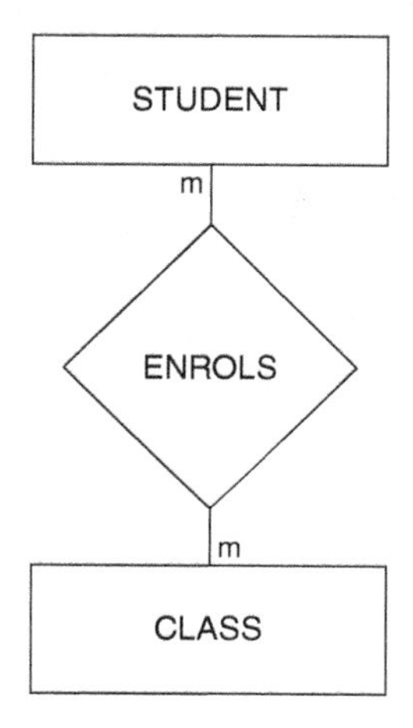

Figure 5.5 An entity-relationship diagram representation of students being enrolled in a class.

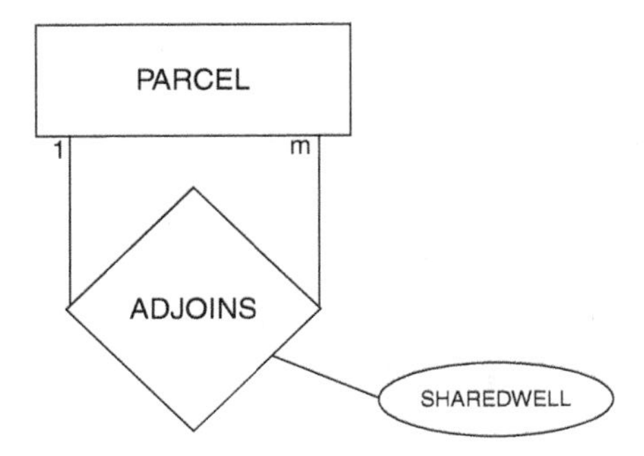

Figure 5.6 A binary recursive spatial relationship with an attribute in an entity-relationship model for parcels.

Relationships also have a *degree* property, which describes the number of entities involved in a relationship. Relationships can be binary, ternary, or n-ary, though typically relationships are binary. A special case of a binary relationship is the recursive binary relationship, where an instance of an entity is related to another instance of the same entity. For example, a land parcel might be related through an *adjoins* relationship to another land parcel (note that this is a spatial relationship) (see Figure 5.6).

Formally in Chen's (1976) notation [20], an entity set E_i is analogous to a table, and an entity e_i is an instance (or row) of an entity set, with properties defined by E_i. A relationship set R_i is a relation among n entities of the entity set

$$\{[e_1, e_2, \ldots, e_n] | e_1 \in E_1, e_2 \in E_2, \ldots e_n \in E_n\} \quad ,$$

whereby each group of entities (e.g., $[e_1, e_2, \ldots, e_n]$) is a relationship. An attribute is a *function* mapping a relationship set or entity set to a defined set of values or a Cartesian product of value sets. For example, in the Nanaimo municipal dataset, the attribute *ADDRESS* of an entity *HOUSE* might map to a subset of the Cartesian product of *HOUSENUMBER* and *STREETNAME*. Additionally, attributes can be defined on relationships. For example, the attribute *SHAREDWELL* might be an attribute on the *ADJOINS* relationship defined between entities in *PARCELS* in a rural area not part of the municipal water system (see Figure 5.6).

While entities and relationships are the core elements of ER modelling, there are numerous extensions and modifications to the original system of Chen [20]. ER modelling is a fundamental step in database design, aimed at capturing the semantics of the information needs of the proposed systems' users and mapping these semantics to a logical database structure that facilitates data integrity, optimized information access and retrieval, and efficient data storage and manipulation. Entities and attributes are not inherent, but context-specific. Much of the process of determining entities, attributes, and relationships is iterative and collaborative, requiring input from domain experts and users. For example, when is "province" an entity and when is it an attribute? One rule of thumb is that if an object requires only itself as an identifier and has no additional descriptions, it should be an attribute, whereas if it has multiple descriptors, it should be an entity. If, for example, "province" is being used as a locational identifier for a business, it would be an attribute, whereas if multiple socioeconomic variables were being compared across provinces, an entity representation would be suitable.

ER modelling in GIS in practice—a worked example

As noted earlier, ER modelling and RDBMS were not originally conceived of with geographic information in mind. However, RDBMS are the most common way of storing geographic information in large organizations, as many of the system requirements for geographic and non-geographic information storage are the same: multi-user access (with multiple data privileges defined through roles), data integrity mechanisms (transaction controls, authentication) with the ability to serve multiple client applications (field data collection software, website, statistical analysis software), and the ability to effectively model the semantics of the real world. The database-centric view of GIS did not gain traction until the release of Oracle 7.3.3 in 1997, which was the first database to have support for spatial data and indexing. In many organizations, penetration of GIS into their information systems is piecemeal—and users are often dealing with a patchwork of legacy systems, file types, processes, and representations—making GIS in the workplace significantly more complex than textbook examples.

Land parcels are a fundamental entity in municipal management information systems. In Nanaimo, BC, land parcels are units of land delineated according to land ownership. The official definition of parcels in British Columbia is "a lot, block or other area in which land is held or into which land is subdivided" (Land Title Act, Ch. 250, RSBC 1996). Land parcels factor into many municipal business processes, one of the most important being property tax assessment. Tax assessments are conducted by another government corporation responsible for tax assessment throughout the province of BC. Assessors carry out property tax assessments, and the municipality is responsible for tax collection. However, tax rates are set by the municipality and are scheduled according to zoning. So for example, in the municipal GIS RDBMS for tax collection, the relationship between zoning and parcels needs to be represented, such as in Figure 5.7.

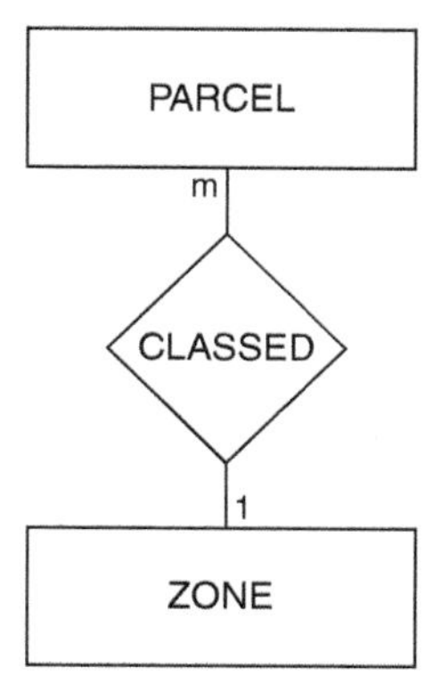

Figure 5.7 A classic ER representation of entities for parcel and zoning data. Cardinality of the relationship indicates that each zone classifies many parcels, but each parcel has only one zoning classification.

Table 5.2 A projection of data stored in the parcel table.

parcelID	HNUMBER	STREETNAME	PLAN	ZONING1	ZONING2	ZONING3	ZONING4
281252	680	TRANS CANADA HWY	11762	PRC1	W2	W1	W3

This representation indicates that each parcel is classified as one zone type. Although this may seem sensible at first, there are cases whereby one parcel has multiple zonings (e.g., an apartment complex with a cafe on the ground floor). In the database derived from the ERD in Figure 5.7, this could not be represented. The shapefile attribute table (Table 5.2) indicates a naive solution to this problem by addition of extra fields (*ZONING2*, *ZONING3*, *ZONING4*). Thus, four zonings are possible per parcel. However, if this tactic were taken during database design, it would require knowing the dimension of this and other relationships beforehand, making the database inflexible to future changes and storing redundant data—poor database design principles. A better alternative would be to allow n zonings per parcel. This could be accomplished simply by altering the cardinality of the relationship such that each parcel is classified as one or more zonings, and each zone is associated with one or more parcels. In keeping with a normalized database design, this is realized in practice through a relationship table that stores only unique combinations of parcels and zonings (*PARCEL-ZONING* in Figure 5.8).

To represent the tax assessment system in a RDBMS, we need to consider the following semantics:

- Multiple legal parcels are possible in one assessment parcel.
- Assessment parcels are often subsets of a single legal parcel (under a common zoning; e.g., mobile home parks).

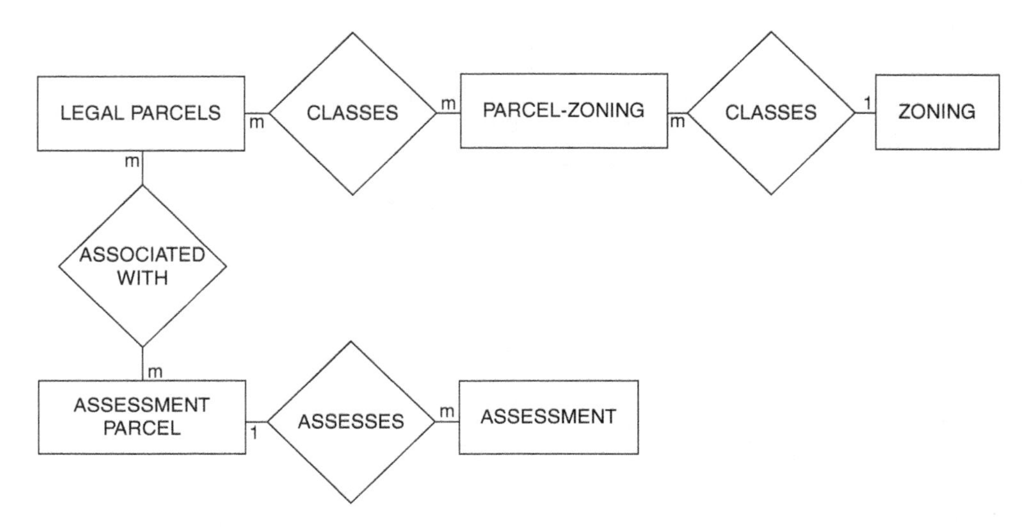

Figure 5.8 A possible ER representation of tax assessment tables in a municipal information system.

- Multiple zonings are possible per legal parcel.
- Tax rates are scheduled according to zonings.
- Legal parcels are zoned by the city.

In the vast majority of cases, assessment parcels and legal parcels will be the same, but we need to be able to design the database to handle all possible (realistic) scenarios that may be encountered. We can formalize these semantics by the following ER constructs:

- Define entities for legal parcels, assessment parcels, zonings, and assessments.
- Enable many-to-many legal parcels to zonings through a relationship table.
- Define tax rate as an attribute of zonings.
- Define a many-to-many relationship between legal parcels and assessment parcels.
- Define a one-to-many relationship between assessment parcels and tax assessments.

To be able to calculate tax bills, we would need to join tables from assessment parcels to zoning to determine the appropriate rate for each assessment parcel, and then multiply by each parcel's area (assuming taxes are per m^2). Once calculated at the assessment parcel level, these assessments could then be used to create tax bills. For situations such as multiple zoning in an apartment complex, we would have overlapping sub-geometries, only differentiated in the z-dimension—difficult to represent in many GIS database systems.

5.2.3 Generating Candidate Tables and Database Normalization

Database normalization is the process of creating tables that store atomic, non-redundant data. Translating a database scheme represented in an ER model to an actual database requires specification of appropriate SQL statements. This process can be automated through the use of computer assisted software engineering (CASE) tools, which allow a graphical ER model to be converted to SQL statements suitable for database creation in commercial and open source RDBMS. The process of normalization occurs iteratively, as originally conceived entities are broken down into constituent subsets of their columns and linked through keys.

A set of rules, called *normal forms*, can help guide the design of a normalized database. Normal forms function as best practices that help a database designer adhere to the principles of sound data modelling to ensure data integrity and reduce redundant storage. *First normal form* (1NF) states there are no repeating columns within a table. This is the most general level of normalization. In the Nanaimo dataset (see Table 5.2 and Figure 5.9), the *parcel* table is not in first normal form because the zoning field is repeated (note that this is a derived table and not representative of the database used by the City of Nanaimo).

A 1NF version of this would be as outlined in Table 5.3 as in Figure 5.10, where the zones are broken out into multiple rows, and the data in other columns are repeated. This form, albeit still problematic, at least supports an arbitrary number of zonings per parcel. However, there are clear problems as values of the PK are replicated over multiple rows, and the PK would now have to be defined on a combination of multiple columns (e.g., *parcelID* and *ZONING*).

parcel
parcelID INT [PK] TYPE VARCHAR(45) HNUMBER SHORT STREETNAME VARCHAR(45) POSTALCODE VARCHAR(7) ACCURACY VARCHAR(45) STATUS VARCHAR(45) AREA DOUBLE PLAN INT [FK] GISLINK INT GID INT ZONING1 VARCHAR(45) ZONE1_DESC VARCHAR(45) ZONING2 VARCHAR(45) ZONE2_DESC VARCHAR(45) ZONING3 VARCHAR(45) ZONE3_DESC VARCHAR(45) ZONING4 VARCHAR(45) ZONE4_DESC VARCHAR(45)

Figure 5.9 The parcel table is in un-normalized form, as the zoning column repeats multiple times.

Table 5.3 Sample of data stored in the parcel table in first normal form.

parcelID	HNUMBER	STREETNAME	PLAN	ZONING
281252	680	TRANS CANADA HWY	11762	PRC1
281252	680	TRANS CANADA HWY	11762	W2
281252	680	TRANS CANADA HWY	11762	W1
281252	680	TRANS CANADA HWY	11762	W3

parcel
parcelID INT [PK] TYPE VARCHAR(45) HNUMBER SHORT STREETNAME VARCHAR(45) POSTALCODE VARCHAR(7) ACCURACY VARCHAR(45) STATUS VARCHAR(45) AREA DOUBLE PLAN INT [FK] GISLINK INT GID INT ZONING VARCHAR(45) ZONE_DESC VARCHAR(45)

Figure 5.10 The parcel table in first normal form, as the zoning column does not repeat multiple times.

Second normal form (2NF) is a higher level of normalization that applies the concept of functional dependencies between columns. Functional dependence between a set of columns A and another set B exists when values for A are uniquely associated with values of B, or rather, each unique value in A has one and only one value in B. Functional dependence is denoted $A \rightarrow B$, where A functionally determines B. When A is the PK, all other columns are functionally dependent on A, since by definition the PK must uniquely identify each row. The problems in the 1NF table in Table 5.3 are clear: the parcel address is repeated for every record. If the address or plan number of the parcel were to change, every record would have to be updated individually, leading to the possibility of inconsistent data. In terms of functional dependencies, we can see that columns *HNUMBER*, *STREETNAME*, and *PLAN* are not functionally dependent on the PK *parcelID*, *ZONING*. Rather, we should break up the table into one that associates zoning to parcel and another that records unique parcel-level information, as in Table 5.4 and Table 5.5,

If multiple sub-geometries were available for the zones within the parcel, these could be stored as a spatial data type (described later in this chapter) in the parcel-zoning table.

Third normal form states that the table must be in second normal form and all non-PK columns are dependent on the PK. Establishing 3NF is largely about removing what are called transitive dependencies, which are dependencies between non-key columns. In the Nanaimo example, this could be storing information about planning areas themselves, which contain multiple parcels (See Figure 5.11). For example, if we wanted to store details about the senior planner associated with each plan that each parcel is a part of, such as in Table 5.6, we would recognize that the column *PLANNER* depends more on *PLAN* than it does on any other column in the table and could therefore be broken out into its own table to achieve 3NF.

The process of database normalization typically seeks to establish design to 3NF. The seemingly simple example of land parcels, zoning, and tax collection illustrates the importance of proper database design practice at the earliest stages. There are

Table 5.4 Sample of data stored for parcel data in second normal form.

parcelID	ZONING
281252	PRC1
281252	W2
281252	W1
281252	W3

Table 5.5 Sample of data stored for parcel data in second normal form.

parcelID	HNUMBER	STREETNAME	PLAN
281252	680	TRANS CANADA HWY	11762

Figure 5.11 The parcels shaded according to *PLAN* number.

higher normal forms, but they tend to be used only in very special circumstances. Also note that there is computational overhead to joining separate tables created by the normalization process. When the size and complexity of relationships in databases grows very large, this computational overhead becomes very important. Recent trends in database storage to accommodate the advent of *big data* have moved away from normal forms and classical RDBMS to more flexible, less-structured data storage designs (NoSQL and key-value databases).

GIS database requirements

The extension of classical RDBMS to geographic information places new demands on what is expected of the database. Many of the requirements were summarized in Hadzilacos and Tryfona [57]. GIS databases must be able to represent an object's location, shape, and orientation. The form of representation is inherently tied to spatial scale. Thus, GIS databases have to be capable of multiple scale representations of geographic objects, such that a GIS database storing information about cities over North America can represent them as point coordinates at small (map) scales, and polygonal areas at large scales. We also need to be able to determine spatial relationships between geographic objects, either on-the-fly through object locations or through explicit storage via topological look-up tables. Spatial scale and spatial relationships are inherently linked, as some relationships are highly sensitive to changes in spatial representation. For example, the distance between two polygons can be calculated in a number of ways, whereas calculating the distance between two point objects is straightforward. As such, inferences made from these spatial relationships are highly contingent on spatial representation.

Table 5.6 Sample of data stored in the parcel table in second normal form.

parcelID	HNUMBER	STREETNAME	PLAN	PLANNER
281252	680	TRANS CANADA HWY	11762	Jim
281239	690	TRANS CANADA HWY	11762	Jim
281223	700	TRANS CANADA HWY	11763	Omar
280713	38	CAMPBELL ST	11016	Cedric

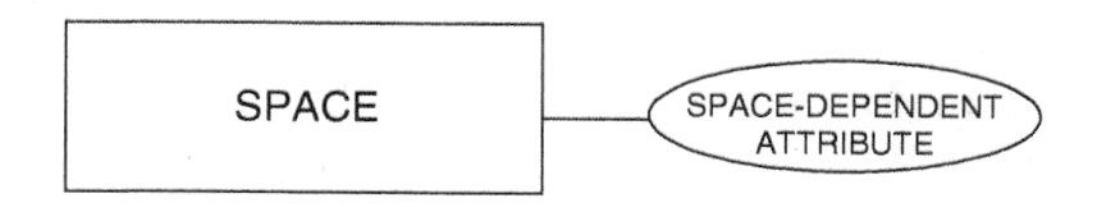

Figure 5.12 An ER representation of spatially continuous fields as space-dependent attributes of a global space entity.

Another desirable feature of a GIS database is the capacity to represent complex geographic objects such as universities, condominiums, or ski resorts as complete entities unto themselves. A complex geographic object can be thought of initially as a union of the constitutive geometries making up the object, but ultimately, we would like to be able to represent these objects including dynamic behaviours and the ability to interact with other objects.

An essential aspect of GIS that hasn't been mentioned in the context of DBMS is spatially continuous phenomena (i.e., fields). The field view of space is inherently non-object-like, and does not fit naturally into the entity-relationship data modelling method. One way around this is to consider spatial fields as properties of a single global object—space (see Figure 5.12). Thus, they are sometimes called space-dependent attributes. For example, a point object representing a meteorological station might contain station elevation as an attribute; however, if the station were relocated 20 km away, that elevation attribute value would no longer hold. A new value would have to be stored, based on the elevation at the new location. While intuitive, this gets at a fundamental idea behind object orientation—object identity. From a database administrator perspective, a space-dependent attribute (i.e., a field) is a function whose domain is space, and range is any set of values.

5.3 Objects and GIS

An object-oriented (OO) approach to data storage and management is an alternative to RDBMS that accords DBMS functionality to objects defined according to object-oriented principles. Object orientation describes an approach to software that emphasizes the development of standalone discrete software objects that interact to form an application. In OO systems, data and application logic acting on data are bundled together into discrete packages. The need to represent, in DBMS, geographic objects (e.g., GIS, CAD/CAM) that were not effectively handled in relational systems was one of the main drivers for the development of OO-DBMS. The dynamism associated with real-world phenomena is much more conducive to an OO representation. For example, a store object can include all the physical geometries associated with the store itself, as well as behaviours associated the store and its information such as order-inventory and hire-employee. Behaviours of an object are actions that change the properties of itself or other objects. The principles of object orientation support a more complex representation of discrete phenomena.

5.3.1 Key Concepts in Object Orientation

An OO-DBMS overcomes shortcomings of RDBMS through its ability to natively represent complex objects. An object (or object class) is any thing in the real world we want to store as a computer representation, and application logic is accomplished through objects passing messages to one another rather than as a sequence of instructions and/or procedures. Thus, an OO system is inherently more dynamic. Message passing (sometimes called interfacing) occurs through the definition of properties (analogous to attributes), methods (that model behaviour), and events. For example, a store object might have a property *address* and a method *sale* that interact with a customer object in an OO-DBMS for a business. A class (or object-class) is defined with a set of properties, methods, and events, and instances of this class are created as objects that exist in memory. Access to contents of objects is wholly restricted to that which is available through its publicly exposed interfaces. The actual data structures of objects are hidden and can therefore not be accessed by generalized languages or functions the way query languages are used in relational databases. However, this independence, in theory at least, allows internal code in an object to be revised without affecting code that uses the object. See Figure 5.13 for a sketch of how this approach might work for our parcels database example.

Three fundamental properties of OO programming languages and OO-DBMS are inheritance, encapsulation, and polymorphism. *Inheritance* means that objects can be defined as a class type that automatically inherits the properties of methods of its parent class. This is a powerful concept that maximizes code reuse and improves maintainability by having objects defined only once. In the municipal tax-assessment example discussed above, the tax assessor may have different assessment protocols for different land tenure types; however, all assessments may have some things in common. Class inheritance can easily model more-specialized objects as sub-objects in an object hierarchy, which implements more specialized properties and behaviours.

Encapsulation is bundling of data and methods together. This bundling makes the internal representation of the data not important to the user, who accesses and manipulates object data through defined methods. This information-hiding aspect of OO means that operations on objects are strictly confined to message passing (i.e., no direct access). In a pure OO language such as Smalltalk, everything is an object,

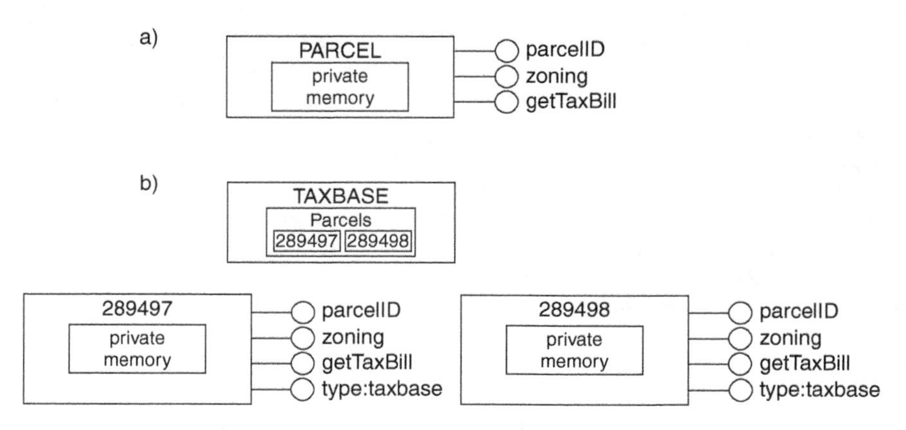

Figure 5.13 An object oriented example of (a) a class and (b) an object hierarchy.

including data types, which in turn inherit from classes, whose class definitions can be changed. In other languages such as Java and C++, a set of foundational data types confines object definitions to assemblages of existing foundational data types. Encapsulation and polymorphism are key concepts that support representation of complex geographic objects in GIS. One example of OO-GIS is Smallworld, which combines an OO programming language and an OO-DBMS to model complex spatial and topological representations.

Polymorphism is the ability of an object to respond differently depending on the context. That is, different objects respond to the same function calls based on their internal composition (i.e., type). Thus, a function that returns an object's location called on a polygon object might return a bounding box, whereas the same method called on a point object would return its x-y coordinates.

OO-DBMS implement OO principles to store, retrieve, and manipulate data. Implementation of OO-DBMS-based GIS has been much more limited than originally predicted. While OO approaches solved some of the problems of RDBMS, they introduced new ones. Bundling of object data and behaviours makes the building of OO-DBMS a fairly extensive programming exercise. Whereas the quest for data independence in the design of the relational model sought to decouple application logic from data storage logic, encapsulation turned this concept on its head. This loss of generality has precluded OO-DBMS or OO-GIS from becoming mainstream technologies. However, incorporation of some of the OO ideas into the relational framework has attempted to achieve the gains of the OO approach while maintaining the RDBMS approach.

A recent example of object-based storage of spatial data is the S4 spatial classes in the statistical programming language R, which is increasingly used for geographic information processing and analysis (i.e., as a GIS). R classes are defined using "slots," which can store any basic data type (numeric vectors, factor vectors, character vectors, scalars, etc.) or store additional custom classes, and are instantiated as objects when a user uses the program for analysis. The spatial classes are called sp classes and include representations for points, lines, polygons, and grids, with or without attribute data. Since R already has class definitions for storing tabular data (called data frames), the spatial classes simply extend the data frame classes using object inheritance. The spatial aspects are stored in spatial classes. Each spatial class—such as SpatialPolygons, SpatialPoints, and SpatialLines—inherits from the base class Spatial, which implements methods common to all descendant spatial classes (i.e., polymorphism). All spatial classes have a slot called "proj4string," which includes spatial reference system information in PROJ.4 format and the bounding box of the spatial coordinates. Individual spatial classes such as the SpatialPoints class include methods for extracting coordinates as a two-column by n-row matrix, while SpatialPolygons includes inherited methods for composite geometric objects of class Polygon, such as ring direction, the area of the polygon, and whether the polygon is a hole or not. Higher-order geometry objects are not built up from lower order objects (i.e., SpatialLines are not groups of SpatialPoints), and topology is not stored.

The key use of object orientation in R spatial objects (and the R language generally) is method over-loading, which means that methods can be implemented to respond differently depending on the type of object that is passed to the function. In the following example we will build up some points and polygons from scratch and pass them to a plot function, then we will use the summary function on them. This behaviour is a

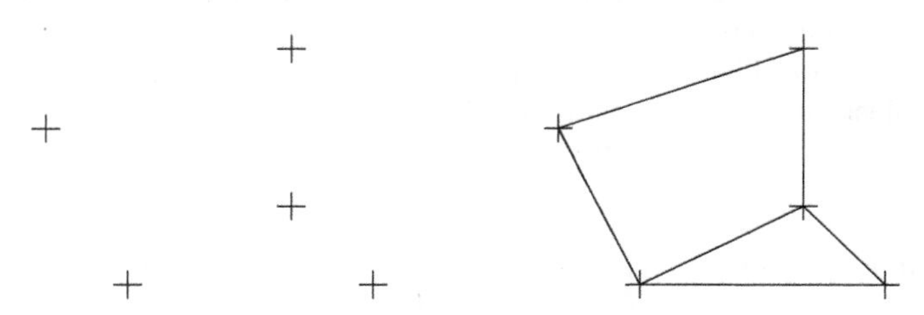

Figure 5.14 Output from sample R code plotting.

very helpful aspect of OO, as the same methods can be used on all sorts of object types. The results of the plotting functions are shown in Figure 5.14.

```
> Sr1.x = c(2,4,4,1,2,5,4,2,5)
> Sr2.y = c(2,3,5,4,2,2,3,2,2)
> SpPt = SpatialPoints(cbind(Sr1.x, Sr2.y))
> plot(SpPt, cex=2, lwd=2)
> Sr1 = Polygon(cbind(c(2,4,4,1,2),c(2,3,5,4,2)))
> Sr2 = Polygon(cbind(c(5,4,2,5),c(2,3,2,2)))
> Srs1 = Polygons(list(Sr1), "s1")
> Srs2 = Polygons(list(Sr2), "s2")
> SpPol = SpatialPolygons(list(Srs1,Srs2), 1:2)
> plot(SpPol, add=TRUE)
> summary(SpPt)
Object of class SpatialPoints
Coordinates:
      min max
Sr1.x   1   5
Sr2.y   2   5
Is projected: NA
proj4string : [NA]
Number of points: 9
> summary(SpPol)
Object of class SpatialPolygons
Coordinates:
  min max
x   1   5
y   2   5
Is projected: NA
proj4string : [NA]
```

The spatial classes illustrated in this example are not true GIS classes because they do not handle attribute information. This is added by objects that extend both Spatial classes and Data Frame classes, such as SpatialPointsDataFrame, and SpatialPolygons-DataFrame classes. Depending on the type of operation required, GIS functions in R will operate on just the spatial data or the spatial data and the attribute data (e.g., spatial join summaries).

5.4 Object-Relational Databases

Object-Relational (OR) DBMS are RDBMS that implement features of OO that enable them to handle complex geographic objects. Thus, O-RDBMS are hybrids. While they maintain tables as the primary stores of data, they also extend classical RDBMS by providing support for user-defined abstract data types, support object class definitions and inheritance, and allow for user-defined methods and behaviour. Modern GIS databases are O-RDBMS, which provide support for GIS objects such as networks, custom topologies, and spatial relationship classes. For example, a stream-network class might be composed of sub-classes streams and rivers, each of which has unique class definitions. However unlike in OO-DBMS, access and manipulation of data are primarily via a query language (e.g., SQL). Further, as hybrids, OR-DBMS are not grounded within a theoretical framework similar to relational or object-oriented approaches but have been developed as extensions to existing systems. For example, query languages have been extended specifically for handling complex geographic objects in OR-DBMS, though implementations tend to be vendor-specific, and a true standard has not emerged.

5.5 Database Storage

GIS data represent a variety of real-world phenomena that are discretized to a small subset of representations within a database.

5.5.1 Data Types

A fundamental building block of almost all databases is data types. Data types define what and how data can be stored at the most basic level or representation inside a database. Database types are specific to individual databases, and some are machine-dependent as well. As RDBMS are extended into O-RDBMS, an increasing number of pre-defined data types are being supported. Typical data types include short integer (2^8 bits representing integer numbers), long integer (2^{64} bits), Boolean (logical true/false), character (for text data), double precision (2^{64} bits representing decimal numbers), and date and time types. Determining data types for representing attributes is a critical step in data modelling.

Numeric data storage: integer and floating point numbers

GIS operations involve extensive numerical operations, which depend on numeric data stored in a binary representation on a hard disk. Numbers are stored as either integers or floating-point representations that approximate real numbers. Integer storage is typically discretized into two data types: 16-bit *short integers* and 32-byte *long integers*. Short integers can represent 2^{16} numbers, either signed (−32,768 to 32,767) or unsigned

(0 to 65,536), and long integers, 2^{32} different numbers, signed (−2.1 to 2.1 billion) or unsigned (0 to 4.2 billion).

Representing real numbers is significantly more complicated because the precision (quantization) of storage is arbitrary. Floating-point data types use a fixed number of digits, called the significand or mantissa, which are scaled by some exponent relative to a base (2 for binary, 10 for decimal numbers). The most widely used floating-point encoding standard is IEEE 754, which defines a finite number as $(-1)^s \times c \times b^q$, where s indicates the sign of the number ($s=1$ negative, $s=0$ positive), c is the coefficient (or significand), b is the base, and q is the exponent. Thus, $(-1)^1 \times 1.430625 \times 2^4 = -22.890$. The term *floating point* refers to the position of the decimal place, which is allowed to float (i.e., move forward or backward) depending on the value of the exponent.

Differences in floating-point representations are based on the size of the coefficient and exponents limited by the amount of computer memory allocated to that type and can vary by hardware and operating system. For example, a double-precision floating-point number occupies two bytes (i.e., 16 bits) of memory. The decimal64 format is a double-floating-point format with 16 digits in the coefficient. The 16-digit coefficient is either encoded as a binary-coded positive integer or a decimal-coded positive integer, though both representations yield the same range of representable magnitudes, with exponent range from −383 to +384. While such precision is sufficient for storing almost all conceivable geographic data, round-off errors can still occur when data are processed and analyzed with geometric algorithms (see Chapter 8 discussion of computer arithmetic). In terms of field definitions for attributes in a GIS, most non-spatial properties that require decimal measures can be represented as double–floating-point numbers.

Character storage: ASCII and Unicode

Storage of character data is determined by two parameters: character encoding and number of allowable characters. Character encoding is a way to map a set of characters (e.g., the English letter alphabet) to a set of binary strings (or codes). The system of encoding is defined by the maximum representable characters allowed by the system. So an 8-bit encoding system stores every character using 8-bit binary strings. The number of representable characters in an 8-bit system is 2^8. Note that binary representation is not the only way to encode data. Other encodings use different bases such as hexadecimal (base of 16) or duodecimal encoding (base of 12).

One of the most widely used character encoding systems is the American Standard Code for Information Interchange (ASCII), which is a 7-bit encoding system (2^7 possible characters). The first 33 characters in the ASCII system are control characters, which are not used for actual character encoding, leaving 95 actual characters that can be represented in the system. Table 5.7 gives some example character encodings using ASCII and the decimal and hexadecimal equivalents. Note that Arabic numbers are included in the characters set, meaning that numbers can be stored as characters using ASCII. In GIS this type of data is common. For example, field measurements of rainfall data are often recorded in millimetres down to one-tenth of a millimetre (e.g., 2.3 mm) and should be stored using floating-point or integer (by adjusting units) data types. However, in practice, very slight amounts of rainfall occur—less than 1 mm, but still

Table 5.7 Character encoding examples in ASCII, decimal, and hexadecimal encoding systems.

Binary	Decimal	Hexadecimal	ASCII
100 0001	65	41	'A'
110 0001	97	61	'a'
011 0111	55	37	'7'
1000111 1001001 1010011	71 73 83	47 49 53	'GIS'

observable—and are often recorded as "Trace" or "T" to denote the presence of rainfall. Data from weather stations recording rainfall this way, if stored in a single column in a database, would have to be stored as character data (at least initially), using a character encoding system. To do numeric computations on the data, numbers stored as characters would have to be converted to their numeric equivalents (although the T's would have to be replaced first). One strategy for dealing with trace amount would be to set it to a very small numeric value (e.g., 0.01).

While 95 characters are sufficient for English language and grammar, there is no way to represent other languages within this system. Consequently other encoding systems have emerged and become much more widely used than ASCII. Unicode is a standard for character encoding that contains over 100,000 characters. There are a variety of character-encoding flavours of Unicode, the most widely used being UTF-8 and UTF-16, which are 8-bit and 16-bit systems, respectively. Importantly, UTF-8 is backward compatible with ASCII such that the code values in UTF-8 are the same as in the ASCII character set but extended to additional characters. UTF-8 is the standard character encoding system used on the Internet.

Spatial data storage: well-known text, well-known binary, GeoJSON, etc.

Database storage of spatial data has become a key feature of O-RDBMS. Storage of spatial data is implemented through a set of standards defining user-defined types for representing simple geographic information. The OpenGIS implementation specification for geographic information [62] describes a common architecture for both binary and text storage of basic geometric primitives (points, lines, polygons). The SQL Geometry Type definitions are organized into a hierarchy as outlined in Figure 5.15. The types represented are independent of the underlying DBMS.

Geometry is a super class associated with a spatial reference system (SRS). Thus, objects defined as type geometry are associated with a SRS. This architecture supports separate SRS between objects in one table, as well as the definition of methods that check for SRS compatibility among geometry objects supplied as arguments. The implementation of SRS associations to geometries is through an SRS Identity (SRSID), which uniquely defines every SRS in the database. The definition of each SRS is stored in a separate table. Sub-types of the geometry type include point, curve, surface, and geometry collection. Geometry collections support multiple geometry objects of different types—a feature that directly supports the storage of complex geographic objects. The representation of the data types in Figure 5.15 can be realized through text or binary constructors.

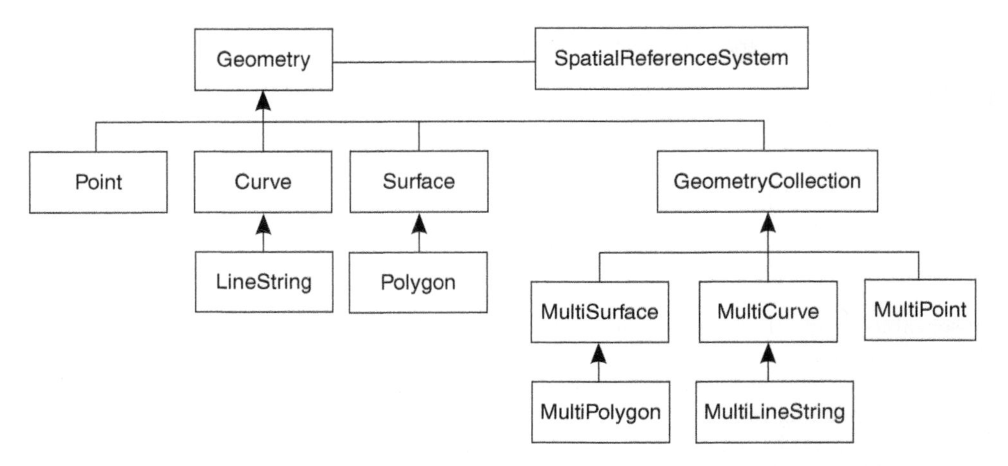

Figure 5.15 Hierarchy as data types for SQL geometry storage as defined in OpenGIS simple features standard.

Web-mapping and web data generally have seen massive growth in the use of JavaScript for web programming and scripting. Coincident with this has been the growth of a JavaScript-based data model, called JavaScript Object Notation (JSON). JSON is a text-based data interchange format increasingly used to transfer data to and between web applications. Objects are defined as groups of name-value pairs called *members*, whose values can be another object, an array, a number, or a string. An object is specified by curly braces; for example, this sample object (provided in the JSON specification)—

```
{
 "Image": {
    "Width":  800,
    "Height": 600,
    "Title":  "View from 15th Floor",
      "Thumbnail": {
         "Url":    "http://www.example.com/image/481989943",
         "Height": 125,
         "Width":  100
      },
      "Animated" : false,
      "IDs": [116, 943, 234, 38793]
    }
  }
```

—specifies an object that has a member Image, which is also an object and includes a member Thumbnail, which is itself another object. The member IDs has an array as its value. Although geographic data can be stored as JSON objects through definition of custom spatial JSON objects, a standard has emerged called GeoJSON, which is JSON encoding of geographic data, including objects for various geometries. A GeoJSON object must have a member with the name "type," which specifies the type of geometry as a string (e.g, Point, MultiPoint, LineString, MultiLineString, Polygon, MultiPolygon, GeometryCollection, Feature, or FeatureCollection) as well as a coordinates "member," which will have a value of a position (defined by one or more arrays of coordinates). Other optional members of a GeoJSON object include coordinate reference system and bounding box information. The point object in the R example above would have the following specification in GeoJSON multipoint object:

```
{ "type": "MultiPoint",
    "coordinates": [ [2, 2], [4, 3], [4, 5], [1, 4],
    [2, 2], [5, 2], [4, 3], [2, 2], [5, 2] ]
}
```

or as a GeoJSON polygon object,

```
{ "type": "Polygon",
    "coordinates": [
      [ [2, 2], [4, 3], [4, 5], [1, 4],
      [2, 2], [5, 2], [4, 3], [2, 2], [5, 2],[2, 2] ]
      ]
   },
```

which includes an extra point to close the polygon at the start location, as is required for GeoJSON polygons. These objects are purely spatial and include no attribute data. As JSON objects, any additional members could be defined to store attribute information, but there are no guarantees that members will exist for all objects. A built-in type for storing attribute is called "features," which has a member called "properties," which has another object as its value. In this way, complex spatial features can be represented using GeoJSON notation. Because JSON is a serialization format (i.e., can be translated to a sequence of bits easily), it can be used to transfer data between systems. GeoJSON is becoming used for real-time geographic information storage and processing, as it can be used to stream bits of information almost continuously updating positional and attribute information. These features make the format very useful for web mapping applications with large and dynamic datasets. For larger datasets, a binary version of JSON is available: binary object notation (BSON).

5.6 Database Access and Manipulation

5.6.1 Structured Query Language (SQL)

Returning to relational schemas, being able to access data is obviously a critical function required of an information system. SQL is the most widely used query language for accessing and manipulating data within DBMS. The language was founded upon Codd's relational algebra, which set out the rules by which data stored in relations could be accessed by generic query language. SQL is the main way that users interact with data stored in a DBMS—through queries that insert, modify, and retrieve data, or queries that alter the database structure itself (sometimes called action queries). SQL operates at the logical design level, using language constructs that treat the data as a set of tables, records, and columns, and are thus independent of underlying physical storage. While numerous variants of SQL exist, the basic constructs of the language for querying data are largely similar across databases. Where extensions have been made to SQL to provide additional functionality (e.g., spatial indices, spatial relationship operators), differences between database implementations tend to be greater.

To retrieve data from a table, a *SELECT* query is called, which specifies the columns, tables, and conditions of the data request. For example,

```
SELECT *
FROM parcel
WHERE  STREETNAME LIKE "TRANS CANADA HWY";
```

This query would return all columns and all records in the table called "parcel" where the value of the *STREETNAME* column is equal to "TRANS CANADA HWY." The special character * indicates that all columns should be returned. Note that in most file-based GIS, SQL queries are implemented through a WHERE clause only, in order to make selections on attribute tables of individual layers. In a RDMBS, SQL is generally implemented through a built-in text editor that connects to the database. The SELECT [columns], FROM [tables], WHERE [conditions] structure can include Boolean operators AND, OR, NOT. For example, to find parcels on the Trans Canada Highway that are not zoned as parkland, we could use the following query:

```
SELECT *
FROM parcel
WHERE  STREETNAME LIKE "TRANS CANADA HWY"
AND ZONING1 NOT LIKE "PRC1";
```

This query would make sense only if there were no multiple-zoned parcels that included parklands; otherwise, parcels where ZONING2 or ZONING3 were zoned as PRC1 could still be returned. This highlights the problem of tables that are not in 1NF. To be sure, the query could be altered to add conditions on ZONING2, ZONING3, and ZONING4 columns. Queries are used for a number of common GIS procedures such as creating selection sets based on attribute criteria, creating data subsets for export, and creating database views (dynamic queries).

Data definition language (DDL) constructs in SQL

DDL constructs in SQL enable the creation, alteration, and deletion of database tables. Commonly used SQL DDL constructs are outlined in Table 5.8.

Data manipulation language (DML) constructs in SQL

DML constructs in SQL enable querying of database tables. Commonly used SQL DML constructs are outlined in Table 5.9.

Operators and keywords in SQL

Operators and keywords in SQL enable Boolean logic, comparison, set operations, and aggregate functions to be implemented in queries. Boolean logic and comparison

Table 5.8 SQL Data Definition Language constructs.

SQL Command	Meaning
CREATE TABLE	define a new table structure and all of its columns
DROP TABLE	delete an existing table and all of its data
ALTER TABLE	add, modify, or delete columns in an existing table
CREATE VIEW	create a new view from an SQL query
DROP VIEW	delete an existing view

Table 5.9 SQL Data Manipulation Language constructs.

SQL	Meaning
SELECT	retrieve data from the database based on a set of criteria
UPDATE	change existing data in the database
INSERT	add new data to an existing database table

operators are implemented within the WHERE clause of SQL queries. Set operations enable multiple queries to be concatenated (when the structure of queries is equivalent). Nested subqueries are another way to implement complex query logic, where a nested SELECT statement is used as a subset for another query. For example, in Table 5.10, a nested subquery on the taxbill table returns a subset of parcelIDs that have a status of outstanding, and these are used to match to the parcelIDs in the parcels table, which is the source of the columns actually returned by the query (i.e., all columns denoted by *).

Joins—querying multiple tables

Table joins are an integral feature of RDBMS and are required to make use of a normalized database. Selecting data from multiple tables without joining results in a Cartesian product, which is the set of all combinations of the two tables. Consider two simple tables, A and B, each with two columns and three rows; see Figure 5.16.

The Cartesian product in Figure 5.16 would be obtained in SQL simply by selecting all records from both tables:

```
SELECT *
FROM A, B;
```

In practice, the results of the full Cartesian product are rarely desired and can become vary large (the product of number of rows in A and B). The Cartesian product is an intermediate step in most queries; it is created and then filtered according to the subsets defined by the join.

Joins between tables use relationships of primary and foreign keys to link together related entities. There are two types of join specification: implicit and explicit. Explicit joins use the JOIN keyword to link tables through keys. Implicit joins are independent of

Table 5.10 SQL operators and keywords.

SQL operators and keywords	Example
AND	SELECT * FROM parcel WHERE AREA >= 2000 **AND** ZONING1 LIKE "R1"
OR	SELECT * FROM parcel WHERE AREA >= 2000 **OR** ZONING1 LIKE "R4"
NOT	SELECT * FROM parcel WHERE AREA >= 2000 **NOT** ZONING1 LIKE "R6"
=, <><, <=, >, >=	SELECT * FROM parcel WHERE x >= 2 AND y <> 7
IN, ANY, ALL, IS NULL, IS NOT NULL, LIKE	SELECT * FROM parcel WHERE parcelID **IN** (SELECT parcelID FROM taxbill WHERE status = 'outstanding')
UNION, INTERSECT, EXCEPT	SELECT * FROM parcel WHERE ZONING1 LIKE 'PRC1' **INTERSECT** SELECT * FROM taxbill WHERE status = 'outstanding'
COUNT, SUM, MIN, MAX, AVG	SELECT **COUNT(*)** FROM parcel
DISTINCT	SELECT **DISTINCT** ZONING1 FROM parcel
GROUP BY	SELECT PLAN, MAX(AREA) FROM parcel **GROUP BY** PLAN
ORDER BY	SELECT PLAN, SUM(AREA) as planArea FROM parcel GROUP BY PLAN **ORDER BY** planArea DESC

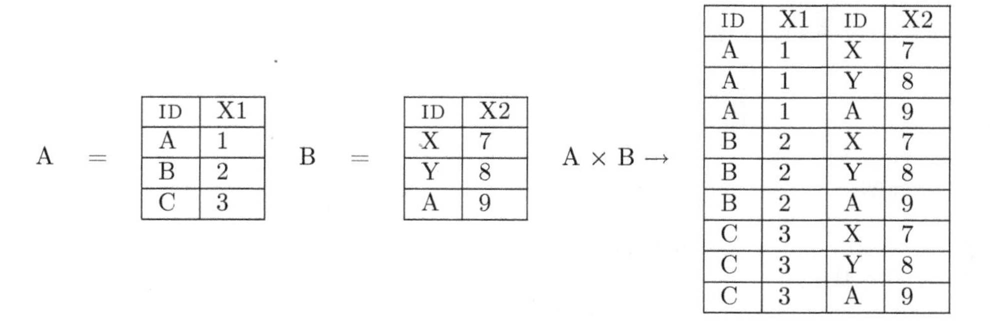

Figure 5.16 Cartesian product of two tables.

A.ID	A.X1	B.ID	B.X2
A	1	A	9

Figure 5.17 Inner join on ID between tables A and B.

database implementations and will be used to highlight basic joining, given two tables with a common field. An inner join between tables A and B returns all rows from A that match a row in B. Matching is based on key relationships specified in the WHERE clause of the query. So returning to the tables A and B in Figure 5.16, an inner join using the common field ID would be specified as

```
SELECT *
FROM A, B
WHERE A.ID = B.ID,
```

which would return the result set in Figure 5.17, which is a subset of the Cartesian product of A and B. Note the syntax of prefixing the table name to distinguish the two identical column names in the two tables.

Inner joins are used to combine data from multiple tables based on matching values of specified fields. In an inner join, only matching records are returned, equivalent to a Boolean AND operator. An outer join is equivalent to a Boolean OR operator. Types of outer joins between sets A AND B include all rows in A OR matching rows in B (LEFT OUTER), all rows in B OR matching rows in A (RIGHT OUTER), and all matching and non-matching rows in A and B (FULL OUTER). Defining table joins to subset the Cartesian product appropriately is important to denormalize data in an RDBMS such that the data can serve client applications such as maps, reports, and models.

5.6.2 Procedures, Triggers, and the Like

Addition of behaviour to RDBSM has been implemented through a series of extensions to SQL. Database triggers are stored functions that are triggered when predefined conditions are met in a table or view. Generally, triggers are used to maintain the integrity of the data across the database. For example, when a parcel is rezoned, a database trigger might also update the tax bills table to reflect the tax rate appropriate for the new zoning classification. Triggers can be designed to maintain the database with DML or DDL. DDL triggers fire when schema-level changes are made, such as the creation, deletion or alteration of a table. In a spatial database, triggers can be used to take advantage of spatial relationships existent in the database. For example, a database trigger might be defined for the creation of new features in a streetlights table, which automatically populate fields for planning areas based on its digitized location. This example of a trigger in PostGIS (written in PL/pgSQL)—

```
CREATE OR REPLACE FUNCTION nanaimo.update_streetlight()
    RETURNS trigger AS $update_streetlight$
BEGIN
  -- Retrieve the name of the planning area
  NEW.planArea = (SELECT planAreaName
```

```
        FROM nanaimo.planningAreas
        WHERE ST_Within(NEW.geom, geom));

        IF NEW.planArea IS NULL THEN
            RAISE EXCEPTION 'outside the city boundaries';
        END IF;

        -- Remember when observation was added/updated
        NEW.date_time := current_timestamp;
        RETURN NEW;
    END;
     $update_streetlight$ LANGUAGE plpgsql;
```

—uses an ST_Within function to find the name of the planning area that the newly digitized streetlight is in. Database triggers can be used to similar effect to enforce user-defined topological constraints between objects/layers.

Stored procedures are similar to database triggers in that they extend SQL and the capability of RDBMS by allowing application logic to be stored directly within the database. A database trigger is actually a special type of stored procedure that is automatically called by an event that occurs in the database. Stored procedures are usually written in database-specific programming languages (e.g., PostGIS - PL/pgSQL, Oracle - PL/SQL, SQL Server - T-SQL) that extend SQL to include declared variable types such as cursors and flow-control statements (IF-THEN, WHILE, CASE) and can call other stored procedures. Stored procedures are used widely in spatial databases to provide information outputs for mapping and analysis. For example, the parcels shapefile on the Nanaimo Open Data website is actually produced from a stored procedure that denormalizes data from multiple tables and combines them into one result set that is output as a shapefile attribute table. The disadvantage of stored procedures is that they are often not transferable between different databases.

5.6.3 Spatial Indexing

A key feature of O-RDBMS is support for abstract data types, and this is integral for storage of geographic information. As described previously, binary data types for vector geometries are the most common form of data typing for representing geographic information. To facilitate fast retrieval of geographic information based on spatial criteria (e.g., return all parcels sharing a border with a water body, find all street lamps within 5 metres of a manhole, or find all habitat patches located downstream from a point source of pollution), some form of spatial indexing is often required. Access to database data is dependent on indexes that provide fast random access to data. Indexes for non-spatial data are relatively straightforward, constructed from a pre-sorting of the data elements, which facilitates fast search and retrieval. However, indexes for spatial data are complicated by the multidimensional space within which spatial objects are defined, the various forms of spatial relationships required to answer spatial queries, and the interdependence between spatial representation and spatial relationships. We now discuss two of the most commonly used spatial indexing approaches.

R-tree indices

An R-tree index uses bounding boxes to index spatial objects in a tree structure. Each node in an R-tree corresponds to an identifier of a spatial object and its minimum bounding rectangle (MBR). The ends of the tree structure are called leafs, which are nodes that correspond to a single spatial feature. Non-leaf nodes contain an MBR and a child pointer, where the child pointer is the address of a lower node in the R-Tree and the MBR covers all MBRs in lower nodes. Thus, for each non-leaf node in the tree, the MBR of each node is the smallest rectangle that contains the MBRs in its child nodes. This data structure provides fast searching of spatial queries using a two-pass system of first using MBRs and then testing only exact spatial relationships for the subset of spatial objects found within the MBR. A graphical representation of an R-tree structure in Figure 5.18 shows the balanced nature of the tree and how the data structure maps to the spatial domain.

For example, a point-in-polygon search (i.e., find any polygons this point intersects) for a point defined at the centroid of the polygon in R8, would proceed from the top level of the tree through R1:R2 → R3:R4:R5 → R8:R9:R10, and then test only on the polygon bounded by R8 with an exact algorithm, which would return TRUE.

Quad-tree indices

A quad-tree index is similar to an R-tree in that space is indexed in a tree structure, and access is accomplished by traversing the nodes of the tree data structure. A quad recursively divides space into contiguous quadrants. Quadrants are stored at the highest level of the tree, and lower levels in the tree are smaller nested quadrants, more closely bounding the stored spatial geometries. Figure 5.19 represents a simple quad-tree structure. A quad-tree is a variable-resolution data structure, such that at each step down, the space is divided into quadrants if the entire area does not contain data. This reduces data storage requirements for large contiguous patches of data. Quad-trees are used for both vector spatial object indexing and continuous raster and image domain data.

5.6.4 Geographic Time-Series

Representation of the temporal domain in GIS databases has been an active area of research in GIScience since the 1980s (see Langran [72] and Peuquet [92]). We'll restrict our discussion to implementation-level techniques for handling spatial data indexed at multiple times that include *identity*, *geometry*, *time* simplified to x, y, t domain data. Such data are very common: temperature and precipitation measurements at a network of meteorological stations, or forest health surveys at permanent sample plots, for example. In an RDBMS, such data are best represented as separate tables: one for x, y data and another for measurements. Often, however, denormalized versions of the data are required for analysis. For example, we may want to calculate a seamless spatial coverage of monthly rainfall from measurements taken at a network of meteorological stations recording rainfall measurements on an hourly basis. Two naive approaches are to represent the data in either long (time indexed as rows) or wide (time indexed as columns) formats. In long format, all non-measurement data (e.g., coordinates, ID) are repeated multiple times, resulting in redundant storage, whereas in wide format (a non-normal form), problems can arise when measurements are not identical for all

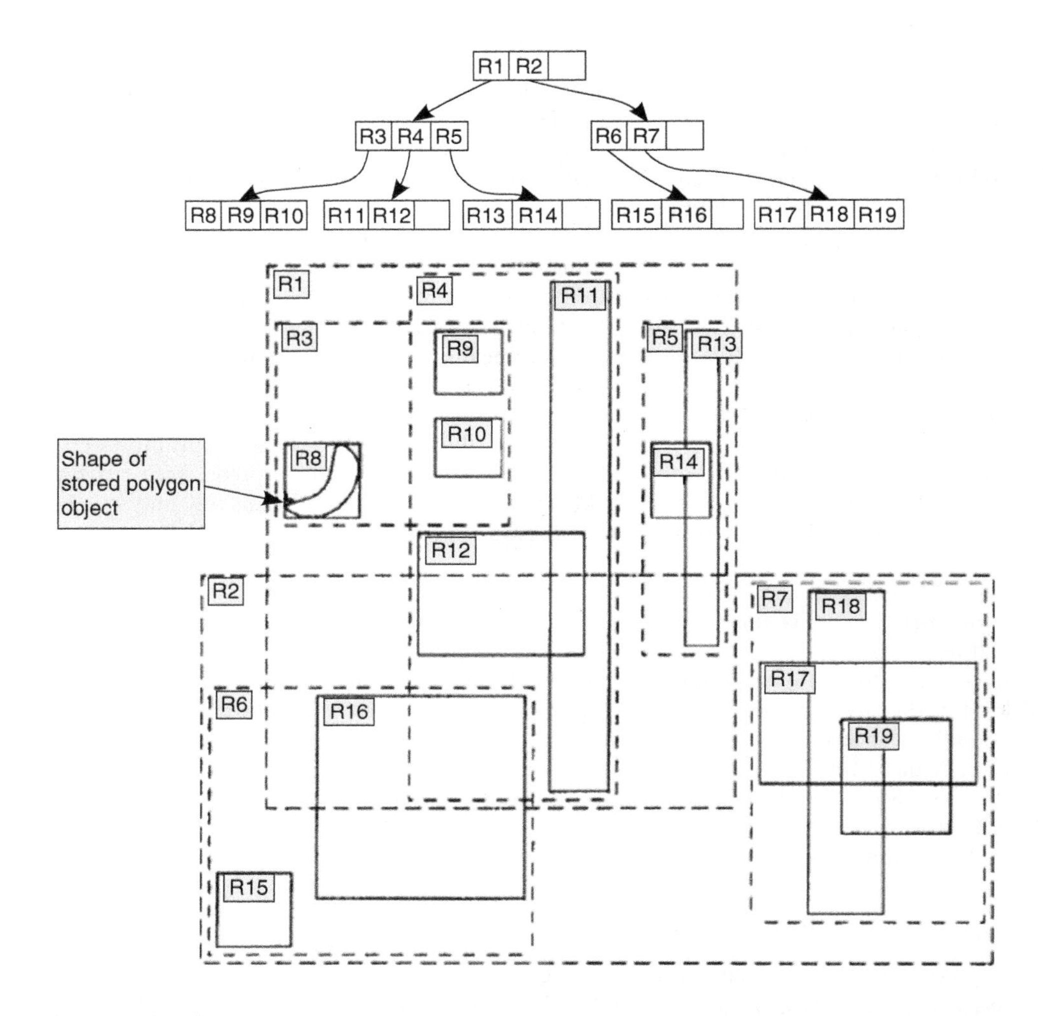

Figure 5.18 Data storage structure for an R-tree spatial index based on minimum bounding rectangles.

Source: Adapted from Guttman[56].

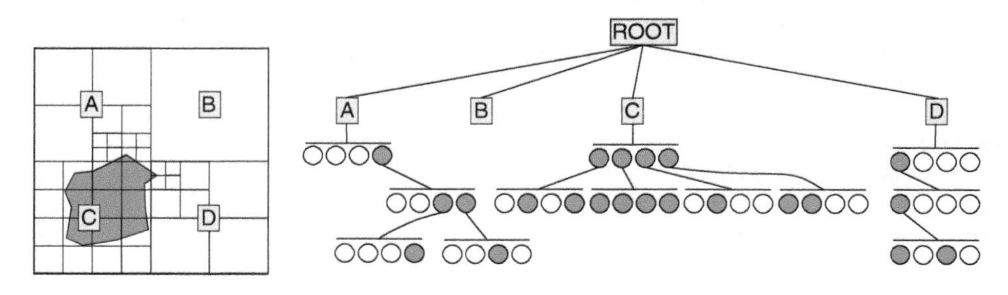

Figure 5.19 Data storage structure for an quad-tree region spatial index.

stations (new stations coming online during the study period, equipment malfunctioning). Pebesma [91] provides a useful description of these space-time representations in the context of space-time data analysis and examples of space-time classes in R.

5.7 Next-generation Databases for GIS

The relational model is founded on principles of structured data (i.e., as tables related through keys). While this is sufficient when details of the information to be stored are known ahead of time, data are increasingly obtained from sources of unstructured data (as documents, images, audio, video, free text). Such data are the result of the new ways people interact with digital devices and the Internet—often called Web 2.0. Sources of Web 2.0 data, such as social networks, cell phone traces, and low-cost environmental sensors, have characteristics not conducive to the relational model: rapidly increasing numbers of users/sources, predominantly unstructured data (no schema), and massive volumes of data distributed across network nodes. New architectures (e.g., document-based, graph-based) now emerging are attempting to address the shortcomings of the relational model, and these will likely be important technologies for GIS and spatial data handling in years to come. Information needs often depend on function—such as search, clustering (e.g., recommendation engines), and networks—that do not lend themselves to transaction-oriented RDBMS.

5.7.1 Document-storage Databases and NoSQL

Document-based databases that adhere to a NoSQL approach to data modelling, storage, and access have emerged. Key principles of NoSQL databases include

- non-relational data models
- flexible replication of data across nodes
- dynamic scaling to handle continuous growth
- application programming interfaces (APIs) for access
- open source

However, the added flexibility of NoSQL comes at a cost of integrity. The relational model guarantees consistent data in a distributed RDBMS by waiting for all instances to accept a request, and when accepted globally, the request is processed. This ensures that all users anywhere in the world see the same data in the database. However, this comes at a performance cost as the number of user and requests increases. A NoSQL database provides less certainty, only guaranteeing consistency after some *reasonable* time period. Thus, there can be times when data are inconsistent across instances, but only for a short period of time. However, the benefit of this "eventual consistency" is much-improved overall system performance.

NoSQL implementations that support geographic data types may provide a storage mechanism for GIS applications that have traditionally been poorly represented (e.g., moving objects) in RDBMS. OO-elements such as object inheritance are part of document-based NoSQL databases; however, due to less structure, querying has to be application-specific. It is worth noting that a document in a "document database" can include binary objects, such as images or files, as well as text data. Some spatial index support for geospatial "documents" has been developed to support spatial querying.

5.7.2 Crowdsourcing: the Undatabase

GIS data and processes are increasingly moving from client-server architectures to Internet-based (i.e., cloud) storage. While initially this represented a shift in access—from application to web—the information being stored is itself increasingly changing, from static representations in a database stored on disk to a dynamic, evolving knowledge base stored in the cloud. The emergence of the Internet as a platform for information generation and sharing has altered the way information is represented and accessed via computer. We might relate the difference between the database and web-based information to that of authoritative and non-authoritative generators of knowledge. For some domain areas, the non-authoritative information sources can be "better" in terms of currency, coverage, and detail than official sources. For example, current events are often talked about on social media, or have Wikipedia entries, before official media outlets have released their stories covering the events. Another example is OpenStreetMap (OSM)—the volunteer-driven project aiming to map the entire world—which provides access to spatial data describing roads, building footprints, and open areas. Because the data are maintained by a global user community, Linus's Law—given enough eyeballs, all bugs are shallow—comes into effect, meaning that maps in OSM are often more up-to-date than official sources. And whereas a DBMS perspective would see OSM as a source of information to be obtained and stored locally, that immediately puts the local version out of date. Similarly, having data accessible via public application programming interfaces (APIs) from online sources of GIS data, such as muncipal open data portals, means that either live or batched exports of the most current data are publicly exposed to application developers. Additionally, the form and structure of online information can be important indicators of contested knowledge bases. Yasseri et al. [125] examined the most-edited Wikipedia articles to investigate socio-spatial disparities in entries, revealing political and religious fissures in entries that would not be evident in one dedicated source of authoritative information. Similarly, contested geographies may be mapped in ways that highlight boundary disputes or environmental damages. What the move toward crowdsourced knowledge bases means for GIS remains to be seen, but the trend toward distributed lightweight applications and large collaborative data sharing networks is unlikely to abate.

Problems

1. Using the parcels shapefile, write down the queries required to find the following subsets of data:
   - all parcels larger than 3000 m^2
   - areas of parcels larger than 3000 m^2 that are zoned as residential
   - full address of all parcels on Banning Court
   - all parcels that are missing a postal code
   - full address of the largest park
2. Use the Internet to find some GIS data for your neighbourhood. Examine the attributes of the data. Create an entity-relationship diagram that includes the data as part of a larger RDBMS.
3. In this chapter, an example is given in Figure 5.8, which represents a model for tax collection in the City of Nanaimo. How could this be represented in an object-oriented database? What would be the advantages and disadvantages of this approach?
4. Examine the OSM data for an area you are familiar with. Examine the attributes associated with features that are mapped. Download OSM data and develop a new schema for this area that reflects poorly represented or missing information. Identify what spatial relationships and indexing schemes would be needed to implement this schema in an object-relational database.
5. Design a database schema for a GIS-based wildfire monitoring application. Discuss each stage of the data abstraction process and comment on any limitations of your design and how they might impact end users of the application.

Chapter 6

Geographic Data Editing

6.1 Introduction

Geographic data editing is required when one wants to make changes to existing geographic data. Although a number of techniques have been developed to facilitate editing GIS data, it is one of the more technically challenging aspects of GIS to automate and often still involves manual and semi-automated workflows. Many of the tools for data editing are similar to graphics programs or CAD software; however, some distinctive characteristics of GIS are worth noting in the context of editing. First, features represented in GIS have a real-world analogue against which accuracy can be measured through positional error metrics (described in Chapter 7). Topology also impacts geographic data editing, as topological data models require special tools for editing—tools that exploit and preserve topological relationships between spatial objects. In this chapter, we review GIS data editing from two broad perspectives: global operations that operate simultaneously across the entire dataset, and local operations that occur only at the object or vertex level.

6.2 Editing Geographic Data Globally

Some editing operations are usually performed on all objects in a dataset. These global operations are typically adjustments made during the dataset compilation process or when adapting a dataset to a new application. For example, when transferring a paper map to a georeferenced raster dataset in GIS, distortions may arise during georeferencing that need to be corrected during post-processing global editing.

6.2.1 Linear Transformation

When data are being mapped from one space (e.g., digitizer units) to another (e.g., map projection space), linear transformation can provide a mapping between source and target coordinate spaces, here both assumed to be 2D planimetric coordinates (i.e., projected data). Transformations require knowing the coordinates of a handful of locations in each coordinate system, termed *control points*, and then fitting a model that relates coordinates in the source coordinate system to coordinates in the target coordinate system. Control points are selected by locating time-invariant features on the source map and the target map (e.g., railroad bridges, built structures, manholes, or fence crossings). Control points should be distributed evenly across the area being transformed so that as many as possible of the distortions between input and output coordinate spaces are modelled.

Given coordinates of a control point denoted x, y, we seek to transform these input coordinates to output coordinate space x', y' and use the estimated transformation parameters to transform all input data to the output domain. Note that these same transformation methods, described below, apply to georeferencing raster imagery.

The general 2D transformation matrix has four parameters,

$$\begin{bmatrix} x' \\ y' \end{bmatrix} = \begin{bmatrix} a & b \\ c & d \end{bmatrix} \begin{bmatrix} x \\ y \end{bmatrix} ,$$

which are used with the x,y coordinates in source map units for each of the control points. These four parameters control changes in scale, translation, and orientation (i.e., via rotation) between input and output spaces.

To see how each of these changes works, consider simple translation of coordinates,

$$\begin{aligned} x' &= x + dx \\ y' &= y + dy \end{aligned} ,$$

where dx and dy are translation parameters. Scaling is similar such that

$$\begin{aligned} x' &= x \cdot \mu_x \\ y' &= y \cdot \mu_y \end{aligned} ,$$

where μ_x and μ_y are scaling parameters. The rotation matrix has been described in the appendix, giving us

$$\begin{aligned} x' &= x \cdot \sin\theta - y \cdot \sin\theta \\ y' &= x \cdot \sin\theta + y \cdot \cos\theta \end{aligned} .$$

Putting all three together (translation, scaling, rotation), we have great flexibility to transform spatial data between spaces. If the parameters in the x and y direction for scaling and translation are equal, we have four parameters: translation, scaling, and two rotations. Usually in transformations we want separate translation parameters but not shearing, so we can use a common μ yielding four parameters: $a = \mu \cdot \cos\theta$, $b = \mu \cdot \sin\theta$, and c, d for translations. We can represent the transformation in the matrix above such that

$$\begin{aligned} x' &= a \cdot x + b \cdot y + c \\ y' &= -b \cdot x + a \cdot y + d \end{aligned} .$$

This is sometimes called the *similarity transformation*, as it scales, rotates, and translates the input data. We can add a term for shearing of the coordinate system:

$$x' = a \cdot x + b \cdot y + c$$
$$y' = d \cdot x + e \cdot y + f$$

In this representation, the meaning of the parameters changes slightly and is better represented in matrix notation. The changes in x and y are now independent of each other, as we have different parameters in the x and y (i.e., shearing). This transformation method is called an *affine transformation*. The similarity transformation is actually a special case of the affine transformation, where $e = a$ and $d = -b$, thereby maintaining equal scale in the x and y directions; as such, the aspect ratio is conserved and shapes are *similar* after transformation (similar to a conformal map projection). Both transformations are available in typical GIS software packages. Figure 6.1 illustrates the transformation from input to output coordinate space.

In Figure 6.2, an affine transformation is used to transform a vector coastline dataset (red line) to a polygon dataset (green polygon with black boundary) that represents more accurate source data. We can see that after transformation the linework matches the polygonal coastline much more closely. The transformation was derived from a series of control points (one shown) along the coastline in order to integrate these two datasets. The new line coastline dataset (blue line) now matches the polygon and includes detail not existing in the polygon dataset.

Warping is the general term used for nonlinear manipulation of an input image. Often, warping employs higher-order polynomial transformations derived from common control points identified in input and output maps. Higher-order polynomial transformations can correct for random distortions such as the pitch and roll in an aircraft or UAV. When higher-order transformations are used, there can be spatial mismatches between source and target spatial coordinates, and resampling algorithms must be used

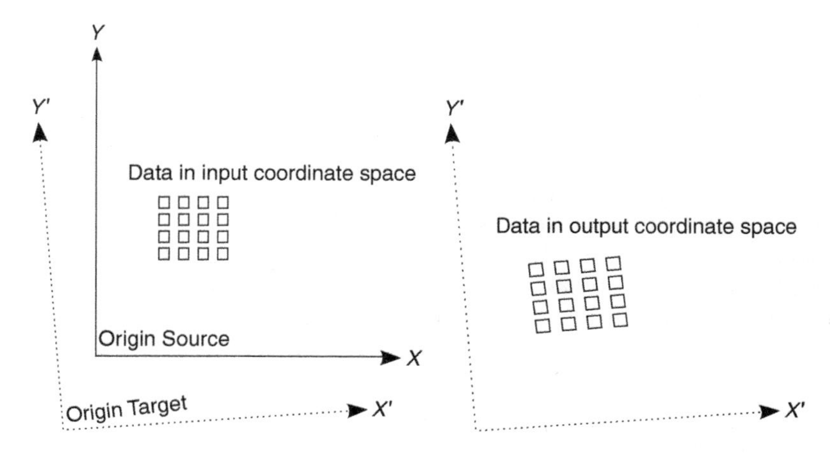

Figure 6.1 Linear transformation from input to output domains.

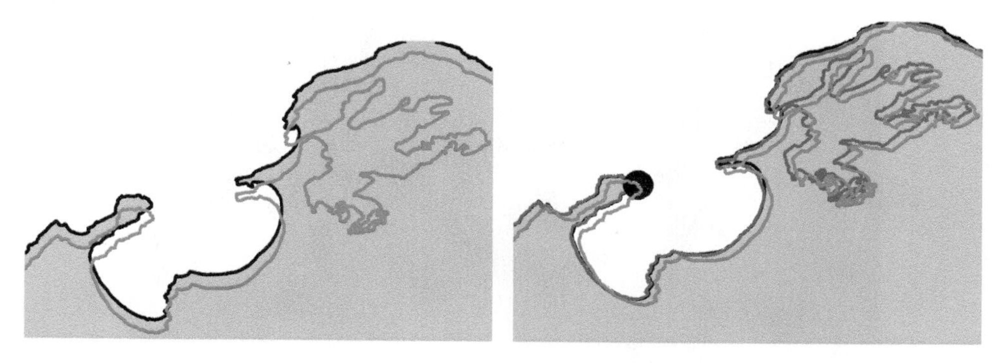

Figure 6.2 Affine transformation of vector dataset, with affine parameters $a = 0.99999$, $b = 0.00003$, $c = -380.18205$, $d = -0.00012$, $e = 0.99972$, $f = 981.46044$; red is before transformation, blue is after transformation.

to reassign input data to the output image space. Examples of resampling algorithms include *nearest neighbour*, where nearest pixels in the input space are used for assigning pixel values in output space, and *cubic convolution*, where a local smoothing function defines output values. New values can be created when using a convolution method for resampling, whereas values can be repeated, creating blocky artefacts when using nearest-neighbour resampling.

6.2.2 Rubbersheeting

While many geometric transformations are global in that spatial adjustments are applied uniformly across the datasets, distortions between source data and target data are often highly variable across the study area (i.e., non-stationary). For example, digitization of historical maps often contends with distortions from variable cartographic generalizations, bleeding of ink in the linework, or the paper itself undergoing nonlinear distortions due to damage or shrinkage. Higher-order (i.e., polynomial) transformations can model some nonlinearity (e.g., x^2, x^3); however, nonlinearity is often highly localized, and different methods are required. *Rubbersheeting* is one way of adjusting map coordinates in a way that allows for different transformations in different parts of the map. Conceptually, the source and target maps are first partitioned, and individual partitions are transformed using unique affine transformations as described above. Thus each partition in S has a matching partition in T. The goal of rubbersheeting is to find a transformation to map the set of S onto the set of T. The name rubbersheeting derives from the visual metaphor of the map as a formable surface being stretched and bent in order to fit the new coordinate space; it is sometimes called spatial adjustment. It is also used after an initial transformation has been performed and further adjustments are required to match spatial features.

One partitioning of input data often used in rubbersheeting is the *Delaunay triangulation* (described in the context of Voronoi Diagrams in Chapter 8), such that when the generator set comprises control points, the circumcircle of each triangle contains no other control points, and the minimum angle of each triangle is maximized. The

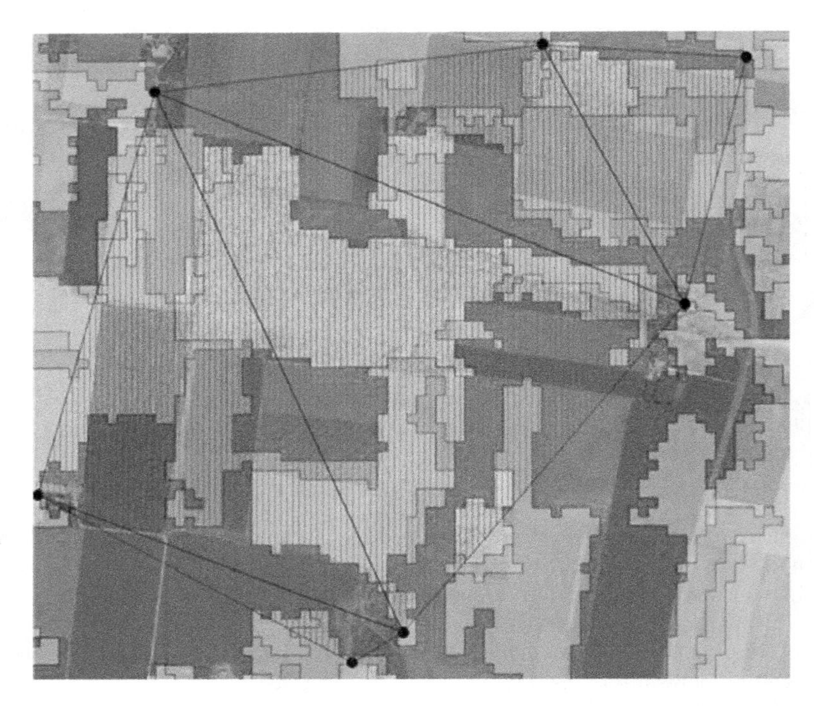

Figure 6.3 Linear transformation using local partitioning defined by Delaunay triangulation of control points.

control points are triangulated, then for each triangle the set of points defining it are transformed with a linear affine transformation (see Figure 6.3). This method is also used often in map conflation (described later in the chapter) to fuse multiple geographic datasets from different sources.

6.2.3 Generalization

Vector GIS data can be created numerous ways: through direct digitizing on a digitizing tablet or on-screen (i.e., "heads-up"), through extraction of GPS-derived data, or through vectorization of raster data (e.g., landcover classes extracted from classification of remotely sensed imagery). The density of nodes and detail with respect to the feature being represented can therefore vary greatly. Many algorithms that operate on geometry have a run-time that scales with vertex density, and therefore it can sometimes be desirable to reduce or normalize the density of vertices in vector linework. Additionally, generalization techniques are used for cartographic mapping such that the geometric representation best suited for a given map scale (i.e., zoom level) is employed. *Typification* is the process of reducing the density of objects that are preserved (or drawn) while maintaining a representative sample. Figure 6.4b shows a case where typification is required, where the contours mapped at one scale are not legible at a smaller scale. An alternate approach to typification (i.e., object removal) is storing multiple

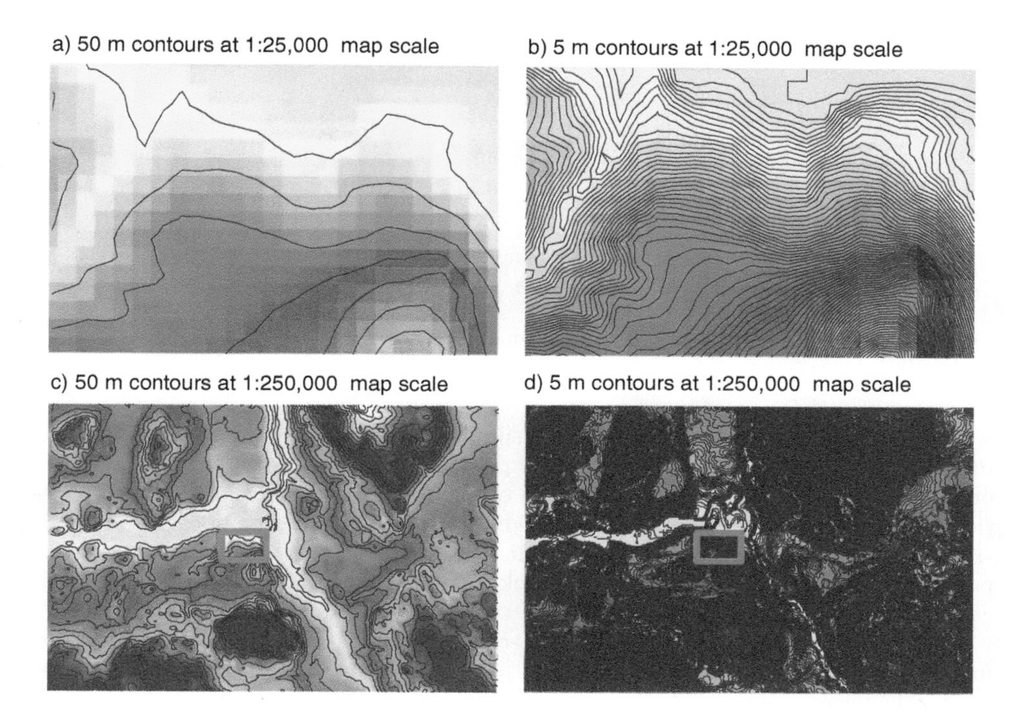

Figure 6.4 Cartographic generalization (typification) of contour data at different map scales.

representations tied to map scale, such as a 5 m contour at 1:25,000 and a 50 m contour at 1:250,000. Web-based maps, for example, often have multiple scale-dependent representations of features such as road networks. As users zoom in, more-detailed representations of the road networks are rendered, and as users zoom out, more generalized representations are rendered (see Figure 6.4).

Whereas the traditional approach to map generalization in a GIS environment has been to store multiple representations of layers at different scales for appropriate mapping, there is potential for on-the-fly generalization to map features dynamically, based on user-supplied zoom levels /map scales [10]. Topological data models can facilitate line and polygon generalization by ensuring that overlaps, gaps (i.e., sliver polygons), and other topological errors do not result from generalization methods run individually on boundaries of shared geometries [60].

Generalization by simplification

Generalization techniques are best exemplified using vector polylines (boundaries of polygons and linear features). Given an ordered set of points $p_i, p_j, \ldots, p_n$ defining a polyline object L, simplification techniques attempt to generalize the geometry at the feature level by removing points from L. The goal of generalization is to minimize the vertices required to represent the feature. Preservation of the representation can be

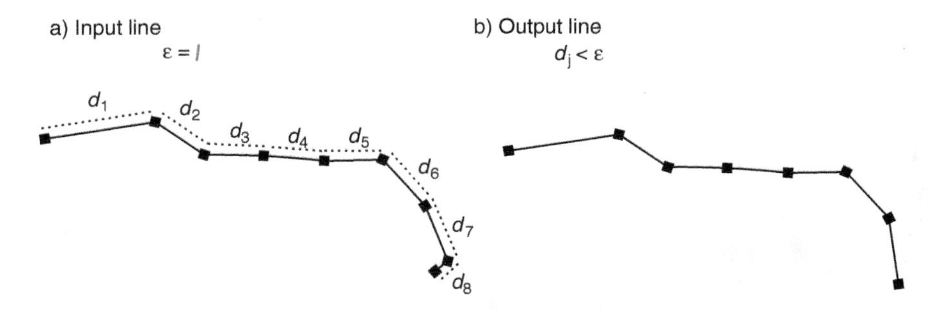

Figure 6.5 Line simplification using the minimum segment length algorithm.

achieved by evaluating a new candidate simplified line, subject to constraints (e.g., measure of shape, topological relationships, proximity relationships with other features, or a lateral offset threshold). The definition of these constraints depends on the application and simplification problem. When used appropriately, line simplification methods simply remove redundant data. For example, GPS-collected data might have an average positional error of ± 7 metres, yet the track-collecting mode may be set to automatically record a point every 1 second. If the GPS were used to map the boundaries of a wetland by a surveyor walking the perimeter, and assuming a walking pace of 5 km/hr, much of the collected data when used for mapping will be redundant, and even misleading in terms of precision. We may wish to simplify the boundary of the wetland by automatically removing vertices in a way that maintains the shape of the mapped feature.

Two criteria are used to determine which vertices are removed in a line simplification method. One approach is to remove vertices subject to some tolerance that defines the minimum allowable distance between vertices along the line. The results of this approach can sometimes be dramatic when curved geometry is being simplified. When the threshold is applied over an entire dataset, it is sometimes called the *grain tolerance*. An even simpler approach samples the line vertices using a systematic sample of every kth point from the input being maintained for the output. Results of this method, while computationally convenient, can produce erratic results.

The second criterion is to iteratively monitor divergence from some measure as candidate points for removal are evaluated. A commonly used line simplification algorithm is the Douglas-Poiker algorithm [34]. In this approach, the measure that is monitored is the maximum perpendicular distance between the original line and the simplified version of the line. Those points that can be removed yet conserve a short distance between the simplified and original geometries are deemed redundant and excluded from the simplified line. In practice, the search for candidate points to remove is recursive over the whole line. Given a set of ordered points and a value for the distance threshold denoted ϵ, first construct a line segment between endpoints and calculate the perpendicular distance between this new line and the furthest point in the original set of vertices (see Figure 6.6a). If the furthest point is within ϵ, all original vertices are removed. If the distance of the furthest point is greater than ϵ, the line is split into two sections defined by beginning and endpoints and the outlier point (Figure 6.6b + c). The same

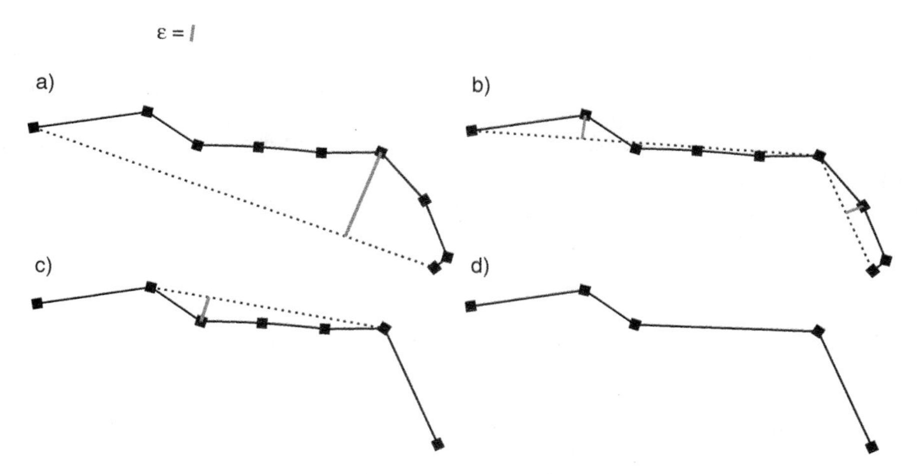

Figure 6.6 Line simplification using the Douglas-Poiker algorithm.

function is then called on each of the sections, and then reassembled (Figure 6.6d). A similar line simplification introduced by Visvalingam and Whyatt [117] uses the area of triangles formed by adjacent vertices to monitor the change in shape with the removal of vertices. Here, adjacent point triples form triangles, and then the vertex removed is the one that causes the smallest change in triangle area when triangles are reformed with adjacent points. This process continues iteratively until all points have been checked.

The linework in Figure 6.7 represents part of the road centreline data from the City of Nanaimo before and after simplification using an ϵ of 10 metres. This line changed

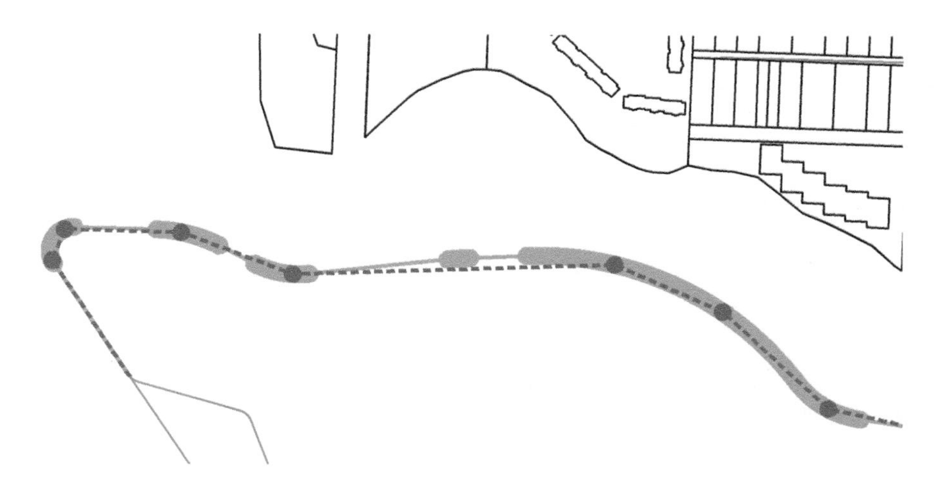

Figure 6.7 Simplification of curved road centreline using 10 m tolerance. Unsimplified line (green) includes many redundant vertices on curved sections, while simplified line (purple dash) retains only seven vertices to represent the road.

from 2162 vertices before simplification to 7 vertices after simplification. The cost of this saving in storage is the smoothness of the curve, which is lost after simplification (dashed line). Another simplification technique uses chords (straight-line segments) iteratively formed by endpoints defined along sections of the line; it compares each chord's length to the length of the original line between those endpoints. The line-chord length deviation is the criterion checked locally and globally against thresholds for removal [75]. Optimal simplification methods have been proposed but are not widely implemented in GIS software due to computational cost of these algorithms.

Generalization by smoothing

Smoothing techniques attempt to smooth out abrupt angular changes in linework. Typically, a simplification method such as Douglas-Poiker is used first, followed by an interpolation using a smoothing function. One smoothing method introduced by Bodanksy et al. [11] uses a distance-weighted average of neighbouring vertices to determine the smoothed line vertices as a first smoothing function. A second pass uses a polynomial approximation over adjacent points of the smoothed points.

Line generalization by smoothing is also possible using the method of Bézier interpolation. Bézier interpolation uses Bézier curves, parametric curves defined through a set of points, which in line smoothing can be a subset of adjacent points on a line. This subset is defined by points $p_0 \ldots p_n$, where n is termed the degree of the curve. When $n = 2$ the Bézier curve is a linear interpolation from p_0 to p_1. A quadratic Bézier curve is thus fit through three points p_0, p_1, p_2. The parameter $t \in [0, 1]$ in a Bézier curve function $B(t)$ defines the location of the function in the curve between p_0 and p_n. The tangent to the curve from p_0 to p_2 intersects at p_1. In addition to the points making up the original line, Bézier curves require points that control the curvature of the smoothed line (these are called *control points* and are different from control points used in transformations).

The general form of the Bézier function $B(t)$ is defined as

$$B(t) = \sum_{i=0}^{n} \binom{n}{i} \cdot (1-t)^{n-i} \cdot t^i \quad ,$$

where $\binom{n}{i}$ is the binomial coefficient:

$$\binom{n}{i} = \frac{n!}{i! \cdot (n-i)!} \quad .$$

Bézier curves are a binomial polynomial form (called Bernstein polynomials). The Bernstein polynomials are

$$B_{i,n}(t) = \binom{n}{i} \cdot (1-t)^{n-i} \cdot t^i \quad .$$

To use this form to interpolate a curve, we add coordinate values as weights to $B(t)$ so that we have a weighted basis function,

$$B(t) = \sum_{i=0}^{n} \binom{n}{i} \cdot (1-t)^{n-i} \cdot t^i \cdot w_i \quad ,$$

where w_i is the coordinate value at the point corresponding to the ith point in the polynomial.

So for a quadratic curve defined by points p_0, p_1, p_2, we can find the new coordinates for a point along the curve as

$$p_x = (1-t)^2 \cdot p_{0_x} + 2 \cdot (1-t) \cdot t \cdot p_{1_x} + t^2 \cdot p_{2_x}$$

$$p_y = (1-t)^2 \cdot p_{0_y} + 2 \cdot (1-t) \cdot t \cdot p_{1_y} + t^2 \cdot p_{2_y}$$

or for the cubic case with two control points for each pair of input points,

$$p = (1-t)^3 \cdot p_0 + 3 \cdot (1-t)^2 \cdot t \cdot p_1 + 3 \cdot (1-t)^2 \cdot t^2 \cdot p_2 + t^3 \cdot p_3 \quad .$$

This with the above, iterating over t from 0 to 1 by a sufficiently small increment, gives us the coordinates needed to fill in the curve between endpoints. Note that in both examples above only the start and endpoints are actual data points, whereas the intermediate points are control points used to draw the curve. We can also see that setting $t = 0$ gives the start point p_0 and setting $t = 1$ gives the endpoint p_2 in the quadratic example. An example of a cubic Bézier curve is given in Figure 6.8, where endpoints are the *real* points, and two control points are in intermediate positions.

However, the issue remains how to determine the control points p_1 in the quadratic case and p_1, p_2 in the cubic case. While there are numerous approaches to this, given that our motivation is line smoothing for GIS data, where we often have long irregular line work, one solution is to break up the line into triplets and do quadratic

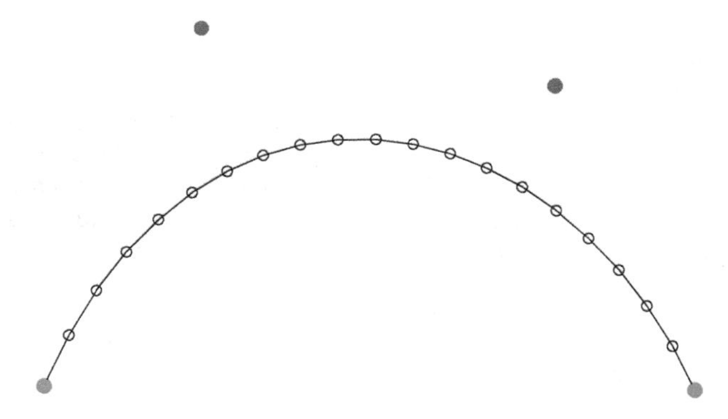

Figure 6.8 Sample cubic Bézier curve between endpoints with intermediate control points.

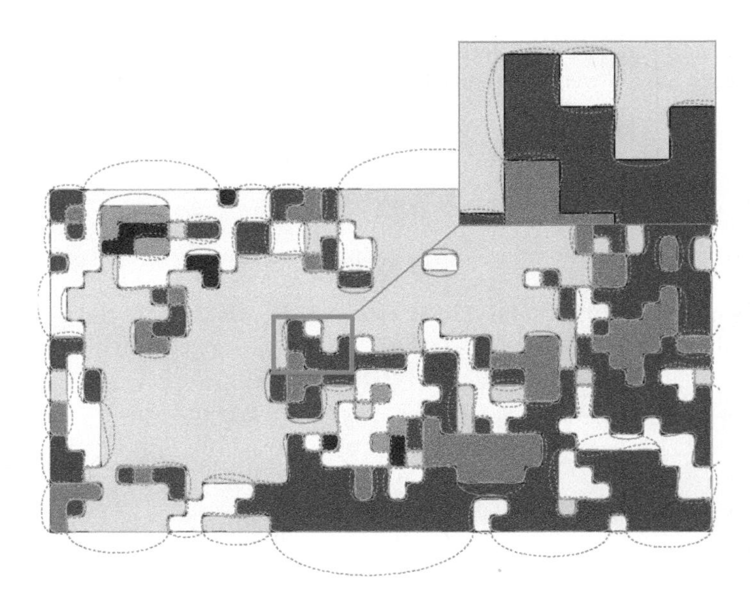

Figure 6.9 Degenerate cubic Bézier curve (dashed line) and weighted averaging smoothing (solid line) on vectorized raster classification polygons.

interpolation. To find the control point in this context, for every sequence of three adjacent points on L, we define p_1 as the midpoint on a line parallel to the line segment connecting p_0 and p_2. The endpoints of the parallel line can be the control points Q_{1i} and Q_{1j} for adjacent quadratic Bézier curves (i.e., a Bézier path). For complex geometry it is typically better to stitch together low-order polynomials than to parameterize with higher polynomials for the whole line.

A very common application of line smoothing is polygons and polylines created from raster-to-vector conversion, such as landcover polygons derived from an image classification. Figure 6.9 shows both Bézier and the Bodanksy et al. [11] weighted averaging methods computed on polygon boundaries derived from a vectorization of the output from an unsupervised classification of a section of the Victoria Landsat dataset. The results indicate the danger of smoothing techniques, as over-smoothing created unrealistic boundaries and, in the case of the Bodanksy et al. method, created significant topological errors. Note that as this was computed on a non-topological data format (a shapefile), smoothing operations operated separately on boundaries of adjacent polygons, which breaks topological relationships in many cases. Figure 6.9 also demonstrates the danger of using smoothing tools, as they can dramatically alter the geometry of data and are more often used when working at very large scales during geographic data editing of specific features of a dataset.

6.3 Editing Feature Geometry

Feature-level geographic data editing is a common task in managing GIS data. To illustrate, we will consider the case of the City of Nanaimo municipal dataset. One of the principal tasks of GIS data managers is to ensure data integrity and develop a set of procedures for detecting and correcting geometric errors in vector GIS data. Topological constraints can be used to automatically enforce data integrity across a GIS database, which can greatly improve data consistency and ensure accurate results during spatial query and overlay operations.

There are numerous tools available for editing vector geometry at the feature/vertex level in current GIS software. Editing vector geometry usually involves invoking an edit mode, adjusting coordinate geometry, and then saving edits back to the file or database. This approach is often time-consuming and prone to error. Before reviewing tools for aiding in the editing of vector geographic data, we will first examine some of the commonly encountered errors.

6.3.1 Detecting Errors

Vector geometry errors can arise for many reasons, though the most common are digitizing errors. Vector geometry errors can be difficult to detect because the errors are often slight, but their impact can be large, resulting in incorrect summaries of geometric properties (e.g., area, length, distance) as well as incorrect results from spatial queries and overlay operations. Figure 6.10 illustrates problems induced into analysis from spatial data editing errors. A digitizing error is a discrepancy between the digitized feature and the representation of the same feature on the source data (e.g., map or air photo). A common way to check for digitizing errors is simply viewing the source and digitized features together and measuring the spatial separation between them. The level of acceptable discrepancy between source and digitized features will vary by application but need not be more precise than the accuracy of the original source data.

Digitizing errors can be classified as those related to spatial accuracy of the digitized feature relative to the source data, or those errors related to the topology of the feature geometry itself. Spatial accuracy errors can only be detected by comparing the digitized data to the underlying source data, typically through visual inspection. Errors can be

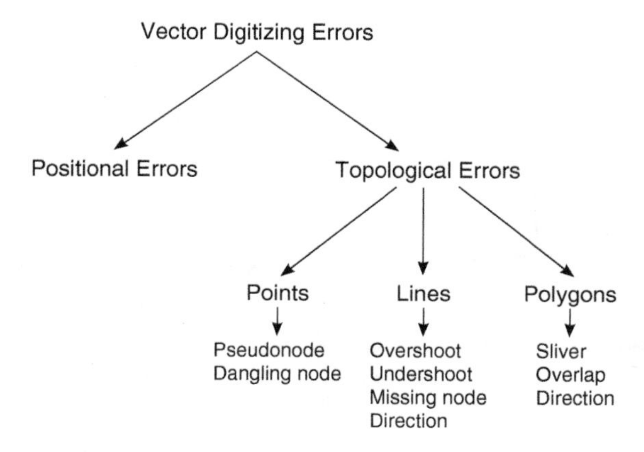

Figure 6.10 Classification of error types associated with digitizing vector geometry.

computed by the separation distance between the digitized feature and the original feature (See Chapter 7 for different types of error metrics).

The most common reason for digitizing errors is fatigue and human error. In the past, digitizing was completely manual, based solely on the operator of a digitizing tablet. Modern GIS software has many tools to aid in the digitizing process. For example, trace tools, which are typically implemented in vector GIS software, trace boundaries of feature geometry to aid in the editing or edge-matching of existing feature geometry. Modern scanners typically have some vectorization algorithms for automatically recognizing some features (e.g., a Sobel filter to detect linear features) in source raster data. Each of these tools can also introduce errors into the vector geometry.

Digitizing errors that are topological in nature can be easier to detect and remedy. Depending on whether the underlying spatial data model stores topology, errors can be automatically detected through implementation of database constraints. The coverage data model for example, explicitly stores topology such that topological rules can be checked and enforced within and between spatial layers through simple queries in attribute tables. A topological error is a malformed geometry based on the geometry definition. The geometry definition is a function of the underlying data model. Typical errors are disconnected linework (e.g., road centrelines that do not connect), overshoots, undershoots, missing nodes in a network, and poorly digitized curves (see Figure 6.11). Because the spatial relationships are known beforehand in a topology, any deviations can be identified immediately. More-recent topological data models allow application-specific topology rules to be built into the database. For example, a topology rule may specify that land parcels cannot overlap or that road centrelines cannot intersect land parcels. If a newly digitized feature violates one of these rules, the error can be flagged and corrected right away. Spatial data models that support vector topology can also support topology editing tools that allow editing of shared geometries. This is particularly helpful for polygon datasets—topological editing of shared boundaries ensures that

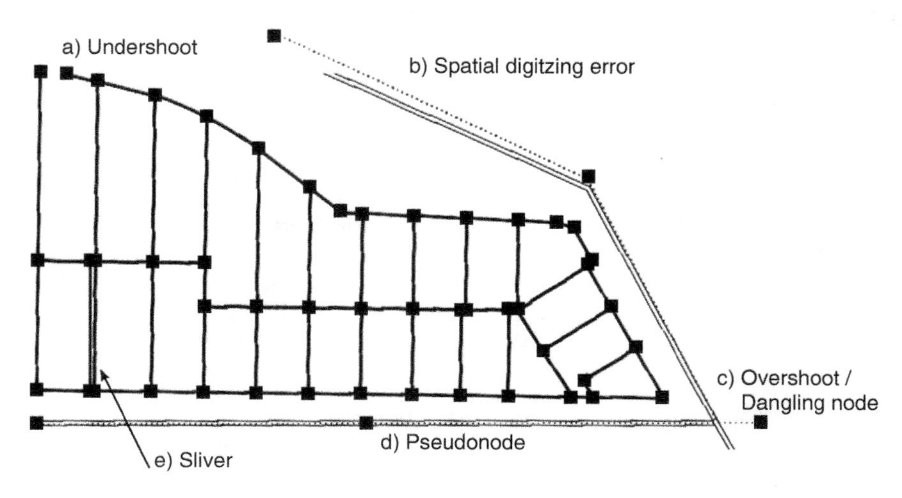

Figure 6.11 Examples of spatial and topological errors in road network data.

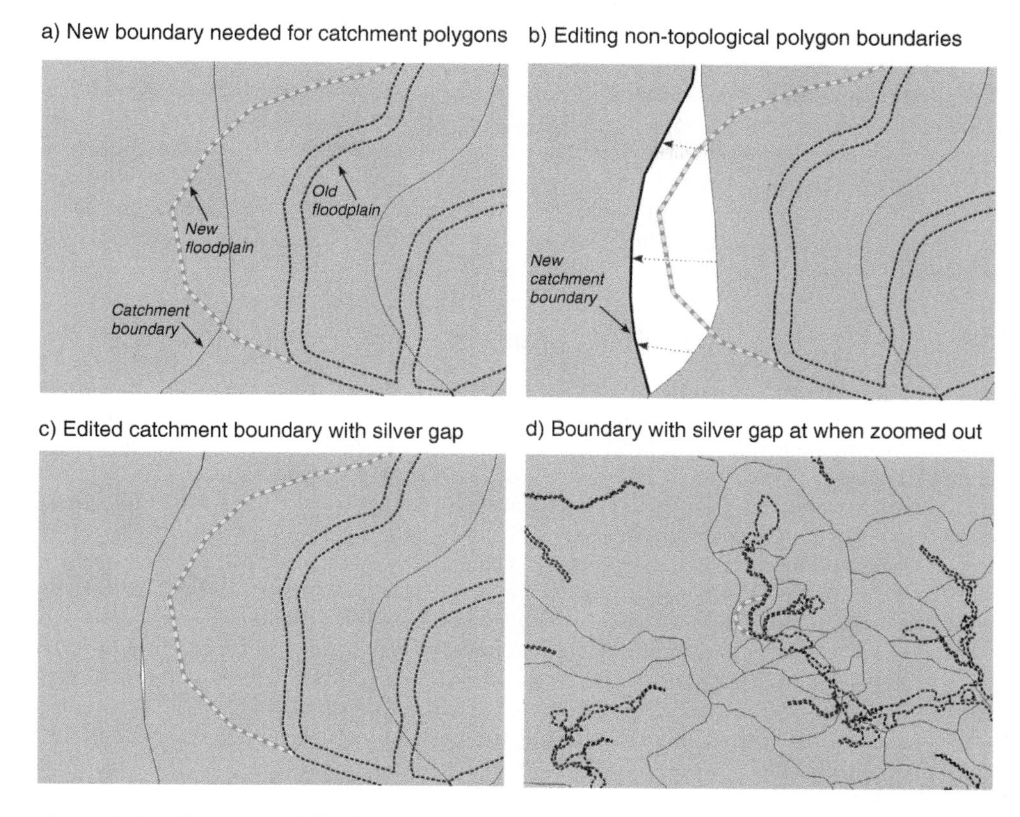

Figure 6.12 Examples of digitizing polygon edges and sliver polygon errors.

edges are matched and that there are no lurking overlaps or gaps in the polygon edges (see Figure 6.12).

The propagation of error in spatial analysis, and particularly overlay operations, is discussed in further detail in Chapter 7. During an editing session in a multi-user environment, feature geometry is checked out of the database for editing (using version control) by a user, and edits are made on a unique version of the data. Versions record changes made to the data. When the data edits are committed back to the database (and the versions are reconciled), the rules are checked and errors can be immediately identified. Multi-user editing using version control is a significant advantage of RDBMS storage for GIS data over file-based storage.

6.3.2 Fuzzy Tolerance and Snapping Features

Many GIS have settings that reduce the likelihood of making digitizer errors. Fuzzy tolerance is the term used to describe the precision of the coordinate space within which the vector geometry is embedded. This can be thought of as a fine grid that coordinates are snapped to during editing operations in order to improve performance and reduce errors. Fuzzy tolerance is also important for reducing inexact intersections resulting from overlay operations.

Snapping is used during editing to adjust existing feature geometry or to add new geometry in a way that snaps to existing geometry. Snapping is identical to fuzzy tolerance except that it is invoked during editing operations rather than as a database-level setting. Another setting available in some GIS is a dangling tolerance, which defines the minimum length of a dangling line (ending with a node unconnected to another line), such that dangling lines smaller than this threshold are automatically removed.

6.3.3 Complex Geometry: Island Polygons, Donut Holes, and Measures

Complex geometries, while often the norm for real-world data, can still cause many problems for spatial analysis of GIS data. Traditional GIS systems have tied representation to "layers" such that each layer was confined to one geometry type. This approach is an outgrowth of the history of how GIS developed as a computer analogue of paper maps, and overlays as digital versions of acetate maps on a light table. While these constraints are no longer needed, most GIS today still represent entities as layers or feature layers defined by a single geometry type.

How geometry is physically stored in a given data format varies, and these variations can impact data editing. When geometries are irregular or complex (e.g., self-intersecting lines, overlapping polygons, multi-part), storage of coordinates is often by convention. For a polygon stored as an interior ring within another polygon, it is required to store the status of that polygon as either an island or a hole. A common technique is to use vertex ordering to flag the status, with clockwise meaning island, counterclockwise hole.

6.3.4 Multi-user Editing and Version Control Systems

Multi-user editing of geographic data has been limited to large-scale RDBMS deployments that employ a database versioning scheme (described above). New techniques developed for versioning allow for structured data editing of versions and multi-user editing in a wider array of deployment scenarios. Version/revision control systems (VCS) allow users to roll back to previous automatically saved versions of a dataset or application. Open source software development projects have led the adoption of revision control, especially for large projects with many developers contributing code. In version-controlled geographic data, usually a complete dataset is checked out and labelled a "working copy," which is then edited locally. Edits made to a working copy are called a branch. These changes are not reflected in the revision control system until the edits have been committed back to the repository. In databases, version control has been used since the earliest multi-user RDBMS and was one of the key factors that led to the adoption of RDBMS in enterprise GIS contexts. Given the new forms of file-based GIS storage today, and large multi-user mapping projects resulting from volunteered geographic information, VCS are evolving to mirror software-based VCS, although this is still in its infancy.

A modern example of VCS for GIS is the multi-user editing environment with literally thousands of users, called OpenStreetMap (OSM). Here, checking out the entire dataset would be unfeasible, but individual features are managed as distinct objects, which are checked out for editing. As such, they tend to be spatially localized features (e.g., a section of an interstate rather than the whole highway). Edits made to a checked-out object are committed back to the OSM database, where all conflicts with existing data must be resolved before the commit will succeed.

Due to the large user community and the capability of rapid geographic data editing, OSM has played an important role in mapping during crises, such as natural disasters and disease outbreaks. As we will see, this check-out, check-in architecture works well for individual editors; however, for bulk edits / uploads through dataset merging, problems arise. One example where web-based data editing has occurred is through online GitHub repositories of geographic data. GitHub's implementation of the VCS system Git (originally created by the Linux operating system development team) has been used for live updating of quickly changing events as new data become available. A good example of this is the 2014 outbreak of the Ebola virus in West Africa, where one of the most comprehensive and up-to-date sources of data on weekly case counts was a GitHub repository (https://github.com/cmrivers/ebola), which received over 2000 views per day at the height of the epidemic. Updating and integrating multiple datasets is generally referred to as data fusion or, in the special case of geographic data, map conflation.

6.4 Map Conflation

As technologies for acquisition of geographic information continue to become smaller, cheaper, and more accessible, there is an increasing need to integrate multiple sources of geographic data. Additionally, GIS data are being used and shared by more applications and users than in the past, making sharing and integration a common computational issue. Map conflation techniques comprise approaches for automated merging of spatially coincident geographic information often obtained from heterogeneous sources. The conflation problem typically includes handling data of varying resolutions, having varying levels of data quality or completeness, and both spatial and attribute information integration. Casado [18] identifies the conflation process as first identifying matching features in each geographic dataset and then finding a transformation that maps features from one dataset to the other. In this way, features can be updated, new features can be added, and existing features can be validated. Factors affecting the conflation process include the spatial resolution of the data and level of generalization, the underlying data formats, and the spatial reference system.

In the case of OSM, conflation means integrating auxiliary geographic data obtained from ordinary citizens, government mapping agencies, and privately held spatial databases with the existing OSM geographic database. A common misconception of OSM is that the data are all contributed by individual mappers using GPS. In fact, the OSM dataset includes many bulk updates of contributed official and private datasets. While conflation tools in OSM are in their infancy, new techniques are being experimented with to add value to user-generated data by incorporating authoritative sources to produce an enhanced overall dataset. Pourabdollah et al. [94] describe a methodology that updates road attributes in OSM by integrating data from the UK Ordinance Survey.

Problems

1. Make a copy of the Nanaimo parcels and buildings datasets. Select a city block as your study area. A new project requires a single polygon dataset with an attribute coded to represent whether each polygon is a building or not. Use geographic editing tools to create this dataset. Map the results and present a snapshot of the attribute table.
2. Using a GPS, collect data in streaming mode while on a run or walk along a trail or sidewalk. Import the data as a line, and try out three different smoothing/generalization techniques. Compare the results in terms of length, number of vertices, and a subjective assessment of how well each line represents your true path.
3. Comment on the quality of the affine transformation in Figure 6.2, and describe how this might have changed if a higher-order transformation were used.
4. Write out the pseudocode to implement the Douglas-Poiker and the Visvalingam-Whyatt line generalization algorithms.
5. Describe how data should be integrated for a crowdsourced mapping project such as OSM. Design a system for implementing an OSM data integration plan in your neighbourhood or city.

Chapter 7

Error and Uncertainty in Geographic Information

7.1 Introduction

Error and uncertainty are important concepts in geographic information handling. This chapter reviews the unique position of maps and geographic data in relation to classical statistical theory and introduces the notion of maps as the outcomes of spatial processes as a conceptual framework for thinking about error in geographic information. We then show how a model can be formulated for error as a random process. We follow with a review of central topics in the understanding of error and uncertainty in geographic data, and we introduce methods for quantifying error in practical geographic information management and analysis.

7.2 Samples and Populations: Statistical Inference in Geographic Analysis

Statistical analysis of maps is problematic. Classical statistical inference is founded on a few basic principles: independent random sampling from a population, numerically characterizing a sample of observations, and making inferences about the population based on measurements of the sample. Under this framework, time is treated as fixed, and space is subsumed within random sampling (i.e., ignored). A geographic example might be random sampling of trees in a forest, measuring the height and diameter of each of the trees in the sample, and making inferences about the forest based on these observations.

Treating each measurement as independent is theoretically convenient; however, it contradicts what we know to be true about the world: it is patchy. *Tobler's Law* captures conceptually what the spatial autocorrelation measures described in Chapter 9 allow us to measure. However, classical inference is founded on the notion that each sample is independent of the others and brings an equal amount of new information into the sample (i.e., it cannot be predicted based on values of nearby samples). In geographic analyses, assumptions of statistical independence are frequently violated unless explicitly controlled for. Similarly, the notion of a sample and inference to a population is also complicated by the fact that in GIS-based analysis of geographic data we often have (or assume to have) a complete mapping of the phenomena being investigated; that is, *our sample is the population.* These theoretical issues lead to practical problems when attempting to quantify and reason about error with geographic data.

When considering error and uncertainty in geographic information, it is helpful first to consider the data-generating process as one with inherent error and uncertainty. We can then reason about the underlying processes by examining the properties and characteristics of the data. At the highest level, we may first want to consider what is represented by the map. Is it a state of reality as measured at a given point in time? Is some measured quantity generalized or aggregated over a region? Do the mapped data represent a physical object like a house, or do they estimate an invisible construct such as fear of crime? As noted above, geographic data are often represented over an entire study area rather than a random sample; this obfuscates the fact that elements of chance are embedded in both the phenomenon being mapped and the measurement process. We can posit a simple model for all measured phenomena as follows:

$$Y = C + E_s + E_m \quad ,$$

where Y is the value observed, C is a model for the deterministic part of the phenomena, E_s is error due to stochasticity in C, and E_m is measurement error. Sophisticated spatial modelling approaches such as state-space models and Bayesian hierarchical methods can estimate C and E_s (and therefore E_m) to fully characterize the process. The degree to which these components are separable depends on the system being represented. While the above holds for all measurements, when Y is GIS data there is often less consideration about the other components contributing to Y (generally, maps are by default trusted more than other types of data). This gives mapped data a veil of authority that is often unwarranted. Few would take the results of a political poll at face value without considering the sample size, the nature of the polling process, the questions asked, and the estimated precision of the results. Yet maps in public use are rarely scrutinized the same way.

Given that Y is the outcome of processes that have stochastic variation, it is natural to turn to statistical inference to characterize the nature of variation in Y. However, as noted above, a key assumption in classical statistics is that samples Y_i are independent random samples from a population of interest, while maps/GIS data often cover an entire population. Researchers have dealt with this contradiction by considering a mapped pattern as a special kind of sample, termed a *sample population*, which is obtained from a hypothetical population. To illustrate, the simple GIS representation model for reality in Figure 7.1 can be used to demonstrate how error and uncertainty can be hidden by spatial representation in GIS. The representation of a complex reality as a series of crisply defined vector objects fails to capture the most obvious feature of the landscape, and we can easily imagine that different interpretations of this landscape would lead to many different types of representations in GIS. The choice of what gets mapped, the spatial data model, and representation depend on the intended use of the

Figure 7.1 A possible representation of reality within GIS.

GIS data. Uncertainty exists at even the highest level of abstraction and therefore can be a source of variability in mapped data. We will briefly review the idea of samples and populations related to maps and GIS data.

Summerfield [109] reviews the arguments geographers employ to taking a statistical approach to sample populations. One idea contends that a sample population is a spatial sample and/or a temporal sample because measurements are obtained for a specific space-time instance when mapped, and statistical inference can inform us about the validity of the findings of analysis of maps in other areas and times. Another argument suggests that, because each observation on a map is recorded imperfectly with some degree of measurement error, the sample population is still a sample of the true unknown distribution of values and that sample-based statistical inference can tell us about the distribution of measurement errors. However, this idea only holds if the distribution of measurement errors is random and independent, and with geographic data this is unlikely to be the case.

An interesting and more plausible idea considers specifically the case of aggregated geographic information. When data are aggregated over arbitrary spatial units, as is common with administrative GIS data such as census and health information, the spatial configuration of these zonal boundaries can be considered one random realization of an infinite number of zonal configurations, and thereby a sample. This takes a well-known problem in geography, termed the modifiable areal unit problem (MAUP)—which describes how statistics calculated over areal units are affected by the scale and configuration of those units—and uses it to construct an inferential argument. One of the earliest demonstrations of the MAUP, by Openshaw and Taylor [87], showed that a huge range of correlation of coefficients is possible from the bivariate analysis of a single dataset if the areal units are randomized. The idea that MAUP links a single map of areal unit data to a population of possible measurements is an intriguing proposition, but as Summerfield [109] points out, this works only if the configuration is in fact random. In reality, administrative boundaries are defined by non-random processes such as ethnic and cultural groupings and physiographic features.

Gould [52] highlighted the major criticisms of applying classical statistical inference to geographic data:

- oversimplified functional relationships (e.g., as linear) between geographic variables
- the sample not being a true sample
- assumption of independence of observations and residuals
- the distribution of the variables and error terms
- the level of significance required, and multiple testing

And while some of these issues are not specific to geography, such as the arbitrariness of selecting a significance level, the critique of Gould [52] is that these issues conspire particularly harshly in geographic research. However, in the time since this critique, significant progress has been made to address many of these issues; for example, with simultaneous autoregressive models for handling spatial autocorrelation in residuals of a regression analysis.

Let us examine some of these issues through an example: the mountain pine beetle aerial survey data obtained for the Morice Timber Supply Area of northwestern British Columbia. As noted earlier, the entire forested area was surveyed each year by helicopter, and locations of attacked pine trees were mapped with GPS. As part of the survey, an estimate of the number of trees in each cluster was noted as an attribute. The "spot infestations" are most pronounced at the early stages of the infestation as beetles disperse a short distance each summer, using pheromones to signal mass attack at a cluster of trees.

Figure 7.2 presents a map of the phenomena over a tract of forest. From the perspective of spatial analysis, there are two ways to approach analyzing the data. The first is to examine the characteristics of the spatial relationships and attribute distributions to learn more about the processes of tree attack and death. For example, we could look at nearest-neighbour distances, or use the K-function to examine the spatial scales of clustering and regularity in the pattern. We could also explore the degree to which infestation intensity is spatially structured by exploring spatial-attribute characteristics jointly using local and global spatial autocorrelation measures (see Chapter 9). The second approach would be to look at spatial and attribute distribution in relation to other geographic variables. This is more typically the domain of GIS-based spatial analysis. For example, we could examine infestation intensity in relation to site factors thought to influence tree defenses and/or beetle dispersal success, such as tree age, size, stand density, or site aspect. In light of this discussion, a key question is, what do these trees represent? Are they a sample population from a hypothetical population of pine trees in the region for which inferences made here can be extrapolated, or new trees planted on this same site decades later? Are inferences made about these trees valid beyond this small study area, and more importantly, if we are doing statistical inference, from where does the randomness derive? Examining some statistics such as the nearest-neighbour distance index (NND) and the global Moran's I, we see evidence of non-randomness in the spatial distribution of attacks but no evidence of spatial clustering of intensity of attacks. If we are explicit about what the null hypothesis being tested is in each of these scenarios—that is, on replication of this analysis with a spatially random process generating tree locations—we would see a pattern as clustered as this one very rarely

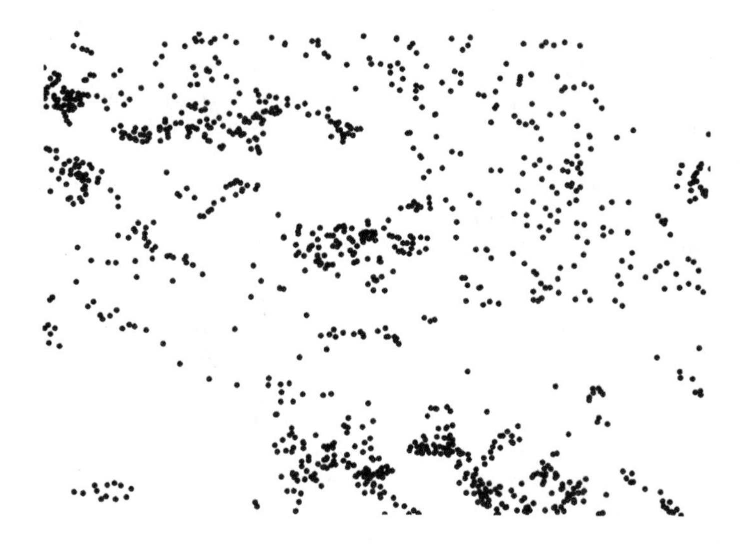

Figure 7.2 A random realization of a spatial process, clusters of pine trees infested with mountain pine beetle. The mean number of infested trees per cluster was 3.21, with standard deviation of 2.51. The nearest-neighbour analysis yielded observed *NND* 206 m, expected *NND* 229 m, Z-score -2.61, p-value 0.008. The Moran's I statistic was 0.003, Z-score 0.48, p-value 0.63.

(less than 1% of the time). Further, for a process where the number of trees infested in a cluster was independent of location, we could see a pattern similar to this one fairly often (63% of the time). Implicit in this interpretation is the idea of replication—that the study could be repeated under identical conditions—something that is almost never possible in geographic studies.

The above discussion highlights the existence of uncertainty at even the theoretical level at which we approach GIS and GIS-based spatial analysis. We see that uncertainty here can lead to errors in inference and decision making. With this in mind, we can now turn to some modern methods—which consider a map as just one single realization of one or more stochastic spatial processes—for handling error in maps as a whole.

7.3 Geographic Patterns as the Outcome of Stochastic Processes

One way of reconciling the idea of sample populations with the ideas of error, uncertainty, and randomness is to view the map as the outcome of single or multiple random spatial processes. The idea here is that if we can learn about the processes generating the data, we can compare the observed data to other data generated synthetically from such a process. In this way, we formulate a model for the data-generating process and consider the likelihood that the observed data could have been realized from this hypothesized process. We therefore consider only the relevance of the stochastic process (and its parameters), not the physical location or time period of the mapped data. Reconsidering the results reported in Figure 7.2, we need not extrapolate these results beyond their relation to the hypothesized stochastic processes, as the link between the process and pattern is sufficient to reason about the pattern probabilistically.

A study by Nelson et al. [84] reports that the spatial accuracy of the heli-GPS survey points is within ±25 m and the attribute accuracy is within ±10 trees. We can therefore expect that repeated realizations of the spatial process of tree infestation measurement—locating the helicopter directly over the centre of the infested cluster, GPS error (multi-path, atmospheric, geometry, etc.), errors induced by fatigue of the surveyor, fog or haze in the atmosphere that could impact ability of surveyor to estimate number of trees—would result in variability within these ranges. We may incorporate these sources of error into the inference procedure used to analyze the pattern of points and attributes. In Nelson et al., a procedure is developed that randomizes attribute and location based on drawing from a uniform distribution of values within the appropriate ranges, estimating a spatial density surface using kernel density estimation, and comparing the error-adjusted surfaces to the naive estimates.

What was found was that the non-error-adjusted surfaces significantly overestimated tree infestation intensity. Figure 7.3 presents the uncorrected, corrected, and difference maps of infestation intensity. The visualization demonstrates that measurement error process is not random but biased upward. Therefore, making decisions based on uncorrected data could lead to incorrect inferences about the nature of the underlying process generating the infestation pattern.

This sort of randomization procedure is an example of *Monte Carlo error modelling* and is one approach for handling spatial and attribute error with geographic data. We can imagine a similar procedure for handling error with raster data. Suppose we have a raster that maps the water table depth in metres above sea level, with accuracy

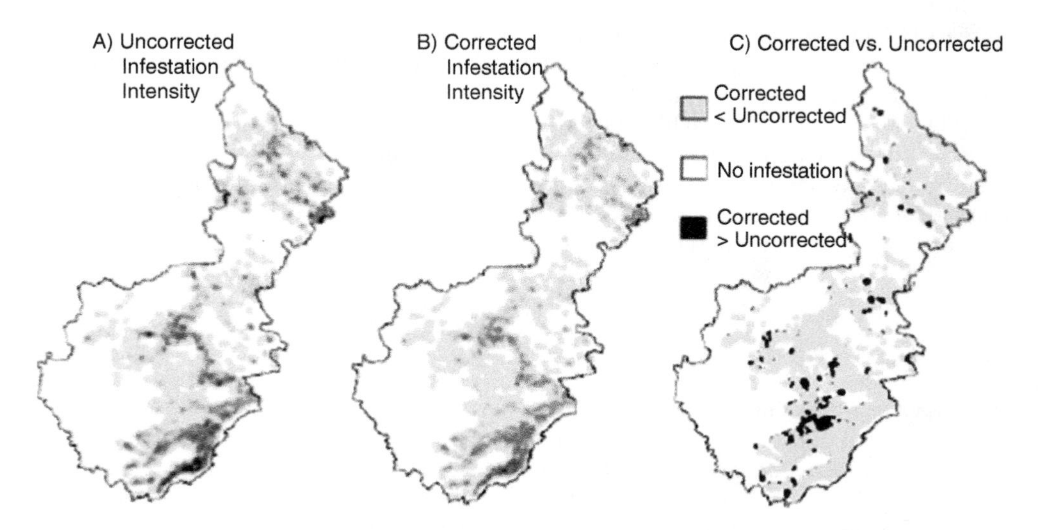

Figure 7.3 Error-adjusted density surface and non-adjusted density surface.

Source: Adapted from Nelson et al. [84].

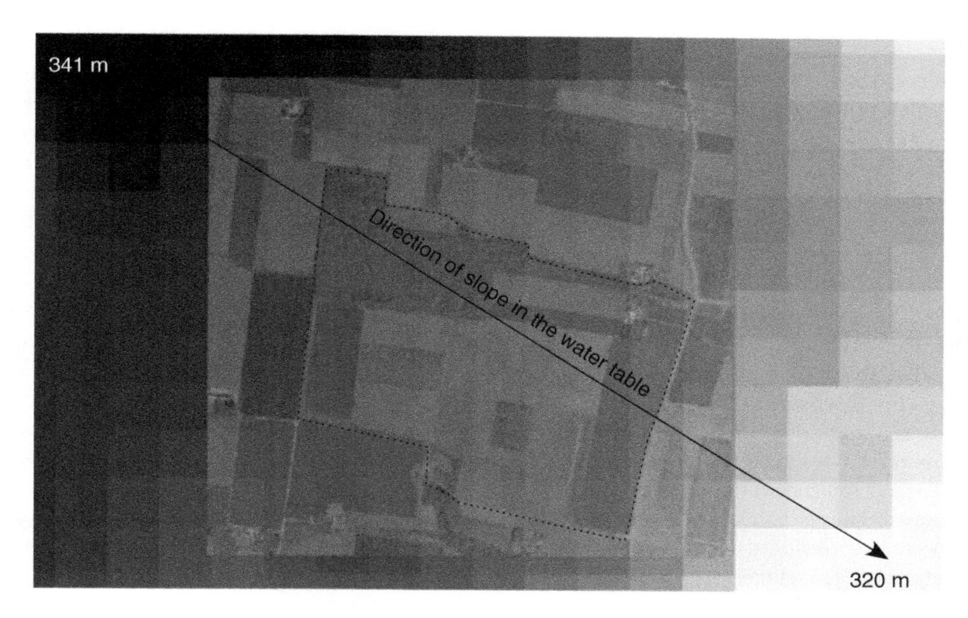

Figure 7.4 Water table raster and high-res airphoto.

of ±5 metres, based on field-validated point samples from wells drilled in the area. We may want to use this dataset to generate an estimate of the water table slope within a specific parcel of land (see Figure 7.4). We could use a Monte Carlo approach to randomly generate new altitude values within ±5 metres of the original values in each pixel (i.e., uniform distribution error model) and recompute the slope from each realization, generating a distribution of slopes within the selected parcel. We can then consider this distribution as a representation of the uncertainty of the true water table slope, analogous to a standard error around a regression slope coefficient.

If error in the raster is large (for example, derived from a sparse sampling of survey points), we would expect the water table slope uncertainty to be large. The assumed error model in the above examples was that of a uniform distribution. However, in spatial data, errors are often correlated due to spatial autocorrelation, and spatial covariance between errors is one way of estimating and correcting for error when true values are not known [47]. In the examples above, a spatial covariance structure could be added to a simulation procedure to generate correlated realizations. This might be especially true in the groundwater example, where knowledge of other variables related to subsurface characteristics likely affect measurement errors. However, this would have to be global, as partial dataset corrections would potentially alter the structure and topological properties of the dataset (see Goodchild [47] for a discussion of an alternative). Note that this sort of randomization error modelling is perfectly reasonable if we think of maps as the outcomes of stochastic processes.

The idea of maps as the outcome of stochastic processes is obfuscated by the fact that we usually have only one mapped pattern or spatial dataset to work with and is easier to envision through a dice rolling analogy. Imagine we have three dice that are rigged so that each rolls a 6 50% of the time and rolls numbers 1–5 10% of the time each. We are asked to predict what the sum of a single roll of all three dice would be. If we knew of the rigging, we would make a higher guess than had we believed the dice were fair. If we turn this question on its head and are given a single dice roll, and are asked to infer whether the dice are fair, it would be very difficult to know if they were rigged.

The total number of outcomes is $6^3 = 216$ for three dice, with sums ranging from 3 to 18. The number of outcomes that lead to each sum are not equal and can be used to determine the probability of each outcome. The highest probability sums for fair dice are 10 and 11, while for the rigged dice the highest probability sum is 13, followed by 18. Inferring whether a rolled sum of 13 is from the rigged set is almost impossible because the probability from a fair set is 9%, while from the rigged set is 14%. Figure 7.5 shows the distribution of a million dice rolls in the fair and rigged settings.

This discussion of dice parallels statistical inference on maps in an important way. If we replicated our 3-dice rolling many times, we would start to see patterns of irregular sums occurring, which might seem suspicious. If we consider the map as an outcome of a stochastic process (such as the sum of a dice roll), all we have to evaluate the process is one single realization. This limits our ability to make inferences about the generating process or processes. It is possible to roll high-valued sums with fair dice, but it is more likely with the rigged dice. Now considering spatial processes, different processes can give rise to similar spatial patterns, and distinguishing process from pattern alone is often possible. For this reason, testing theory about spatial processes with maps (i.e., patterns) has typically been restricted to extremely simple spatial outcomes: is the pattern clustered, is it random, or is it regular? Unfortunately, scientific significance and theory typically require much more nuanced questions about generating stochastic

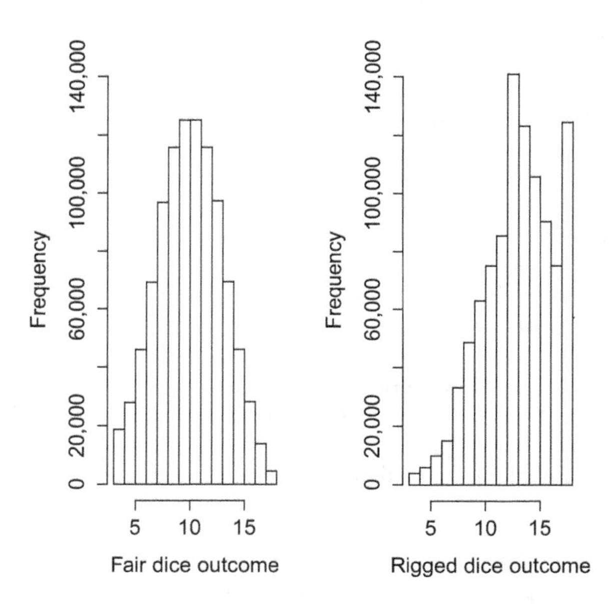

Figure 7.5 Patterns and processes in fair and rigged dice: distribution of the sum of a roll of three dice when fair and rigged for higher values.

processes. With analysis of spatial patterns and processes, we can reason about the probability of a pattern arising from a process, but we can never establish a causal link.

A final point is that there is inherent masking of underlying uncertainty associated with stochastic processes when forced into GIS data structures. Further, this masking is compounded by the typical workflow of GIS analysis: integrating data from heterogeneous sources, indexed by spatial location, to derive new data, maps, and statistical summaries. We will discuss this issue in more detail toward the end of the chapter, but we will first outline the major types and measures of error and uncertainty in geographic information.

7.4 Spatial Data Error, Uncertainty, and Spatial Data Quality

Given a patchy state of nature modelled with various GIS representations (vector polygon, raster surface, etc.), each representation may include various forms of error and/or uncertainty. Note that error and uncertainty are present at each stage of the abstraction process—from filtering out the world into conceptual models of space to representing these conceptual models within spatial data models, to the actual numerical representation of the geometric structures within the computer. Figure 7.6 outlines how error and uncertainty increase with levels of abstraction. Each step in this process is susceptible to error and can impact data quality. To control for spatial, attribute error and data uncertainty, we need to define what each of these terms means.

Error can be defined as the difference between the measured value and the true state of nature. In the general measurement equation (see Section 7.2), the error is the difference between the observed value Y and the value given by the model of the process or phenomena C, so $E = E_s + E_m$. Error is unavoidable in GIS data because no method of measurement is perfect. Error can be absolute: the true difference between a measured value and its correct value, but usually error is relative to some other source

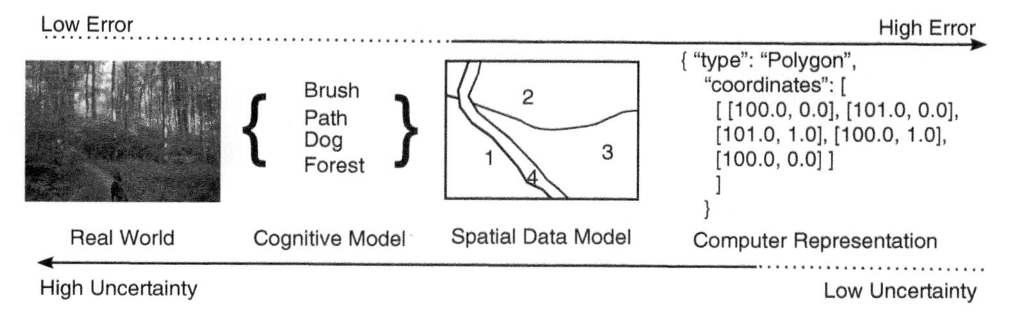

Figure 7.6 Levels of abstraction in geographic information handling and their role in error and uncertainty. Dashed lines indicate a less significance issue, and bold lines indicate greater significance. In the real world, true values for geographic phenomena are not subject to error, whereas they are subject to uncertainty (i.e., vague geographic concepts). Uncertainty is less significant at the computer representation level, as the numeric representations are fairly stable and consistent, but their association and meaning in the real world are not.

of data, which itself has some (lower) degree of error (i.e., Y is unknown). Errors are measured on the scale and support of the data as $e_i = y_i - \hat{y}_i$. Individual errors e_i can be summarized in different ways to characterize the overall degree of error, the distribution of errors, and the characteristics (such as spatial patterns) of the errors. The average error is called the *accuracy.*

A concept related to error but that measures the dispersion in error is called *precision.* This is best described in relation to measurement error. Where a measurement device such as a GPS at a fixed location (i.e., true value has no location variation) records repeated location measurements (i.e., GPS points) that are tightly centred on the true location, this GPS would be said to be highly accurate (i.e., low error) and highly precise (i.e., low variability in error). But if the spread of GPS points were tightly centred around a distant location, the GPS would have low accuracy and high precision. Accuracy and precision can best be understood as summary measures of error.

While error is a concrete measurable quantity, uncertainty is fuzzier. *Uncertainty* can be defined as stochastic in nature, where repeated measurements or samples draw a range of values that define the amount of uncertainty. This is called *statistical uncertainty. Systematic uncertainty* is where variability is the result of missing data or a lack of knowledge about the system being represented. In the water table example in Figure 7.4, calculating the hydraulic gradient might vary based on the errors made during digitizing, errors in the interpolation process, the density and spatial arrangement of measurements, and other factors. But not knowing if a mineral deposit between sampled locations interrupts water flow would be subject to systematic uncertainty associated with the lack of knowledge about areas not sampled. Different contextual factors could therefore influence the degree of systematic uncertainty. Repeated samples (in the same locations) will not tell us anything about systematic uncertainty, whereas it would be informative of stochastic uncertainty.

GIS analysis of spatial and attribute error is part of the accuracy assessment step in a broader data quality evaluation. Accuracy assessment involves characterizing the overall degree of error: how well some spatial or attribute data (e.g., a model or reality) corresponds with reality, which can be generalized to the expected degree of error in a given dataset. First, we will examine positional accuracy (i.e., derived from spatial error) measures for vector data and raster data, respectively. Typically, positional accuracy measures are expressed in linear units of the projection, such as metres. Second, we will examine attribute errors and measures of attribute error for different measurement scales.

7.4.1 Vector Positional Accuracy

In spatial accuracy assessment, positional accuracy metrics can be computed to assess the difference between a GIS representation and reality. Of course this implies knowledge of the true location, which is typically only approximated from measurement using high-accuracy GPS and surveying techniques or existing geodetic control data. Accuracy assessment is therefore relative, comparing an entity in a GIS database to another representation obtained from a higher-accuracy data source typically called the *reference data.* Depending on the representation, differences can be conceptualized as differences in location, size, shape, or orientation. The measurement of spatial error

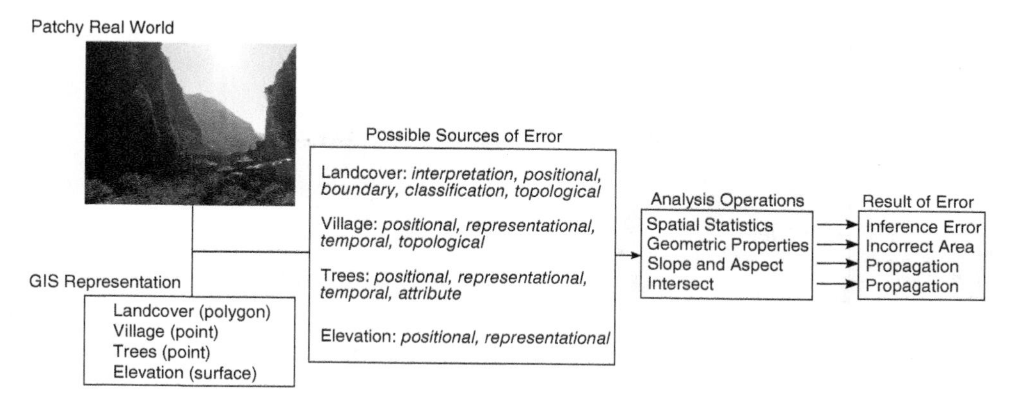

Figure 7.7 Reality, geographic abstraction, and sources of measurement of error in GIS data.

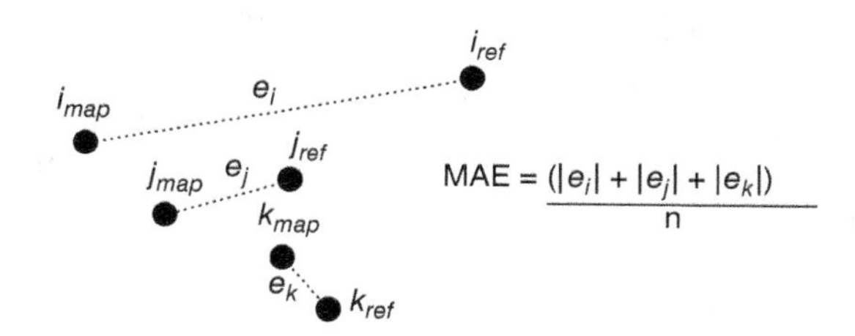

Figure 7.8 Visual representation of mean absolute error calculation from three points.

is confounded with the choice of representation of a real world object (see Figure 7.7). A stop sign in a municipal GIS database would likely be digitized as a point object, which is zero-dimensional, whereas in reality the stop sign encompasses three dimensions in space (four if including time). Representational issues are subsumed within the choice of spatial data model and not considered part of spatial error assessment.

For point data, spatial (i.e., positional) error is simply the distance between the point as represented in the GIS database and the point's location in the reference data (see Figure 7.8). For a single point i, we can determine the error e_i as the *distance* between the reference point (x_{ref}, y_{ref}) and the mapped point (x_{map}, y_{map}). Represented as a positive distance, this is called the *absolute error*. For multiple point locations on a map (e.g., all road intersections in the study area), it is more common to consider the average of point-wise error by subtracting reference coordinates from target coordinates. To cancel out the sign of the difference, we compute the root mean square error (RMSE), which summarizes errors as

$$RMSE = \sqrt{\frac{\sum_{i=1}^{n} (x_{\text{map}_i} - x_{\text{ref}_i})^2 + (y_{\text{map}_i} - y_{\text{ref}_i})^2}{n}},$$

which for points is the average of the point-wise spatial errors. While this references horizontal positional accuracy, the method also applies to vertical positional accuracy,

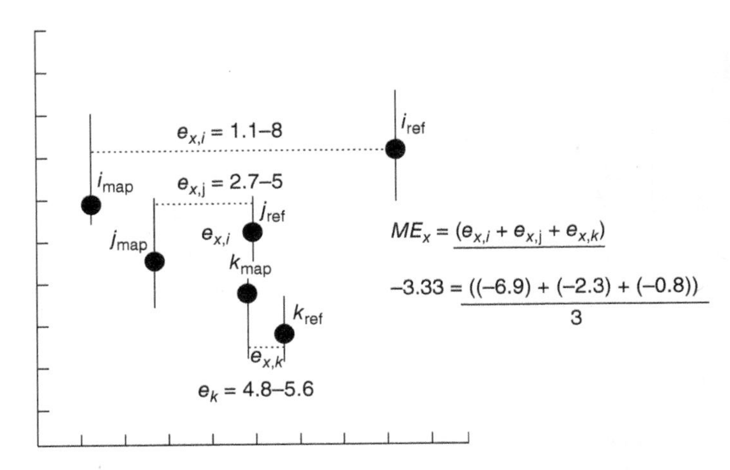

Figure 7.9 Visual representation of mean spatial bias in x-direction from three points.

reported as RMSE in linear units. Like any average, the RMSE is sensitive to outliers, and other central tendency measures such as the root median square error and mean absolute error are possible.

Another component of accuracy assessment is to measure systematic error or bias. This can be done by simply not squaring the differences and therefore not taking the square root in the RMSE formula, so that signs of errors no longer cancel each other out. Bias is often measured separately for each direction, such as

$$ME_x = \frac{\sum x_{\text{map}_i} - x_{\text{ref}_i}}{n} ,$$

which will indicate spatial bias in the x-dimension if it exists (due to a datum shift, for example). Such information may help to determine the source of the errors.

For line vector data, positional accuracy can be measured naively by computing point-wise RMSE at vertices. This approach is only useful when vertices are similar, such as at fence posts for a digitized fence line. In this way, the line is simply treated as a series of points. However, there is often no concordance in vertices between mapped linear features and reality/reference data (i.e., vertices have no physical analogues or are not significant landmarks so may be chosen at different locations). Landmark points can be learned by considering distinctive properties of the lines, such as acute angles formed by adjacent vertices. A related method is to sample the line vertices to common points identifiable on both the reference and target maps and then use point-wise RMSE.

Straightforward extensions of Euclidean distance for line and polygonal objects would be summary distance measures such as the minimum distance, maximum distance, or the centroid distance between objects in the GIS and the reference map (see Figure 7.11). This is the basis for the Hausdorff distance, which is covered in the following section.

Another approach to compare whole lines involves buffering the representation of the reference line and the mapped line, and computing the area of intersection to the

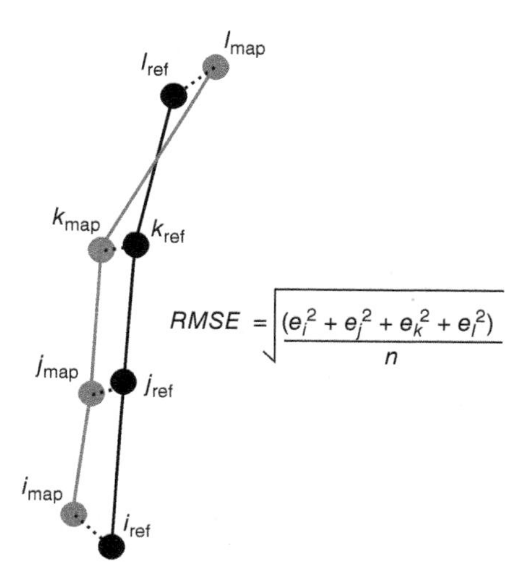

Figure 7.10 Visual representation of RMSE between lines, using vertex matching.

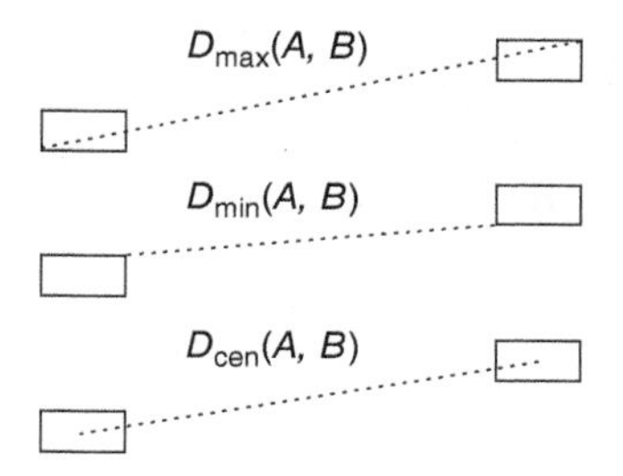

Figure 7.11 Summary distance measures for polygons.

area of the line buffer [115]. Done iteratively, the proportion of overlap at different buffer distances indicates the degree of positional agreement between the two linear features. An alternate method simply measures the proportion of the line length that intersects the buffer at a user-defined threshold [51] (d is buffer distance, B is buffer in Figure 7.12). A threshold amount of intersection can also be used and the buffer distance reported as precision, for example, when 95% of the test line's length is contained within the buffer of the reference line.

Another dimension of positional accuracy with complex geometry is shape. Do the two lines look the same? Do they have the same curvature or angular structure? Trajectory analysis offers methods to compare linear features in terms of shape. These methods are usually employed as analysis tools rather than for characterizing error. For example, it may be of interest to identify shapes of fishing vessel trajectories associated with different fishing behaviours such as trawling or long-line fishing. For accuracy assessment, shape-based methods can reveal similar linear representations even when locational differences exist. Orientation of lines can be detected through point-wise measures, but measuring the orientation of the overall line may be more revealing.

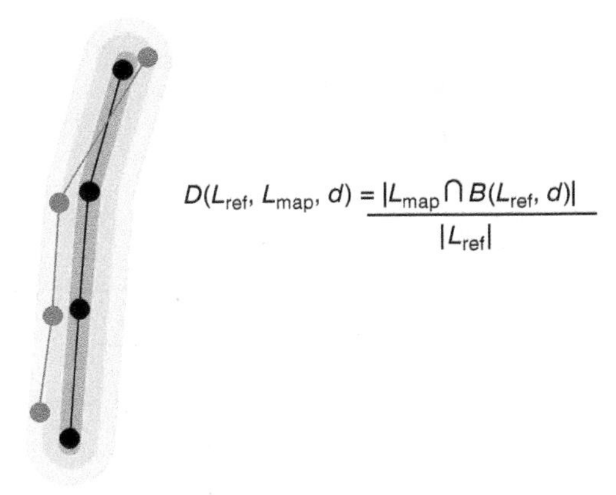

Figure 7.12 Visual representation of Goodchild and Hunter buffering algorithm for distance between lines [51].

Given a polyline L as a set of ordered points $p_i \ldots p_n$, where points meet at a set of angles $\theta_i \ldots \theta_n$, we can compute the average orientation (see Chapter 8) and compare orientation to determine orientation error between reference and map lines. Such an approach could also reveal situations where digitizing errors "flipped" the expected orientation of the line, as line direction often carries with it meaning associated with flows (e.g., traffic, trade, water).

While the above methods cover point–point and line–line, distance relations are more uncertain between point–line and polygon–polygon. Recall that a distance metric must satisfy four properties:

- non-negativity: $d(x_{\text{map}}, x_{\text{ref}}) \geq 0$
- identity: $d(x_{\text{map}}, x_{\text{ref}}) = 0$ iff $x_{\text{map}} = x_{\text{ref}}$
- symmetry: $d(x_{\text{map}}, x_{\text{ref}}) = d(x_{\text{ref}}, x_{\text{map}})$
- triangle inequality: $d(x_{\text{ref}}, x_{\text{map1}}) \leq d(x_{\text{ref}}, x_{\text{map2}}) + d(x_{\text{map2}}, x_{\text{map1}})$

For polygonal representations, distance can be computed by centroid distance (see Figure 7.11), although this is often a poor choice when polygons are irregular or overlapping. Given polygon A and polygon B, which are each defined by points denoted p_i, we can define the Hausdorff distance between A and B as the largest of the shortest distances from points in A to points in B, and vice versa. So going from A to B, we have

$$h(A, B)_1 = \max_{p_a \in A} \{\min_{p_b \in B} \{d(p_a, p_b)\}\}$$

and from B to A,

$$h(B, A)_2 = \max_{p_b \in B} \{\min_{p_a \in A} \{d(p_a, p_b)\}\} \quad ,$$

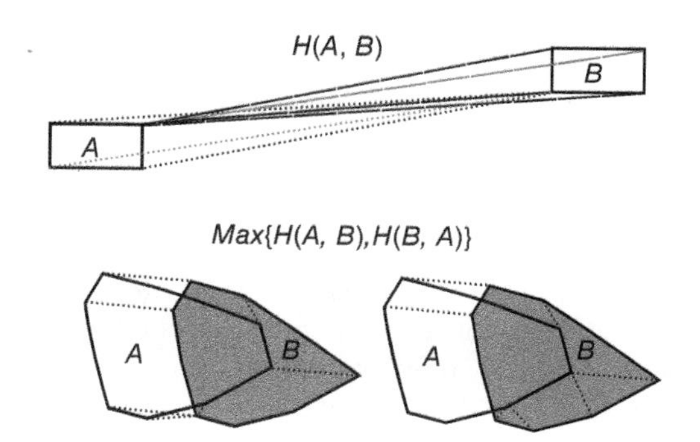

Figure 7.13 Hausdorff distance computed over two sets of polygons. Dashed lines indicate set of distances for which the maximum is the Hausdorff distance between A and B.

where $\max\{h(A,B)_1, h(B,A)_2\}$ gives the Hausdorff distance between the two polygons. The Hausdorff distance is a metric similar to Euclidean distance but works with point sets. An example is given in Figure 7.13, which shows the distances that would be compared for calculating $H(A,B)$ for a simple scenario and for overlapping polygons.

Additionally, error in derived metrics such as polygon area and perimeter, and in shape measures such as perimeter/area ratio, can be used to compare positional accuracy. It is good GIS practice to routinely calculate geometric properties when manipulating spatial data.

7.4.2 Raster Positional Accuracy

Positional accuracy of raster data is more difficult to measure directly. Normally the spatial accuracy of raster data is determined at the time of creation as part of georeferencing (see Chapter 2), and accuracy is derived for known locations (control points) and used to estimate error over the entire dataset (RMSE). Since raster data are usually created through computational methods (e.g., satellite sensors, rasterization of vector data) and space is continuously represented, positional errors are more likely to be systematic than random. Uncertainty in the positioning of features represented in raster depends on the spatial resolution and is an important consideration when deciding on the spatial data model.

7.4.3 Attribute Accuracy

Measuring attribute accuracy in geographic information depends on the measurement scale of the variable and its numeric representation in the computer. Attribute accuracy is therefore based on specifying a model for error that is appropriate for the data. In the simple binary case, the error is therefore $e_i \in 0, 1$ and can be summarized in various ways to determine measures of accuracy. An example of binary error is classification error in raster data, where a grid cell has been incorrectly labelled relative to ground control or a higher-resolution raster dataset. For example, a landuse classification may classify a pixel as urban when in reality it is forest. With such a set up (and all

Table 7.1 2 × 2 Contingency Table.

	Reference Map	
Test Map	a	b
	c	d

attribute accuracy tasks), we can represent the comparison of test and reference values in a *contingency table*, where cells of the table present counts of the cells with correctly (i.e., diagonal) and incorrectly (i.e., off-diagonal) classified raster cells.

In the case of a binary classification (e.g., forest–non-forest) the diagonal cells in Table 7.1 would correspond to forest correctly classified as forest and non-forest correctly classified as non-forest. Elements a and d are called true positives and true negatives, as they correspond to correct predictions from the model or classification. Elements b and c are called false positives (non-forest labelled as forest) and false negatives (forest labelled as non-forest), respectively. The objective in any classification or prediction exercise is to maximize a and d while minimizing b and c. Errors b are sometimes called Type II errors, while c are known as Type I errors in classical statistics. Note that the same set-up could result from any binary attribute value, such as the predictions from a logistic regression model. Switching notation from Table 7.1 such that $a = TP$, $b = FP$, $c = FN$, $d = TN$, we can define the true positive rate (also called sensitivity or recall) as

$$TPR = TP/(TP + FN) \quad ,$$

whereas the true negative rate (also called specificity) is

$$TNR = TN/(FP + TN) \quad .$$

Another important summary statistic is the positive predictive value (also called precision), which describes the proportion of positive predictions that are correctly classified,

$$PPV = TP/(TP + FP) \quad ,$$

and finally, the total classification accuracy is

$$TCA = TP + TN/(TP + FP + TN + FN) \quad ,$$

which is simply the proportion of correctly classified predictions.

Returning to our forest–non-forest binary classification example, we can compute each of the above classification statistics over a sample classification. A sample of the Victoria dataset is given in Figure 7.14. After an unsupervised classification of this image with two classes (forest, non-forest), we have a map where each pixel contains a prediction of whether that pixel represents forest-cover or not.

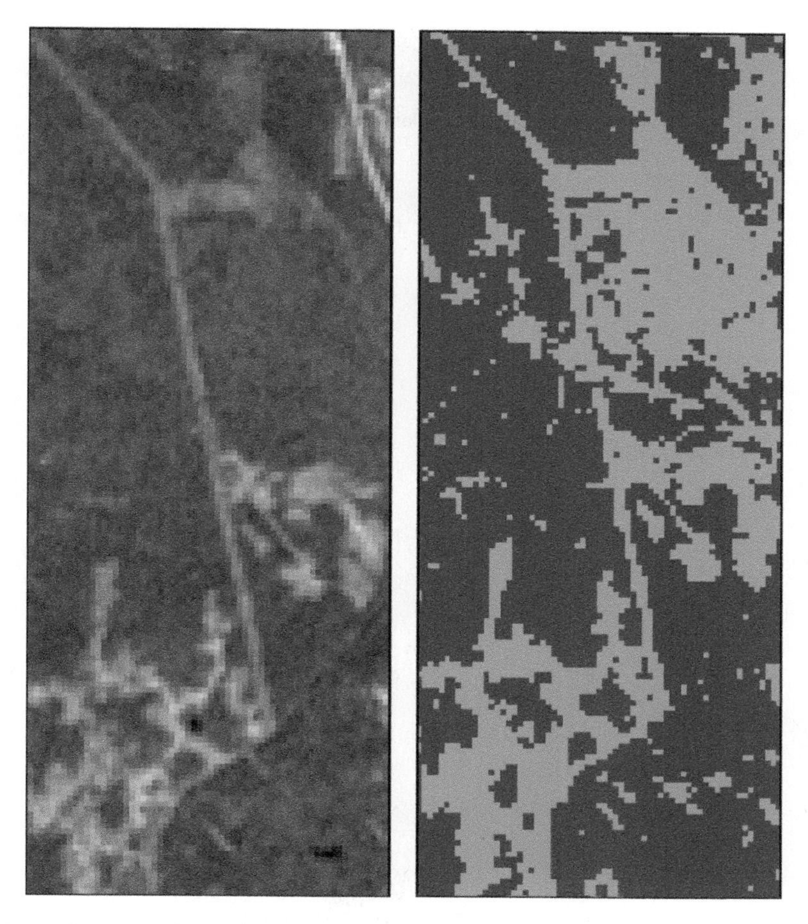

Figure 7.14 Section of Landsat imagery near Victoria, BC, and result of isocluster unsupervised classification.

To assess the accuracy of the classified raster dataset, we need to establish the true status for a subset of representative pixels in scene. Before proceeding, we can inspect the classification visually and see that the linear feature indicating a road was classified as non-forest. However, in the northeast part of the image there are a lot of predicted non-forest pixels that appear to be forest cover in the Landsat image. The reason for this misclassification is likely because that section of forest is younger and therefore has less biomass and different spectral properties from older, more-mature forest areas. We now randomly select 20 sites in the study area for which we obtain validation data and characterize errors in Tables 7.2 and 7.3 and Figure 7.15.

Continuing on, we take the error results reported in Table 7.2 and summarize them into a table of the form of Table 7.1, from which we can derive our desired error statistics. Enumerating the elements of the contingency table yields Table 7.3, which

Table 7.2 Classification accuracy for 20 random locations in Victoria study area.

ID	Ground	Classified	Result
1	NF	F	FP
2	F	F	TP
3	F	NF	FN
4	F	F	TP
5	NF	F	FP
6	NF	F	FP
7	F	NF	FN
8	F	F	TP
9	NF	F	FP
10	F	NF	FN
11	F	F	TP
12	NF	NF	TN
13	F	F	TP
14	F	F	TP
15	F	F	TP
16	NF	NF	TN
17	NF	F	FP
18	F	F	TP
19	F	F	TP
20	F	NF	FN

Table 7.3 2×2 contingency table for forest–non-forest classification.

		Reference Data		
		Forest	Non-forest	
Test Data	Forest	9	5	14
	Non-forest	4	2	6
		13	7	

gives the following summary error statistics in Table 7.4. We can see that in total, only 55% of the tested field sites were correctly classified by the unsupervised classification. However, the forest areas classified better than the non-forest areas, as the true positive rate was 69%. This means that if you were to visit a pixel labelled as forest, on average it would be correct 69% of the time.

Table 7.5 extends this to $k \times k$, where k is the number of map classes. Here we collected more sample field data to improve the sample size to 60, and extended the map legend to Young Forest, Mature Forest, Road, and Clearing. With $k \times k$

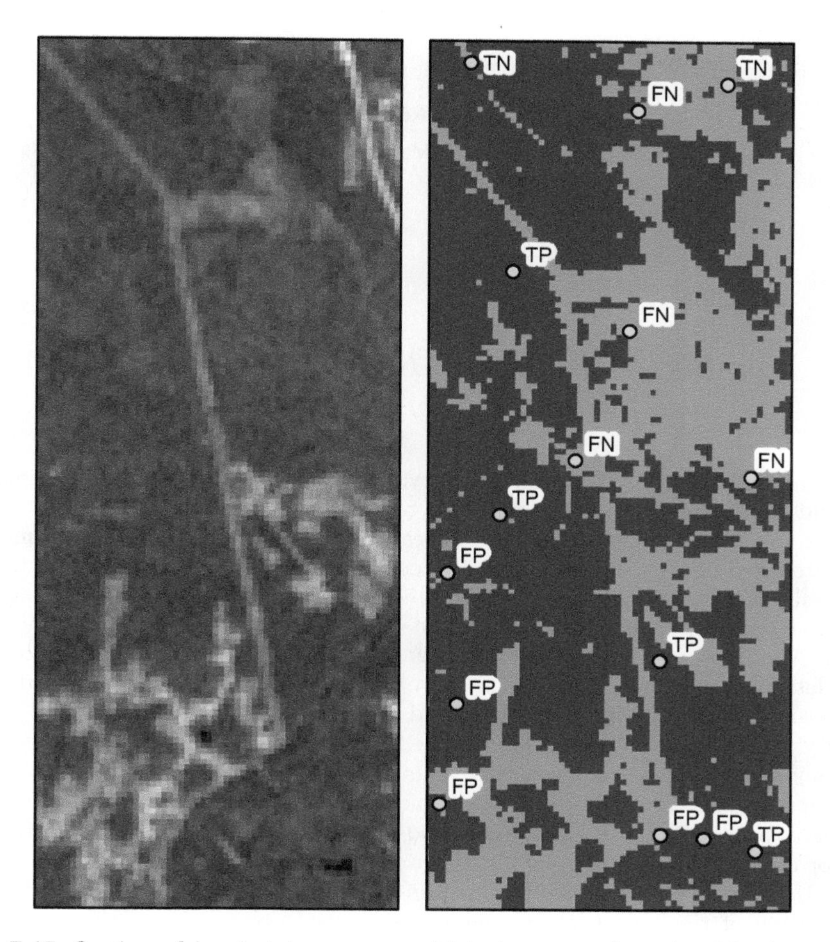

Figure 7.15 Section of Landsat imagery near Victoria, BC, and result of isocluster unsupervised classification with reference sites and classification errors (15 of 20 validation sites shown).

Source: Upper Saddle River, NJ: Prentice Hall

Table 7.4 Summary error statistics for binary classification of forest–non-forest.

Metric	Value
TPR	0.69
TNR	0.28
PPV	0.64
TCA	0.55

Table 7.5 $k \times k$ contingency table for land cover classification.

		Reference Data					
		Young Forest	Mature Forest	Road	Clearing	**Totals**	User's
Test Data	Young Forest	4	2	0	1	7	57%
	Mature Forest	1	15	1	3	20	75%
	Road	3	0	11	6	20	55%
	Clearing	4	1	2	6	13	46%
	Totals	12	18	14	16	60	TCA
	Producer's	33%	83%	79%	38%		60 %

tables—especially for classification accuracy assessments—it is typical to compute the marginal statistics, which across rows are called User's Accuracy measures and down columns are called Producer's Accuracy measures. These names refer to who might be most interested in the associated type of accuracy. A user of a map derived from an image classification, using the map to identify field sites, would want to know the accuracy of the mapped classes (i.e., what proportion of mature forest pixels are actually mature forest). A map producer may want to know what proportion of the mature forest sites were classified correctly. This nomenclature is common in remote sensing studies, but the measures are identical to sensitivity and specificity described above. In a $k \times k$ contingency table (Table 7.5), we compute the marginal proportions along rows as User's Accuracy and down the corresponding columns for Producer's Accuracy. A final way to summarize this is the more interpretable *commission errors* and *omission errors*, defined as (100 − Producer's Accuracy) and (100 − User's Accuracy), respectively.

For an overall assessment of classification accuracy, TCA seems the most obvious choice. However, TCA can be high even when the classification is no better than random. A limitation of the binary error measures described above is that they are greatly affected by the relative proportions of map classes. For example, if 90% of a study area is forested, a random classifier would still correctly classify a large number of pixels as forest. We can account for the expected proportion of cells that would be correctly classified due to chance, given the composition of the map classes. For example, chance is likely playing a role in the high TPR in Table 7.4, as forest is much more common than non-forest.

The Kappa coefficient is an index of agreement that accounts for the random element in quantifying classification accuracy [25]. The observed fraction of agreement is corrected for the expected fraction of agreement if the cells had been randomly reshuffled. The general formula for Kappa is

$$K = \frac{p_0 - p_e}{1 - p_e} \quad ,$$

where p_0 is the observed proportion of correct classifications (i.e., sum of diagonal elements in the contingency table divided by total number of cells in the map) and p_e

is the expected proportion of correctly classified map units based on chance. The p_e is determined by the marginal sums of the contingency table,

$$p_e = \sum_{i=1}^{k} n_{i+} \cdot n_{+i} \quad ,$$

where n_{i+} is a column proportion for column i and n_{+i} is a row proportion for column i. Returning to the example of Table 7.3, the Kappa index would be

$$-0.023 = \frac{11/20 - [(13/20 \cdot 14/20) + (7/20 \cdot 6/20)]}{1 - [(13/20 \cdot 14/20) + (7/20 \cdot 6/20)]} \quad ,$$

which, since the K value is very close to zero, indicates that even though over 50% of the cells in the map were correctly classified, this was not greater than what would be expected given a random classification. The next step would then be to try to improve the classification accuracy by possibly splitting the map classes into more categories (e.g., Table 7.5) or to try alternate classification algorithms. Additional considerations may be made for cases where samples used in the contingency table are not independent, a more advanced topic covered by Foody [40].

Attribute error can also be at higher measurement scales, and while all error can be converted to binary, there is a loss of information when doing so, and greater insight into the magnitude of errors is possible when preserving full information. For example, if we have multiple forest classes (young forest, mature forest), we may consider misclassifying a pixel as the wrong type of forest to be less of an error than misclassifying a young forest pixel as farm land. The appropriate treatment of ordinal classification errors is far less developed than for binary errors, although some techniques do exist. A measure of ordinal errors should capture the degree of success in preserving the ordering of the predicted classes when compared to the validation data, and should maintain the property that $e_i = (p_i a - q_i b) \leq e_j = (p_j a - q_j d)$.

One approach to handling ordinal data is to examine the ranks of classified and validation data though a correlation statistic. A rank order correlation statistic called Spearman's rank correlation coefficient is based on ordering classification and validation data vectors p and q and simply computing the regular correlation coefficient over the numeric ranks as

$$\rho = \frac{\sum_{i=1}^{N}(p_i - \bar{p}) \cdot (q_i - \bar{q})}{\sqrt{\sum_{i=1}^{N}(p_i - \bar{p})^2 \cdot \sum(q_i - \bar{q})^2}} \quad ,$$

where N is the number of validation sites and ρ has the same interpretation as Pearson's correlation coefficient. Since the numbers used for the ranks are used directly in the calculation, the magnitude of correlation can be affected by the number of mapped classes, which may not be desirable. An alternate method is Kendall's τ rank correlation, which counts pairwise orderings of predicted and validation pairs. Given an ordering of P such that $p_i \leq p_j \leq p_k \ldots \leq p_n$ and pairs p_i, q_i are considered jointly, a pair is called *concordant* if the relative rank of p_i and q_i agree (i.e., $p_i > p_j$ and $q_i > q_j$). If the ranks are different, the pair is called *discordant*. If the ranks are tied ($p_i = p_j$ or $q_i = q_j$), they are counted separately for p and q. The Kendall τ is defined as

$$\tau = \frac{c - d}{\sqrt{(c + d + e_q) \cdot (c + d + e_p)}},$$

where c is the number of concordant pairs, d is the number of discordant pairs, e_t is cases of ties in the validation pairs (e.g., q), and e_p is cases of ties in the predicted pairs (e.g., p). The interpretation of τ is the same as ρ, on the scale of -1 to $+1$.

Interval and ratio level attribute errors can be quantified with the standard error statistics and diagnostics previously described, such as the mean absolute error, root mean squared error, scatterplots, and Pearson's R correlation coefficient.

7.4.4 Completeness

At the level of a dataset, data *completeness* is an aggregate measure of spatial data quality that summarizes the degree to which a test dataset covers the same area as another reference dataset. Measuring completeness depends on identifying matching features in datasets and enumerating the proportion that exist in the test dataset relative to the reference dataset. In theory, completeness could apply to attribute as well as spatial dimensions. Further, like all aspects of data quality, the determination of what constitutes "complete" data depends on the intended use of the dataset—the *fitness for use* concept.

Matching features in two datasets can be based on spatial concordance, matching unique identifiers, or, for attributed street networks, address matching. Spatial concordance can be determined through a spatial join or buffering approach, while matching identifiers can be achieved through database joins via linking IDs in the attribute tables. Address matching typically involves geocoding to geographic coordinates using a common geocoder, then assessing spatial concordance. A common application of evaluating spatial completeness is data quality evaluation of volunteered geographic information (VGI) (e.g., Haklay [59], Jackson et al. [65]). A simple measure in this context is the proportion of matching features by enumerating the number of features, found in both datasets (intersection)—those in reference but not in test dataset (reference complement) and those in the test but not the reference (test complement). After features have been matched, an assessment of their spatial accuracy is usually carried out, which depends on the form of spatial representation (i.e., using spatial accuracy measures described above).

7.5 Error Propagation in Geographic Analysis

Geographic analysis and processing tools often create new data from existing datasets (see Chapter 10). One of the principal powers of using GIS as the basis for analysis is the idea of space as an indexing system for integrating disparate data sources. As geoprocessing operations—such as Buffering, Intersection, Union, and Clip on vector datasets, or spatial filtering routines on raster datasets—are used to derive new geographic data from existing datasets, the error and uncertainty within each input dataset can propagate and accumulate through the analysis. The result of this error propagation is that the geographic data products used for decision making may have much greater uncertainties and error in them than the original datasets.

To illustrate, the example in Figure 7.16 gives a simple two-map overlay of soils and landcover. We assume that each came from independent processing techniques, the landcover map, with a user's accuracy of 89%, and the soil map, with a user's accuracy of 70%. The probability that any joint soil-landcover combination will be correct is $0.89 \cdot 0.70 = 0.623$, or 62.3%. Subsequent processing of data with other datasets, such as elevation, slope, soil moisture, and temperature, would quickly add error to resultant datasets. Propagation of error is a classical issue of data quality specific to GIS. To deal with error propagation, careful tracking and/or estimating of the error propagation is required, which can be done through Monte Carlo or numerical methods [63].

We discussed previously a Monte Carlo approach to spatial uncertainty in Nelson et al. [84], an older and still-useful technique that uses analytical methods. We'll consider the case of deriving new attribute values U from existing attributes $A_i \dots A_m$. Following Burrough and McDonnell [17], given a function f mapping $a_i \dots a_m$ to new output U,

$$U = f(A_i, A_j, A_k, \dots, A_m) \quad ,$$

classical error analysis gives us the following tools:

$$s_U = \sqrt{\sum_{i=1}^{m} \left(\frac{\delta U}{\delta A_i}\right)^2 \cdot s_{A_i}^2} \quad ,$$

when all A are independent, so the relative contribution of each attribute to the total error can be estimated. In the case of correlated inputs, as is typical with geographic

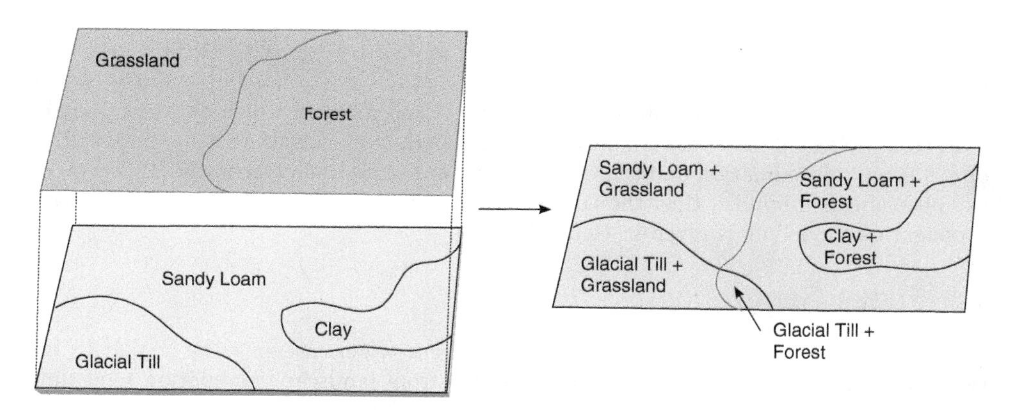

Figure 7.16 Example of error propagation in GIS-based analysis (adapted from Lo and Young [77]).

data, we add an additional term onto the above, the notable term being ρ_{ij}, or the correlation between A_i and A_j,

$$\left[\sum_{i=1}^{m}\sum_{j=1}^{m}\left(\frac{\delta U}{\delta A_i}\right)\cdot\left(\frac{\delta U}{\delta A_j}\right)\cdot s_{A_i}\cdot s_{A_j}\cdot\rho_{ij}\right] \quad .$$

The effect of propagation therefore varies based on the type of function f. For sums of two inputs where both A_i and A_j are independent, the error as quantified in s_U simplifies to

$$s_U = \sqrt{s_{A_i} + s_{A_j}} \quad .$$

However, to generalize the above to any continuously differentiable function $g(\cdot)$, we approximate the function using the Taylor series method [63]. Given an output map U derived from some analytical function or geoprocessing operation $g(\cdot)$ mapping input attributes (or raster cell values) $A_1, A_2, \ldots, A_n$, we can define a model for the error in each A_i such that we represent the deterministic estimate of the data as one component of A_i, which we represent as b_i, and the error component V_i, such that we have $A_i = b_i + V_i$ (note that V is analogous to $E_s + E_m$ in the general model that started the chapter). Thus each of the inputs to $g(\cdot)$ contributes some error through V_i into U. A Taylor-series expansion to approximate $g(\cdot)$ around $\overline{b}$ is

$$U = g(\overline{b}) + \sum_{i=1}^{m}(A_i - b_i)\cdot g_i'(\overline{b}) \quad ,$$

where $g_i'(\cdot)$ is the first derivative of $g(\cdot)$ with respect to the i-th argument. The expected value of the function is therefore the mean of U, and the variance is

$$s_U = \sum_{i=1}^{m}\sum_{j=1}^{m} s_{A_i}\cdot s_{A_j}\cdot\rho_{ij}\cdot g_i'(\overline{b})\cdot g_j'(\overline{b}) \quad ,$$

which only requires the variance and covariance of $A_i \ldots A_m$ and importantly can be used to decompose the contributions of error to U and identify which datasets perhaps need to be collected at a finer resolution or improved. Note that the above linearization of $g(\cdot)$ is an approximation and therefore incurs approximation error itself. Higher-order expansions can minimize this, though are not covered here. A full discussion of error propagation modelling is given in Heuvelink [63].

7.6 Fuzzy Geography

A special case of uncertainty in GIS data can occur when assessing the similarity between two categorical maps, which often results from landscape simulation modelling or evaluating a map derived from a new data source (e.g., a citizen science project). While the binary attribute (e.g., classification) error measures described earlier can be used to quantify error between two categorical raster maps (or binary attributes), there

Table 7.6 Fuzzy vector representation for map categories in landcover classification example.

Map Class	Vector			
Young Forest	(1,	0.6,	0,	0)
Mature Forest	(0.6,	1,	0,	0)
Road	(0,	0,	1,	0)
Clearing	(0,	0,	0,	1)

are some important considerations specific to handling geographic data. In particular, a slight spatial misalignment can create hugely inflated error statistics when measures are computed on a cell-by-cell basis. More generally, a human observer comparing two maps will recognize similarities and differences, beyond discrepancies in cell-by-cell values, due to local anomalies, global trends, and logical consistency, among other observations [58]. An extreme example would be the comparison of two checkerboard maps, one starting with white in the upper left cell, the other with black in the upper left cell. A human observer would immediately recognize the similarities in pattern and structure, whereas a cell-based error comparison using a contingency table would find the maps highly dissimilar (i.e., total disagreement between the two maps).

One way to handle the above problem is to allow uncertainty into the definition of categories and locations on the map. Hagen [58] describes the use of a "fuzzy" representation whereby map categories can have degrees of membership in other categories (fuzzy membership). Degrees of membership are represented as a vector $V_{\text{cat}} = [\mu_{\text{cat}_1}, \mu_{\text{cat}_2}, \ldots, \mu_{\text{cat}_n}]$ for all n categories represented in the map. Returning to our forest classification example, we may represent the degrees of membership differently for map classes that are more similar or different; for example, the Young Forest class having partial membership in the Mature Forest class and vice versa (see Table 7.6). In this way, by formulating the fuzzy relations between map categories, we can handle ordinal map categories and nominal map categories.

The same concept can be applied to location, such that the map category of a cell depends partially on the categories of cells in its neighbourhood, with some definition of the scale and functional form of distance decay in the neighbouring cells (e.g., exponential, linear, Gaussian). This may be useful where uncertainty in boundaries are common between map categories (e.g., forest to marsh). The comparison of maps therefore jointly considers uncertainty (fuzziness) in both category and location to simulate human judgement in comparing maps.

7.7 Topological Error

When topology is enforced in geographic data, it can be exploited to detect and correct for errors. Ubeda and Egenhofer [116] describe how topological errors in geographic data can be due to the data structure, either where the data do not have topological structure (e.g., the spaghetti model) or where the data do not respect the structure that exists (e.g., unclosed polygon, self-intersecting polylines). A secondary source of topological errors may be due to the semantics of the features being represented in the data. For example, a line representing a road centreline intersecting a polygon representing a building footprint.

Recalling the 9-intersection model for point-set topology, we can use these relations to enforce or detect topological errors. One way to accomplish this is through topological constraints, which are rules that encode topological relations and thus flag violations to these rules. As the 9-intersection model provides topology for point, line, and area relations based on intersection of boundary (b), interior (i), and exterior (e), many possible topological constraints can be created. Given the 3×3 matrix of topological relations between feature A and feature B as

$$\begin{bmatrix} A_b \cap B_b & A_b \cap B_i & A_b \cap B_e \\ A_i \cap B_b & A_i \cap B_i & A_i \cap B_e \\ A_e \cap B_b & A_e \cap B_i & A_e \cap B_e \end{bmatrix},$$

the number of relations realized can be grouped based on the types (and dimensions) of the objects being considered. For example, where A is a point and B is a point, there are only two possible topological relations, whereas where A is a line and B is a polygon, there are 19 possible topological relations. Topological relations can be grouped into similar categories as well; for example, where A is a line and B is a polygon, all topological relations where the interior of A intersects the interior and exterior of B could be grouped together as *cross*. The next step would be to encode semantic relations as an extra layer of constraints onto the topological constraints. This type of topology is emerging in modern spatial database systems in order to build application logic into the geographic data at the database level. This can have significant benefits for preventing data entry errors and maintaining topological integrity across multiple datasets during spatial data editing or when representing dynamic spatial objects.

Problems

1. Select a sub-study area in the mountain pine beetle dataset that captures about 100 points. Add a new attribute and randomly assign values 1–50 such that each label appears twice. Compute the distance between all matching labels, compute the RMSE, and produce a histogram of the point-wise distances.
2. Compute an aspect surface for the digital elevation model that is part of the Grande River dataset. Extract all south-facing cells as a new polygon dataset. Develop a procedure to estimate the variability in the derived south-facing slopes given an uncertainty in the input DEM of 20 metres.
3. Explain how the Hausdorff distance differs from the centroid distance and the minimum distance, and give an example of how it could be used for analysis of polygonal data.
4. Explain how error and uncertainty modelling differs between raster and vector data. What are the advantages and disadvantages of each?
5. Design a methodology for measuring and modelling error in three-dimensional GIS data, representing buildings and interior spaces of a university campus, that will be used for an emergency routing application.

Chapter 8

GeoComputation

8.1 Introduction

In this chapter we delve into some topics that one might normally encounter in a computer science text, but because they are fundamental to understanding the theoretical basis for GIS, we spend some time here discussing the relevant ideas and details. Understanding the theory allows one to make reasonable and sensible decisions while problem solving and is thus useful in addition to being interesting in its own right. The first section of the chapter is devoted to algorithms, the second to computational issues.

8.2 Algorithms

An algorithm is a formalized set of steps in a process to achieve some goal or calculation. By formalized, we mean describable unambiguously in a standard language. Often the standard language of implementation is a computer language, and the communication of the algorithm may be just the native code or so-called pseudocode. Pseudocode is a hybrid between natural language and computer code meant to describe the algorithm without reference to a particular computer language. In this section we describe in some detail a few algorithms used quite frequently in GIS practice. Thinking about solving problems with an algorithmic approach is often foreign to many GIS students but a worthwhile addition to your toolkit of ideas. Note that in several cases we describe what we feel is the most intuitive methodology, not perhaps the most efficient or widely used.

8.2.1 Line Segment Intersection

In rigorous computational geometry, integer coordinates are used wherever possible. This avoids the pitfalls associated with the introduction of floating point representations of real numbers, as discussed later in this chapter, and provides for the possibility of proving the correctness of the results. However, line segment intersection is a problem in which floating point calculations are unavoidable. Furthermore, line segment intersection is a fundamental operation in GIS; for example, in merging datasets and in spatial intersection queries. The discussion in this section is adapted from O'Rourke [88]. We now give a sketch of one popular solution approach.

Let endpoints of the two segments be given by

$$\text{first segment: } \underline{a} = \{x_a, y_a\} \;,\; \underline{b} = \{x_b, y_b\}$$
$$\text{second segment: } \underline{c} = \{x_c, y_c\} \;,\; \underline{d} = \{x_d, y_d\} \quad .$$

So line segment $\underline{ab}$ is contained in line L_{ab}, and line segment $\underline{cd}$ is contained in line L_{cd}. Then we create a parameterized form of the two line segments as follows:

$$\begin{aligned} &\text{Let} \\ &\underline{ab} = \underline{b} - \underline{a} \\ &\underline{cd} = \underline{d} - \underline{c} \quad , \\ &\text{then let} \\ &\underline{p}(s) = \underline{a} + s \cdot \underline{ab} \\ &\underline{q}(t) = \underline{c} + t \cdot \underline{cd} \quad . \end{aligned}$$

This gives a parameterization of the two lines L_{ab} and L_{cd}, the two parameters being s and t, respectively. If we restrict s and t to lie in the closed interval $[0, 1]$, the above equations model all the points in their respective line segments. Note that with the parameter values $s = 0$ or $t = 0$, this gives the starting point of the respective lines, whereas parameter values $s = 1$ or $t = 1$ gives the respective endpoints of the lines. Now, to find the intersections we set the two parameterized line equations equal to each other and solve for the parameters:

$$\underline{p}(s) = \underline{q}(t) \quad .$$

Note, this gives two equations and two unknowns as follows:

$$\begin{aligned} x_a + s \cdot x_{ab} &= x_c + t \cdot x_{cd} \\ y_a + s \cdot y_{ab} &= y_c + t \cdot y_{cd} \quad . \end{aligned}$$

To solve for t, for example, we get

$$\begin{aligned} s \cdot x_{ab} &= x_c + t \cdot x_{cd} - x_a \\ s \cdot y_{ab} &= y_c + t \cdot y_{cd} - y_a \\ s &= \frac{x_c + t \cdot x_{cd} - x_a}{x_{ab}} \\ s &= \frac{y_c + t \cdot y_{cd} - y_a}{y_{ab}} \quad . \end{aligned}$$

So setting these last two equations equal to eliminate s, we get

$$\frac{x_c + t \cdot x_{cd} - x_a}{x_{ab}} = \frac{y_c + t \cdot y_{cd} - y_a}{y_{ab}}$$

$$x_c \cdot y_{ab} + t \cdot x_{cd} \cdot y_{ab} - x_a \cdot y_{ab} = y_c \cdot x_{ab} + t \cdot y_{cd} \cdot x_{ab} - y_a \cdot x_{ab}$$

$$t \cdot (x_{cd} \cdot y_{ab} - y_{cd} \cdot x_{ab}) = y_c \cdot x_{ab} - y_a \cdot x_{ab} - x_c \cdot y_{ab} + x_a \cdot y_{ab}$$

$$t = \frac{y_c \cdot x_{ab} - y_a \cdot x_{ab} - x_c \cdot y_{ab} + x_a \cdot y_{ab}}{x_{cd} \cdot y_{ab} - y_{cd} \cdot x_{ab}} \quad .$$

Next, using the definitions of $\underline{ab}$ and $\underline{cd}$, we get

$$\begin{aligned}
t &= \frac{y_c \cdot (x_b - x_a) - y_a \cdot (x_b - x_a) - x_c \cdot (y_b - y_a) + x_a \cdot (y_b - y_a)}{(x_d - x_c) \cdot (y_b - y_a) - (y_d - y_c) \cdot (x_b - x_a)} \\
&= \frac{y_c \cdot x_b - y_c \cdot x_a - y_a \cdot x_b + y_a \cdot x_a - x_c \cdot y_b + x_c \cdot y_a + x_a \cdot y_b - x_a \cdot y_a}{x_d \cdot y_b - x_c \cdot y_b - x_d \cdot y_a + x_c \cdot y_a - (y_d \cdot x_b - y_c \cdot x_b - y_d \cdot x_a + y_c \cdot x_a)} \\
&= \frac{y_c \cdot x_b - y_c \cdot x_a - y_a \cdot x_b + y_a \cdot x_a - x_c \cdot y_b + x_c \cdot y_a + x_a \cdot y_b - x_a \cdot y_a}{x_d \cdot y_b - x_c \cdot y_b - x_d \cdot y_a + x_c \cdot y_a - y_d \cdot x_b + y_c \cdot x_b + y_d \cdot x_a - y_c \cdot x_a} \\
&= \frac{x_b \cdot y_c - x_a \cdot y_c - x_b \cdot y_a + x_a \cdot y_a - x_c \cdot y_b + x_c \cdot y_a + x_a \cdot y_b - x_a \cdot y_a}{x_d \cdot y_b - x_c \cdot y_b - x_d \cdot y_a + x_c \cdot y_a - x_b \cdot y_d + x_b \cdot y_c + x_a \cdot y_d - x_a \cdot y_c} \\
&= \frac{x_a \cdot (y_b - y_c) + x_b \cdot (y_c - y_a) + x_c \cdot (y_a - y_b)}{x_a \cdot (y_d - y_c) + x_b \cdot (y_c - y_d) + x_c \cdot (y_a - y_b) + x_d \cdot (y_b - y_a)} \quad .
\end{aligned}$$

If we let the denominator equal D,

$$D = x_a \cdot (y_d - y_c) + x_b \cdot (y_c - y_d) + x_c \cdot (y_a - y_b) + x_d \cdot (y_b - y_a) \quad ,$$

we get

$$t = \frac{x_a \cdot (y_b - y_c) + x_b \cdot (y_c - y_a) + x_c \cdot (y_a - y_b)}{D} \quad .$$

The intersection point is then

$$x_{q(t)} = x_c + t \cdot x_{cd}$$

$$y_{q(t)} = y_c + t \cdot y_{cd} \quad .$$

Note that we could also have solved for s and used the equation for $p(s)$, as the intersection solution, if it exists, is on both lines. However, it is useful as shown below to calculate both s and t. Following similar algebra, we get

$$s = \frac{x_a \cdot (y_d - y_c) + x_c \cdot (y_a - y_d) + x_d \cdot (y_c - y_a)}{D} \quad .$$

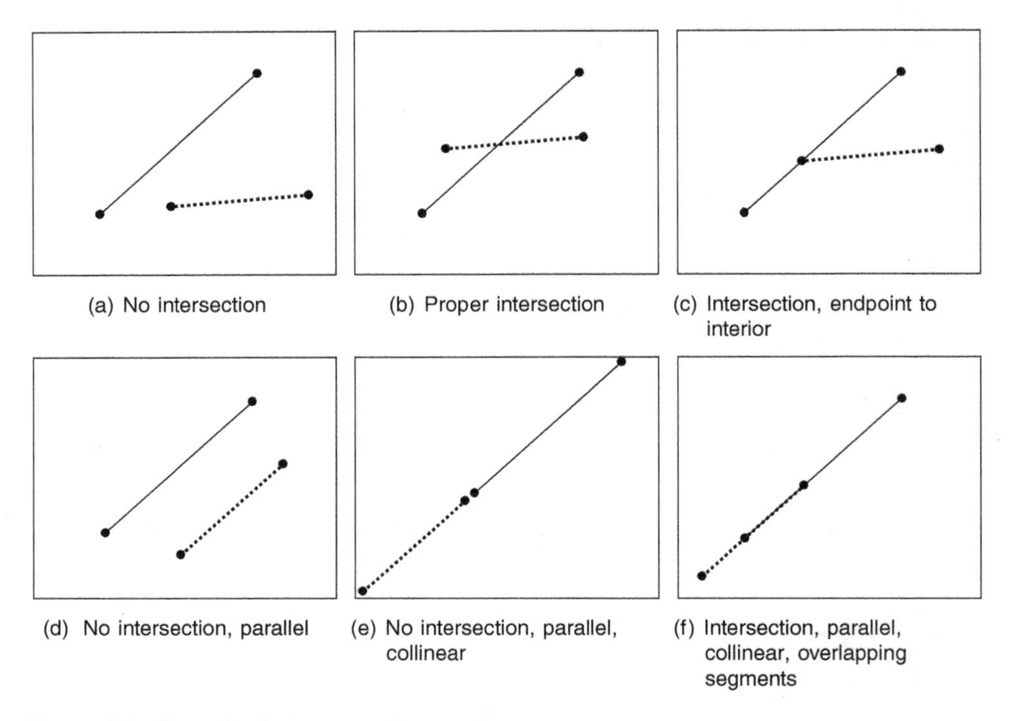

Figure 8.1 Cases for line segment intersection.

It turns out that if $D=0$, the lines are parallel. Also, there are several types of intersection that have to be handled by the solution, best illustrated by Figure 8.1. The calculations we performed above can be used to differentiate the six cases as follows (see Figure 8.2).

You will see in the next section hints of why it is useful to distinguish the types of intersections of two line segments. Further, notice that if we we relax the restrictions on the range of the parameters s and t, we can use the parametric equations to represent rays and lines. Also, the parametric form generalizes nicely to higher-dimensional settings. Finally, note that normally we do not need to globally test for pairwise intersections of all line segments, as we usually use a plane sweep algorithm (as described in Chapter 4) or take advantage of problem-specific information to reduce the number of segments to be tested for intersection.

8.2.2 Point in Polygon

One basic spatial query operation is determining whether a point lies in a polygon. For instance, show all the store locations (if stored as point data) in a prescribed neighbourhood of a city. The methodology for performing this search is presented below, as it at first seems a rather counterintuitive idea. To determine if a given point lies in a polygon, you construct a ray (in practice a line segment) from the point in an arbitrary direction away (in practice horizontally) from the point to a reasonable distance, perhaps to

If $D = 0$ (the two lines are parallel)
  If $\underline{a}$, $\underline{b}$, $\underline{c}$ are not collinear
    Case d)
  If $a_x \neq b_x$ ($\underline{ab}$ not vertical if it is test y values for betweenness)
    If ((c_x between a_x and b_x)
      or (d_x between a_x and b_x)
      or (a_x between c_x and d_x)
      or (b_x between c_x and d_x)
      Case f)
    Else
      Case e)
Else
  If ((the numerator of $s = 0$ or the numerator of $s = D$)
    or (the numerator of $t = 0$ or the numerator of $t = D$))
    Case c)
  If (($0 \leq s \leq 1$) and ($0 \leq t \leq 1$))
    Case b)
  Else
    Case a)

Figure 8.2 Test cases for line segment intersection.

just beyond the bounding rectangle of the polygon. Then you count the number of times the ray intersects the boundary of the polygon. If it intersects an even number of times the point is outside the polygon. If it has an odd number of intersections the point is within the polygon (see Figure 8.3). Note that this method works for concave and convex polygons. A few special cases must be dealt with to ensure unambiguous results, Figure 8.4 outlines these cases and the procedures to deal with them. Note that the method we have just described deals inconsistently with points that lie on the boundary of the polygon, but choices can be made to deal with this—keeping the polygon closed, for example; that is, points on the boundary of the polygon belong to

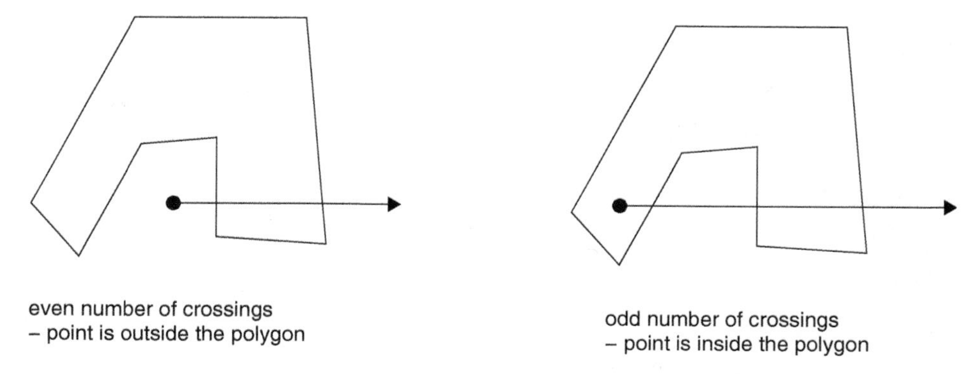

Figure 8.3 The ray-shooting method for the point in polygon query.

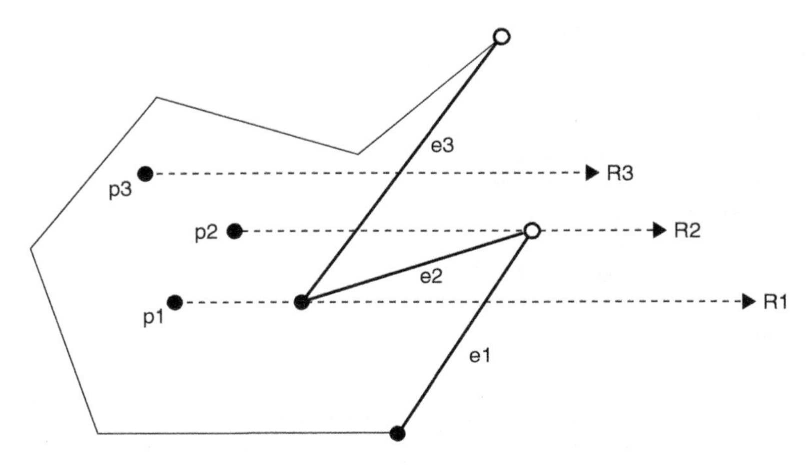

For edge e to count as a crossing, one of e's endpoints must be strictly above ray R and the other endpoint on or below R. An edge that is colinear with ray R is not counted as a crossing.

Example p1 generates R1: edges e1, e2, and e3 are counted as crossings: 3 is odd, so p1 is inside the polygon.

Example p2 generates R2: edges e1 and e2 are not counted as crossings, e3 is counted as a crossing: 1 is odd, so p2 is inside the polygon.

Example p3 generates R3: edge e3 is counted as a crossing: 1 is odd, so p3 is inside the polygon.

Figure 8.4 Special cases in the ray-shooting method.

the polygon. Historically, however, some GIS data models prescribe edges as occuring once but belonging to two adjacent polygons. In this case it maybe preferable to have the point assigned to only one polygon or the other, as the above procedure does.

8.2.3 Convex Hulls

A convex hull of a finite set of points S in the plane is the enclosing convex polygon P with the smallest area or perimeter. We can define convexity in the plane:

$$\text{Set } S \text{ is convex in the plane if} \\ \underline{x} \in S \text{ and } \underline{y} \in S \text{ implies the line segment } \underline{xy} \subseteq S \quad .$$

Note that this is one of many ways to define or describe a convex hull. According to O'Rourke [88], convex hulls and Voronoi diagrams (described in the next section) are the two most important geometric structures in computational geometry. Computational geometry is the engine underlying much GIS functionality. A convex hull is used, for example, to model the home range of a species whose location is stored as a set of points. One method for calculating a convex hull is the Quickhull algorithm. In Figure 8.6 below, we give the pseudocode for the algorithm. It is important to note that this is a recursive function; that is, it calls itself. These types of functions are very useful.

We start by finding two distinct extreme points. The rightmost lowest, P_R, and leftmost highest, P_L, are extreme points. Extreme points are points on the convex hull with interior angles strictly less than π (180°). Note that we can discover these two starting points without calculating the entire convex hull, so no circular reasoning is involved here. These two points form a line segment that splits the convex hull into an upper and lower portion. The final convex hull is created by the set $\{P_L\}$ $\cup$ *Quickhull*($\{P_L\}$,$\{P_R\}$,$\{S_1\}$) $\cup$ *Quickhull*($\{P_R\}$,$\{P_L\}$,$\{S_2\}$) $\cup$ $\{P_R\}$, where S_1 and S_2 are the set of points of S strictly above and below $\underline{P_L P_R}$.

Referring to the algorithm pseudocode, we note that there are two operations we need to describe a little more closely. First, finding a point in S that is the maximum distance from $\underline{ab}$ is done by finding the slope of a line perpendicular to $\underline{ab}$. This slope is simply the negative reciprocal of the slope of line segment $\underline{ab}$. Then we find the perpendicular line segments ending at points in S and choosing the maximum among these. Second, finding points strictly to the right of the given line segment (e.g., $\underline{ac}$) requires the notion of signed area. Given triangle $T(a, b, c)$ in general position in the plane (see Figure 8.5), its area can be determined via the determinant of its vertex coordinates as follows:

$$\begin{aligned}
2 \cdot A(T(a,b,c)) &= \begin{vmatrix} x_a & y_a & 1 \\ x_b & y_b & 1 \\ x_c & y_c & 1 \end{vmatrix} \\
&= x_a \cdot (y_b \cdot 1 - 1 \cdot y_c) - x_b \cdot (y_a \cdot 1 - 1 \cdot y_c) + x_c \cdot (y_a \cdot 1 - 1 \cdot y_b) \\
&= x_a \cdot (y_b - y_c) - x_b \cdot (y_a - y_c) + x_c \cdot (y_a - y_b) \\
&= x_a \cdot y_b - x_a \cdot y_c - x_b \cdot y_a + x_b \cdot y_c + x_c \cdot y_a - x_c \cdot y_b \\
&= x_a \cdot y_b - x_a \cdot y_c - x_b \cdot y_a + x_b \cdot y_c + x_c \cdot y_a - x_c \cdot y_b + 0 \\
&= x_a \cdot y_b - x_a \cdot y_c - x_b \cdot y_a + x_b \cdot y_c + x_c \cdot y_a - x_c \cdot y_b + (x_a \cdot y_a - x_a \cdot y_a) \\
&= x_b \cdot y_c - x_a \cdot y_c - x_b \cdot y_a + x_a \cdot y_a - \\
&\quad (x_c \cdot y_b - x_a \cdot y_b - x_c \cdot y_a + x_a \cdot y_a) \\
&= (x_b - x_a) \cdot (y_c - y_a) - (x_c - x_a) \cdot (y_b - y_a) \quad .
\end{aligned}$$

Now we get the following interpretation of the results:

If $2 \cdot A(T(a,b,c)) > 0$, then

c is left of $\underline{ab}$ and c is counterclockwise ordered from $\underline{ab}$.

If $2 \cdot A(T(a,b,c)) == 0$, then

c is collinear with $\underline{ab}$.

If $2 \cdot A(T(a,b,c)) < 0$, then

c is right of $\underline{ab}$ and c is clockwise ordered from $\underline{ab}$.

Figure 8.7 shows the method applied to subset of mountain pine beetle outbreak point data. Note in this case that there are no points in S below $\underline{P_L P_R}$.

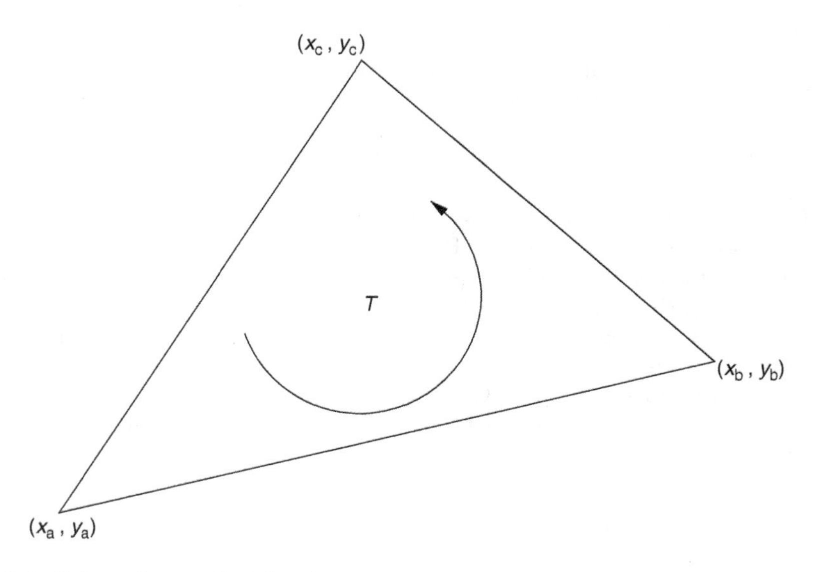

Figure 8.5 Oriented area of a triangle.

```
function Quickhull(a,b,S)
        if S == ∅ then
                return ()
        else
                {c} ← point in S with maximum distance from ab
                {A} ← points in S strictly right of ac
                {B} ← points in S strictly right of cb
                return Quickhull(a, c, A) ∪ {c} ∪ Quickhull(c, b, B)
```

Figure 8.6 Pseudocode for the Quickhull algorithm.

8.2.4 Voronoi Diagrams

A planar Voronoi diagram is a transformation from zero-, one-, or two-dimensional objects to a two-dimensional tessellation of the plane. It partitions the plane into areas whose component points are closer to each of their respective seed objects than any of the other input seed objects. Here we consider zero-dimensional or point objects as the seeds. Formally this is a planar ordinary Voronoi diagram (see Okabe et al. [85]). We let the generator set $\underline{P}$ of seed points be

$$\underline{P} = \{\underline{p}_1, \underline{p}_2, \ldots, \underline{p}_n\} \quad ,$$

where

$$\underline{p}_i \text{ is an ordered pair of coordinates and } 2 \leq n < \infty \quad .$$

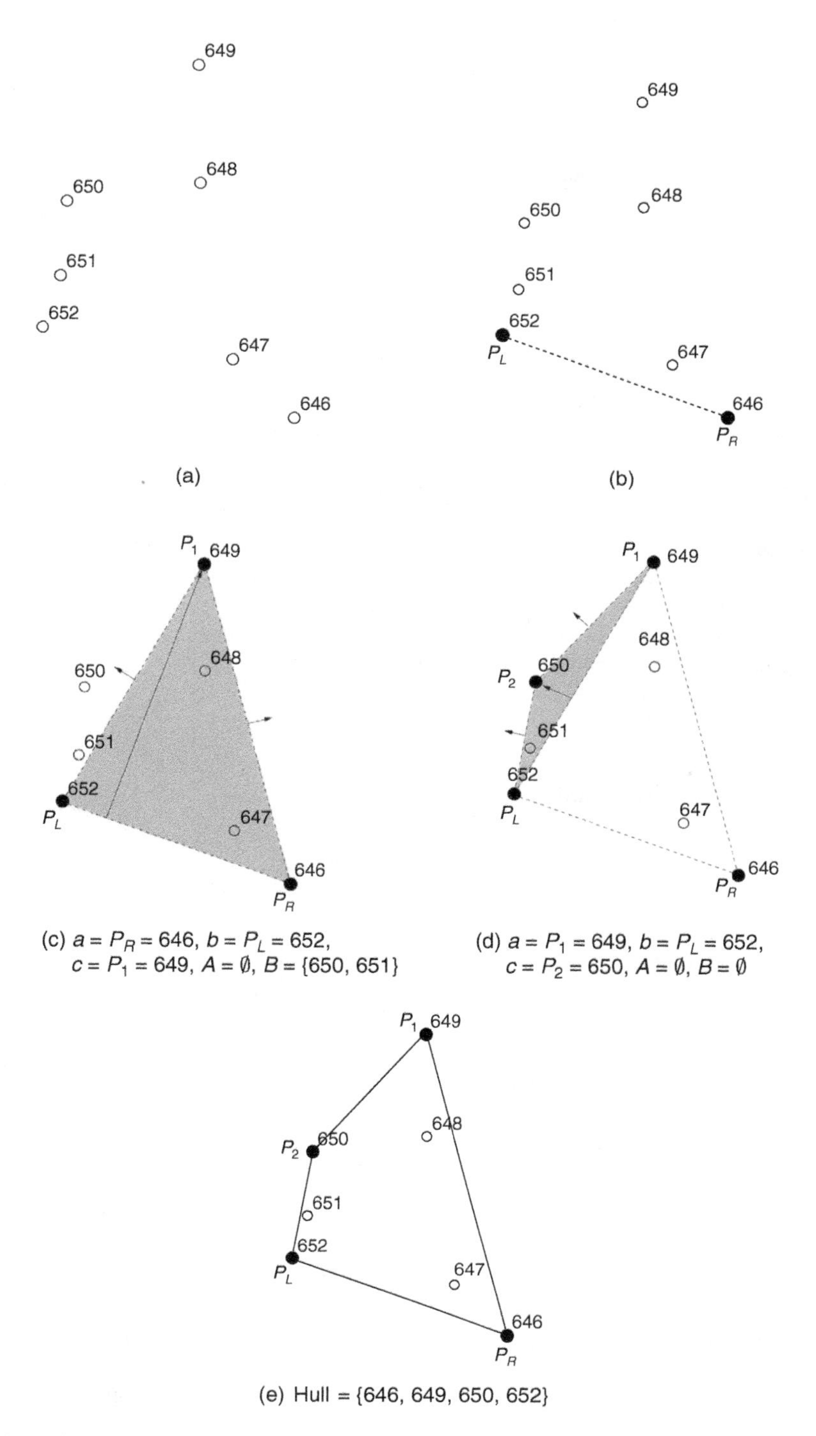

Figure 8.7 Creating a convex hull, using the Quickhull algorithm with a sample of mountain pine beetle data.

The Voronoi cell or polygon $V(\underline{p}_i)$ for a given point $\underline{p}_i$ is given by

$$V(\underline{p}_i) = \left\{ \underline{x} \;\middle|\; \| \underline{x} - \underline{x}_i \| \leq \| \underline{x} - \underline{x}_j \| \text{ for } j \neq i \, , \; i, j \in \{1, 2, \ldots, n\} \right\} \quad .$$

The Voronoi diagram $\mathcal{V}$ generated by $\underline{P}$ is then given by

$$\mathcal{V} = \{V(\underline{p}_1), V(\underline{p}_2), \ldots, V(\underline{p}_n)\} \quad ,$$

where

$$2 \leq n < \infty \quad .$$

One method for creating the planar ordinary Voronoi diagram is an iterative approach. We take an arbitrary first two points, then construct the perpendicular bisector between them. This division of the plane has the property that all parts of the plane on the side of the line of each seed point are closest to that point and not the other. We continue to add points and construct pairwise perpendicular bisectors, being careful to properly resolve the length of the boundary segments created by the perpendicular bisectors. The process is shown graphically in Figure 8.8. Note that using the notation from the section on line intersection above, the midpoint (the bisector) of a line in parametric form is given by setting the parameter, s, equal to 0.5. Further, the slope of a line perpendicular to a given line is the negative reciprocal of the slope of the given line. So if the original line was our segment $\underline{a}$ from above whose slope is

$$m_a = \frac{y_b - y_a}{x_b - x_a} \quad ,$$

then the slope of the perpendicular line is

$$m_{pa} = -\left(\frac{x_b - x_a}{y_b - y_a}\right) = \frac{x_a - x_b}{y_a - y_b} \quad .$$

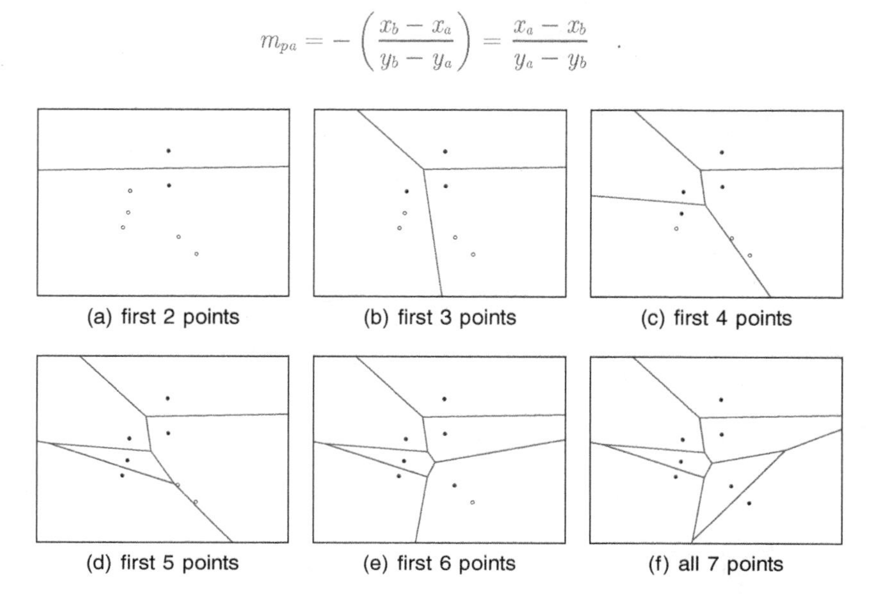

Figure 8.8 Creating a Voronoi diagram, using a sample of mountain pine beetle data.

So the perpendicular bisector is the line going through

$$(x_a + 0.5 \cdot x_{ab} \ , \ y_a + 0.5 \cdot y_{ab}) \quad ,$$

with slope m_{pa}. Voronoi diagrams have been used in many contexts (see Okabe et al. [85]) and can be generalized to higher dimensions and in other ways such as using weighted distances.

8.2.5 Shortest Path Dijkstra's Algorithm

Geographic data can often be abstracted as a one-dimensional network. The most common instance is representing a road system as a network of interconnecting line

```
input: An edge weighted directed graph, G = {V, E},
       a set of edge weights, w, and
       a single source vertex, s
output: A shortest path from s to all the other vertices in G in the form of
       a set of attributes to the vertices which store the label of the
       predecessor vertex in the shortest path to that vertex, π

initialize:
for each vertex v ∈ V[G]
       d[v] = ∞ (a set of attributes to the vertices which give the current
                 cumulative distance from the source vertex s to vertex v)
       π[v] = NIL (a set of pointers to the previous vertex in the shortest path)
end for
d[s] = 0

S = ∅ (a set of vertices whose final shortest path weights from the source s have
       already been determined)
Q = V[G]

while Q ≠ ∅ do
       u = min{Q}
       Q = Q − {u}
       S = S ∪ {u}
       for each vertex v adjacent to u do
              if d[v] > d[u] + w(u, v) then
                     d[v] = d[u] + w(u, v)
                     π[v] = u
              end if
       end for
end while
```

Figure 8.9 Pseudocode for Dijkstra's algorithm.

(a) Location of Nanaimo bar outlets and potential routes between them

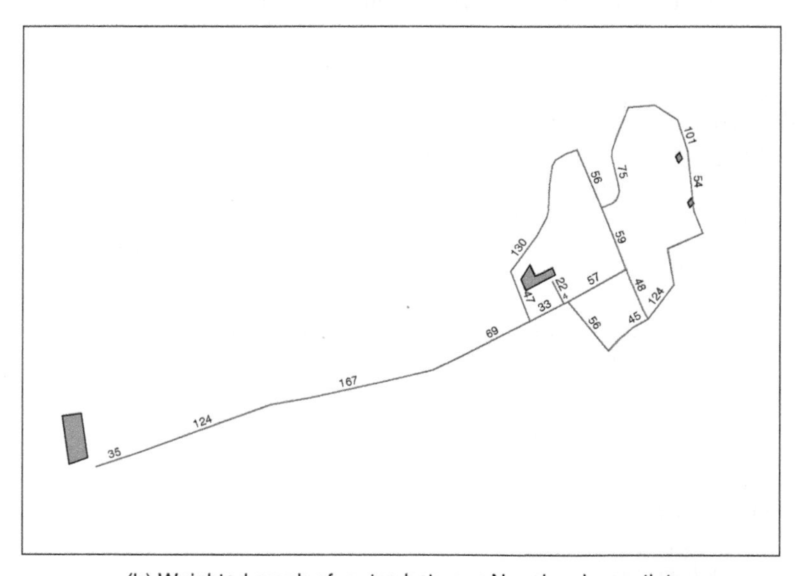

(b) Weighted graph of routes between Nanaimo bar outlets

Figure 8.10 Nanaimo bar outlets shortest path example, Nanaimo, BC.

segments. A common abstraction of such networks is as a mathematical graph (see Appendix II). With such a model in place we can do analysis such as the shortest path through a network. We now describe one of the popular algorithms for doing such analysis, Dijkstra's shortest path algorithm, and illustrate its use with an example of a shortest path between four shops in Nanaimo, BC.

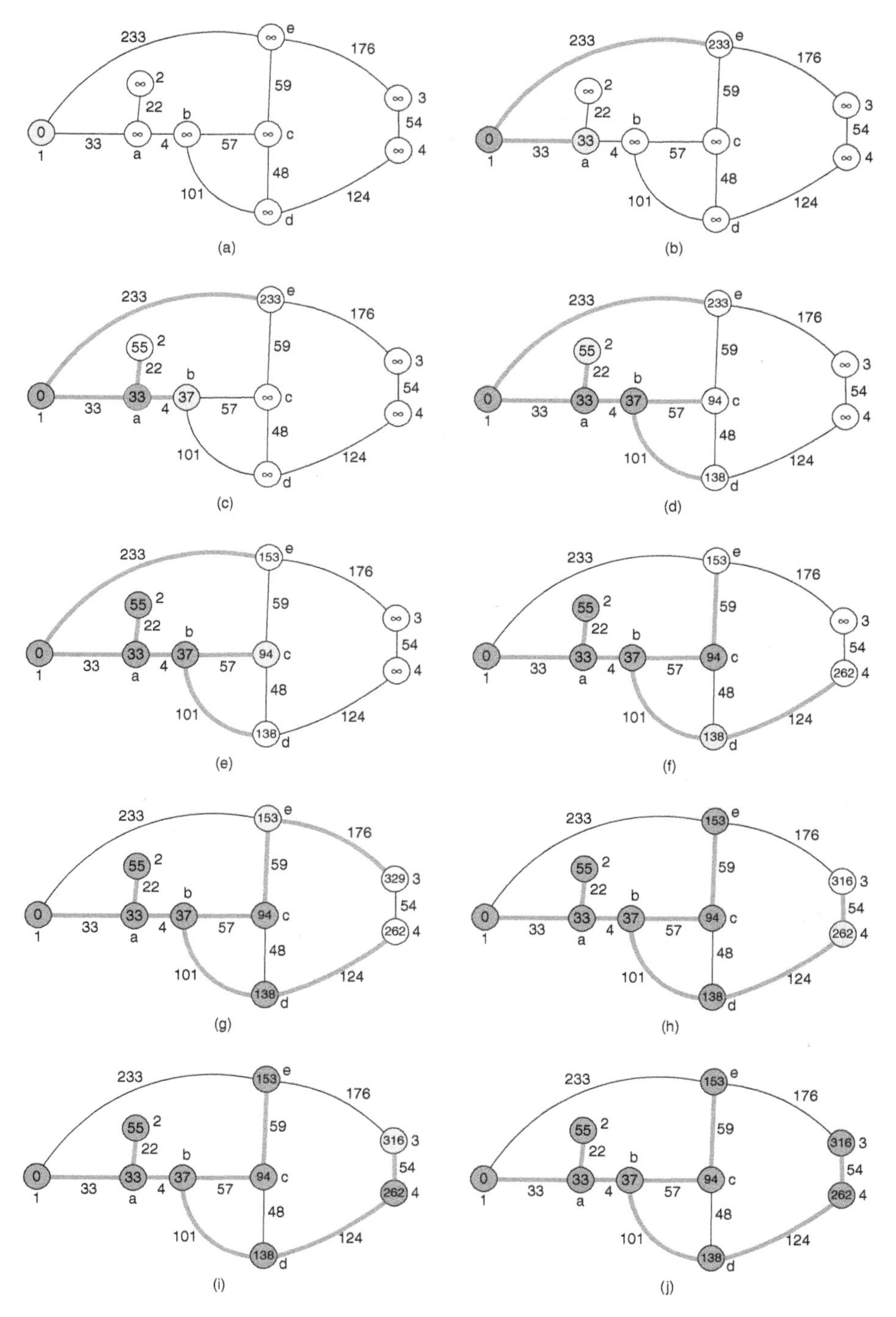

Figure 8.11 Shortest path Dijkstra algorithm applied to routes between Nanaimo bar outlets.

Dijkstra's algorithm works on a directed edge-weighted graph, with all the edge weights non-negative. The algorithm returns the shortest path from a designated source to all the other vertices in the graph. In Figure 8.9, we give the pseudocode for the algorithm. The pseudocode is based on discussion in Cormen et al. [27]. Note that the while loop iterates exactly $|V|$ times, so we calculate the shortest distance to all vertices.

We now give a small worked example. Nanaimo, BC, is a community on the east coast of Vancouver Island. Assume we are in the town sampling Nanaimo bars along the Nanaimo bar trail. We are at one location, and since we are trying to find the best bar and are in a bit of a hurry we want to find the shortest routes to three other outlets that sell the bars. Figure 8.10 (a) presents the input data of the location of the parcels of the outlets and a single line street network of a reasonable subset of the potential shortest routes between them. We abstract the locations of the outlets to vertices on the street network and add a few walkway segments as shown in Figure 8.10 (b). We then abstract this modified street network as a bidirectional weighted graph, with the edges merged between intersections, and edge weights being the resultant summed street segment distances. Note that there is only one path from the first location for part of the way, so we collapse those edges to a vertex on the graph. This means less work in the shortest-path calculations and we can always add the collapsed distance back to get true absolute distance in our final result. We show the operation of Dijkstra's algorithm graphically in Figure 8.11 (a)–(j). The final panel of this figure shows the shortest-path solution.

8.3 Computational Issues

8.3.1 Finite Precision Storage and Overflow Errors and Rounding Errors

We look now at how two of the most-often-used types of numbers, the integers and the reals, are represented in computers and the ramifications of this representation for computational geometry and thus GIS. The discussion here is based on notes provided by Dr P.H. Calamai.

Fixed-point integer representation

We emphasized in Chapter 4 that in GIS we do not work in $\mathbb{R}^2$ but rather in at best $\mathbb{Q}^2$. Similarly, computer arithmetic, which we use in computational geometry, is not equivalent to exact arithmetic. In a computer, which is a finite-state machine, we are generally stuck with a fixed range of possible values to work with. For instance, an integer may be represented as

$$\begin{aligned} x &= \pm\,(d_t d_{t-1} \cdots d_1)_\beta \\ &= \pm\big(d_1 \cdot \beta^0 + d_2 \cdot \beta^1 + \cdots + d_t \cdot \beta^{t-1}\big) \quad , \end{aligned}$$

where, $d_i \in \mathbb{I}$, $0 \le d_i \le \beta - 1$

$\mathbb{I}$ is the set of integers

$i \in \mathbb{I}$

$t =$ the number of digits or precision

β is the base of the number system.

For example, with $\beta = 10$ and $t = 3$,

$$\begin{aligned} 352 &= 2 \cdot 10^0 + 5 \cdot 10^1 + 3 \cdot 10^2 \\ &= 2 \cdot 1 + 5 \cdot 10 + 3 \cdot 100 \\ &= 2 + 50 + 300 \quad . \end{aligned}$$

With fixed-point integer representation, only integers within the range

$$[-(\beta^t - 1)\ ,\ (\beta^t - 1)]$$

are available. If calculations or data input yield integers above or below this range, overflow errors are generated.

For example, with $\beta = 10$ and $t = 3$ as in the previous example,

$$\begin{aligned} &[-(10^3 - 1), (10^3 - 1)] \\ &= [-(1000 - 1), (1000 - 1)] \\ &= [-999{,}999] \quad . \end{aligned}$$

Normed-floating point real representation

Normed-floating point numbers are a subset F of the real numbers. F consists of all numbers of the form

$$f = \pm .d_1 d_2 d_3 \cdots d_t \cdot \beta^e \ \text{ and zero,}$$

where

$$\begin{aligned} & 0 \leq d_i < \beta,\ d_1 \neq 0,\ L \leq e \leq U \\ & \mathbb{I} \text{ is the set of integers} \\ & i \in \mathbb{I} \\ \beta \equiv\ & \text{base of the number system} \\ t \equiv\ & \text{precision} \\ [L,\ U] \equiv\ & \text{is the exponent } e \text{ range.} \end{aligned}$$

For example, here is the set F for $\beta = 2,\ t = 3,\ L = 0$, and $U = 2$, but listing only the positive values (noting that the subscript 2 indicates a base 2 or binary number, and denormalizing refers to eliminating the base to a power term by shifting the decimal place):

$$\begin{aligned} 0.100 \cdot 2^0 &= 0.100_2 \ \text{(denormalize)} \\ &= 1 \cdot 2^{-1} + 0 \cdot 2^{-2} + 0 \cdot 2^{-3} \ \text{(as decimal)} \\ &= 2^{-1} \\ &= 0.5 \end{aligned}$$

$$\begin{aligned}0.100 \cdot 2^1 &= 1.00_2 \quad \text{(denormalize)}\\ &= 1 \cdot 2^0 + 0 \cdot 2^{-1} + 0 \cdot 2^{-2} \quad \text{(as decimal)}\\ &= 2^0\\ &= 1\end{aligned}$$

$$\begin{aligned}0.100 \cdot 2^2 &= 10.0_2 \quad \text{(denormalize)}\\ &= 1 \cdot 2^1 + 0 \cdot 2^0 + 0 \cdot 2^{-1} \quad \text{(as decimal)}\\ &= 2^1\\ &= 2\end{aligned}$$

$$\begin{aligned}0.101 \cdot 2^0 &= 0.101_2 \quad \text{(denormalize)}\\ &= 1 \cdot 2^{-1} + 0 \cdot 2^{-2} + 1 \cdot 2^{-3} \quad \text{(as decimal)}\\ &= 2^{-1} + 2^{-3}\\ &= 0.5 + 0.125\\ &= 0.625\end{aligned}$$

and the rest,

$$\begin{array}{lll} & 0.101 \cdot 2^1 = 1.25 & 0.101 \cdot 2^2 = 2.5\\ 0.110 \cdot 2^0 = 0.75 & 0.110 \cdot 2^1 = 1.5 & 0.110 \cdot 2^2 = 3\\ 0.111 \cdot 2^0 = 0.875 & 0.111 \cdot 2^1 = 1.75 & 0.111 \cdot 2^2 = 3.5 \quad .\end{array}$$

In Figure 8.12, note that the floating-point numbers are unevenly spaced, denser in some areas than others, and that underflows and overflows are possible. Some modern graphics and numerical computation libraries have variable-precision routines to take advantage of the denser parts of the floating-point set. However, note that $\mathrm{F} \neq \mathbb{R}$, so there are many numbers that do not exist in F that might be required of calculations but are not representable. In other words, we do not have a robust computing platform if we blindly assume we are in $\mathbb{R}$ (see for example Kettner et al. [68]).

To be even more specific, the number of elements in the finite set of *machine representable* real numbers (with β, t, U, and L defined as above) is given by

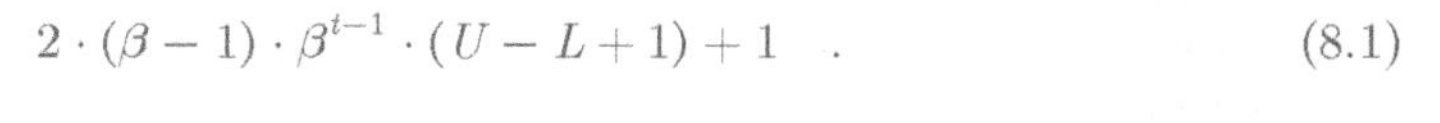

$$2 \cdot (\beta - 1) \cdot \beta^{t-1} \cdot (U - L + 1) + 1 \quad . \tag{8.1}$$

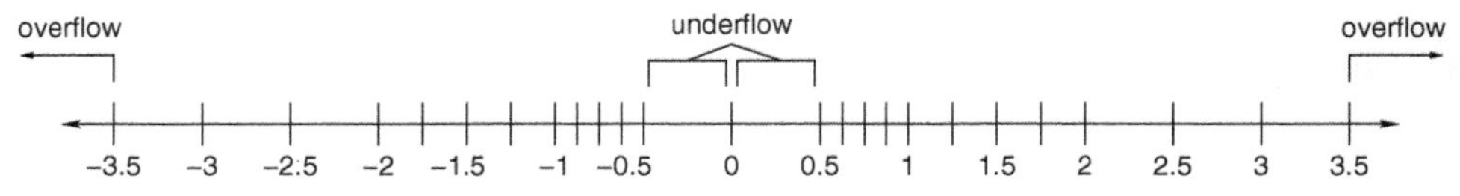

Figure 8.12 The set F for $\beta = 2, t = 3, L = 0,$ and $U = 2$ (source: adapted from Golub and Van Loan [46, p. 60]).

In this equation, the leading 2 takes into account the positive and negative values possible for each number. Continuing the explanation, $(\beta - 1) \cdot \beta^{t-1}$ counts how many numbers are possible at each scale interval (remembering that no leading zeros are possible in the normalized form). The next term, $(U - L + 1)$, counts the number of scale intervals, and the final 1 takes into account zero.

From the above example, $\beta = 2, t = 3, , L = 0$, and $U = 2$, we can calculate how many machine representable real numbers are available:

$$\begin{aligned} &2 \cdot (2-1) \cdot 2^{3-1} \cdot (2-0+1)+1 \\ &= 25 \quad . \end{aligned}$$

That is, ± 12 and zero.

To give a typical example of numeric formats in common use, we now consider a double-precision floating-point format referred to as decimal64. In this case, $\beta = 10, t = 16, L = -383$, and $U = 384$, and we can calculate how many machine representable real numbers are available:

$$\begin{aligned} &2 \cdot (10-1) \cdot 10^{16-1} \cdot (384-(-383)+1)+1 \\ &\doteq 1.3842 \cdot 10^{19} \quad . \end{aligned}$$

This is very much less than you might expect from naively looking at the range of the numbers available, which is of the order 10^{384}, and overwhelmingly less than the true number of reals available within this range, which is of course infinite.

Round-off errors and loss of significant digits

If we have fixed precision in our computer calculations, we can accumulate small errors over a series of calculations such that we may end up with an inaccurate final answer. For example, if we are working with 3-digit precision, $0.821 \cdot 0.447 = 0.366987$ becomes either 0.366 or 0.367, depending on the computer. Round-off errors generally lead to a slow degradation of accuracies in the least-significant bits of the digital representation of numbers in calculations.

Catastrophic cancellation or loss of significant digits occurs to the left end of the number by the subtraction of two nearly equal quantities. For example,

$$56.345611 - 56.345322 = 0.000289$$

(recall that zeros to the right of the decimal point only count as significant if recorded to the right of a number). Here we started with two numbers with eight significant digits and end up with a number with three significant digits, a loss of five significant digits. This issue can sometimes be avoided by rearranging the terms of a calculation. This type of error is referred to as numeric instability.

Here is a familiar example—finding the roots of a quadratic expression. Recall that the solutions of

$$a \cdot x^2 + b \cdot x + c = 0$$

are

$$x_1 = \frac{-b + \sqrt{b^2 - 4 \cdot a \cdot c}}{2 \cdot a} \quad \text{and} \quad x_2 = \frac{-b - \sqrt{b^2 - 4 \cdot a \cdot c}}{2 \cdot a} \quad .$$

If $b^2 \gg 4 \cdot a \cdot c$ (that is, b^2 is much bigger than $4 \cdot a \cdot c$) and $b < 0$, then there is a possibility for cancellation of significant digits in x_2 as follows:

$$\begin{aligned} & -(-b) - \sqrt{b^2 - 4 \cdot a \cdot c} \\ & = b - (b - \mathit{very\_small\_value}) \\ & \approx 0 \quad . \end{aligned}$$

In this case we can algebraically manipulate the solutions to avoid the computational issue:

$$x_1 \text{ is as above, but we let } x_2 = \frac{c}{a \cdot x_1} \quad .$$

Similarly, if $b < 0$, we keep x_2 the same as normal but let $x_1 = \dfrac{c}{a \cdot x_2}$. To demonstrate that these are in fact valid solutions, we show the equivalence for the first case to our original well-known form, and the second case may be verified in a similar fashion.

$$\begin{aligned} \frac{c}{a \cdot x_1} &= \frac{c}{a} \cdot \frac{2 \cdot a}{-b + \sqrt{b^2 - 4 \cdot a \cdot c}} \\ &= \frac{2 \cdot c}{-b + \sqrt{b^2 - 4 \cdot a \cdot c}} \\ &= \frac{2 \cdot c}{-b + \sqrt{b^2 - 4 \cdot a \cdot c}} \cdot \frac{-b - \sqrt{b^2 - 4 \cdot a \cdot c}}{-b - \sqrt{b^2 - 4 \cdot a \cdot c}} \\ &= \frac{2 \cdot c \cdot \left(-b - \sqrt{b^2 - 4 \cdot a \cdot c}\right)}{b^2 - (b^2 - 4 \cdot a \cdot c)} \\ &= \frac{2 \cdot c \cdot \left(-b - \sqrt{b^2 - 4 \cdot a \cdot c}\right)}{b^2 - b^2 + 4 \cdot a \cdot c} \\ &= \frac{2 \cdot c \cdot \left(-b - \sqrt{b^2 - 4 \cdot a \cdot c}\right)}{4 \cdot a \cdot c} \\ &= \frac{-b - \sqrt{b^2 - 4 \cdot a \cdot c}}{2 \cdot a} \\ &= x_2 \quad . \end{aligned}$$

8.3.2 Supercomputing

Large datasets and real-time applications can demand a large number of calculations per second. Increasingly, this need is being met by various parallel processing hardware and software. The text by Healey et al. [61] describes in considerable detail some of the major GIS operations adopted for parallel computing environments. For example, our previous algorithm for Dijkstra's shortest-path algorithm is shown to be amenable to parallelization by two types of problem domain partitioning. A recent trend for numerical computing is to utilize large banks of graphics processing units (GPUs) bundled onto extension boards or dedicated servers to create desktop or workstation supercomputers. The GPUs have fast on-board arithmetic processing units developed for interactive graphics that can be harnessed for numerical computing. Typically, modern general purpose GPU-based (GPGPU) hardware contains thousands of processing nodes. Libraries of open source software are being developed to work with these new platforms in specific problem domains. Other approaches to large data include cloud computing, where virtual computers are created with upward of a million networked computers that contribute CPU cycles to process large problems, and large shared-memory computing, where entire databases may be stored in primary memory for extremely fast query and update. All of these advances have direct relevance for GIS, as datasets are getting larger and sources of spatially aware data are rapidly expanding.

Problems

1. Calculate the intersection point of the two line segments given by $line1 : \underline{a} = \{1, 6\}$ and $\underline{b} = \{6, 4\}$ and $line2 : \underline{c} = \{2, 2\}$ and $\underline{d} = \{5, 5\}$.
2. Describe a few reasons why we prefer to do arithmetic on computers using integers.
3. How many machine representable real numbers are there if our floating point model has base $\beta = 10$, $t = 16$ digits of precision, and the exponent range $[L, U] = [-383{,}384]$? What are biggest and smallest machine representable numbers in this model? How many real numbers are in this range?
4. If we used Dijkstra's algorithm to calculate the shortest path from the first store to the second store using the Nanaimo example data from this chapter, how could we speed up the calculation?
5. It is possible to use higher-dimensional features as seeds for generating Voronoi diagrams. Explain how you might use an existing GIS point Voronoi function to create a Voronoi diagram with polylines as seeds.

Chapter 9

Geographic Measures

9.1 Introduction

Much of the material in the next two sections of this chapter should be review from an introductory geography statistics course, and the remaining material might be covered in an introductory spatial analysis course. We discuss the following topics: scales of measurement, central tendency and dispersion in one and two dimensions, and measuring geographic patterns. We will move quite quickly through the one-dimensional measures, as most of them will be familiar, and spend a little more time on the two-dimensional measures.

9.2 Scales of Measurement

Scale of measurement, sometimes referred to as level of measurement, is a characteristic of data and is an indication of what operations and comparisons are appropriate. For example, it can suggest what measure of central tendency we should use. Traditionally, there is a hierarchy of four scales: nominal, ordinal, interval, and ratio. Some authors have suggested that this categorization may need to be expanded in light of new types of digital data currently in widespread use, such as images, video, and sound.

9.2.1 Nominal

Nominal data consist of qualitative or categorical classes. The categorization must be exhaustive and mutually exclusive. For example, a variable collecting the colour of an observed phenomenon might consist of the values {red, green, and other}. A common geographic example of nominal data is soil class (see Figure 9.1).

9.2.2 Ordinal

Ordinal datasets have all the characteristics of nominal data, with the additional constraint that the values can be rank ordered. For example, flood risk $\in$ {minor, moderate, severe}. A geographic example of ordinal data is soil capability for agriculture (see Figure 9.2). In this scale, "1" is the best rating, "5" the worst.

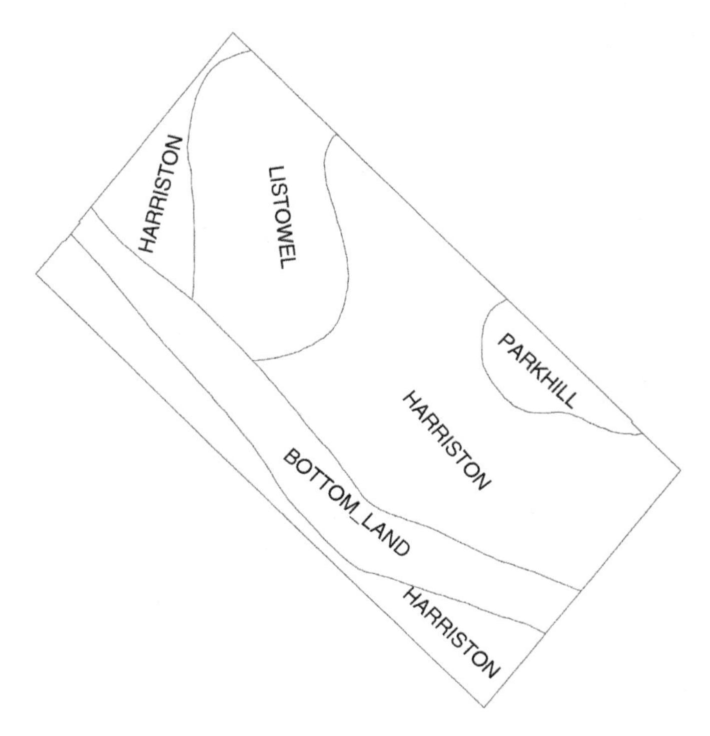

Figure 9.1 An example of nominal data: soil class.

9.2.3 Interval

Datasets are characterized as interval if they have the all characteristics of ordinal data with the further restriction that the interval between values is meaningful (can be measured), but their zero is arbitrary. Consider, for example, temperature in degrees F or degrees C.

9.2.4 Ratio

Ratio datasets have all the properties of interval datasets with the further constraint that their zero has physical meaning. Consider, for example, precipitation in millimetres, length in metres, and temperature in Kelvin.

9.3 Central Tendency and Dispersion (1D and 2D)

The shape of a dataset is one way to characterize a set of observations. Commonly one talks first about the central tendency and dispersion of a dataset. These terms are reasonably self-explanatory, but their meaning should become clearer as the various ways to measure them are outlined below. We discuss some of the measures of these two characterizations in the next sections. We begin with the one-dimensional cases, which

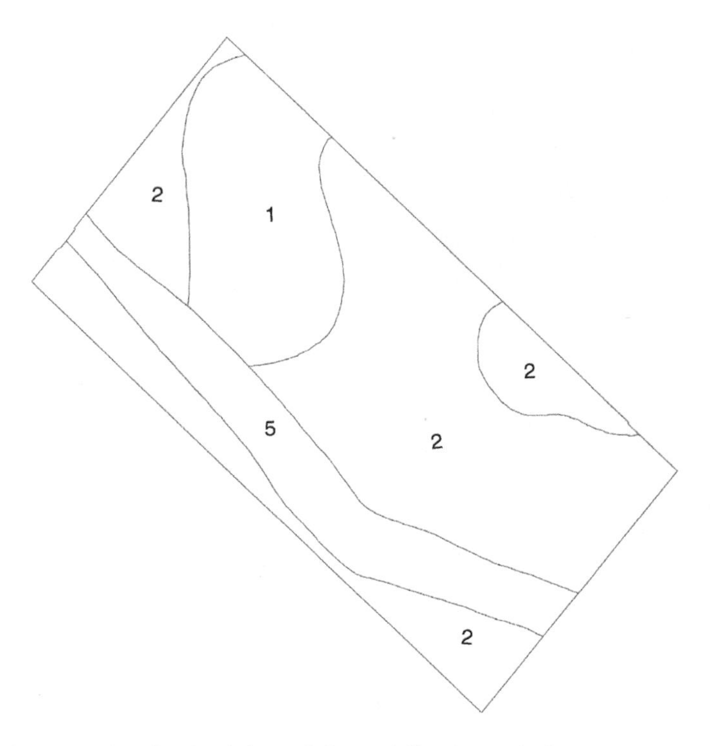

Figure 9.2 An example of ordinal data: Soil capability for agriculture.

are probably already familiar, and follow this with an introduction to two-dimensional measures. Note that these statistics are the sample moments; that is, they are estimated from the data by the formulas given below. This is in constrast to population statistics, which are either calculated over a whole population or derived from theoretical considerations. We will revisit this idea when we discuss expected value and Monte Carlo testing later in this chapter.

9.3.1 Central Tendency 1D

In this section we introduce three measures of central tendency in decreasing order of familiarity but increasing scope of applicability: mean, median, and mode.

Mean

The sample mean may be calculated only for interval and ratio data levels. It is calculated using the familiar formula

$$\overline{x} = \frac{\sum_{i=1}^{n} x_i}{n} ,$$

where x_i are data values and n is the number of data values. Please see Appendix I, section 2, if you are unfamiliar with the summation notation in the above formula.

Weighted mean

The sample weighted mean may be calculated only for interval and ratio data levels. It is used for grouped data or when there are reasons to allow observations to have varying degrees of influence on summary statistics. It is calculated using the following variation on the formula for the sample mean:

$$\overline{x}_w = \frac{\sum_{i=1}^{n} w_i \cdot x_i}{\sum_{i=1}^{n} w_i},$$

where x_i are data values, w_i are the observation weights or class frequencies, and n is the number of data values.

An example of interpolation of height value on a TIN facet is given below (see Figure 9.3). Note that in this application the areas of the triangles are the weights, and

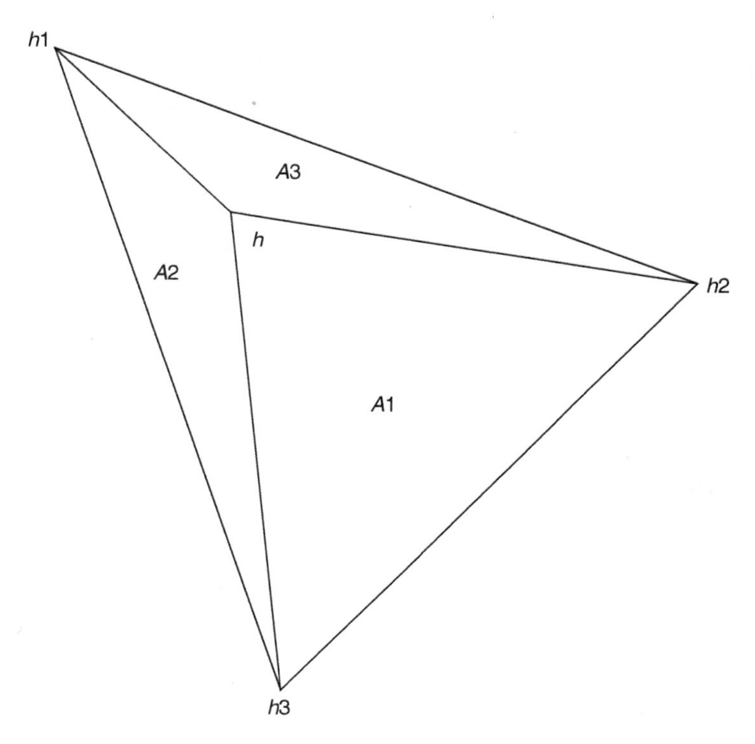

Figure 9.3 Estimating height at a location on the facet of a TIN.

the solution is set up so that the triangle areas are labelled to correspond to the points opposite their bases. This means that if the location of the point height to be estimated is nearer a given point, the corresponding triangle area is greater, giving that point a greater weight in the weighted mean calculation.

$$h = \overline{h}_w = \frac{\sum_{i=1}^{3} A_i \cdot h_i}{\sum_{i=1}^{3} A_i} \quad .$$

Median

The median may be calculated for ordinal, interval, and ratio data levels. The median is the middle value of a sorted set of data values for an odd number of data values or the average of the two middle values for an even number of values. As a measure of central tendency, the median is more robust against outliers, extreme high or low values, than the mean. The median is also known as the 50th percentile. In general, the pth percentile of a sorted dataset may be estimated using the following formulas:

$$\hat{y}_p = y_{(ip(m))} + fp(m) \cdot \{y_{(ip(m)+1)} - y_{(ip(m))}\} \quad ,$$

where

$y_{(\cdot)}$ = the sorted observation at the index position given by the value calculated in brackets

$m = n \cdot p + \frac{1}{2}$

n = the number of observations

$ip(m)$ = the integer part of m

$fp(m)$ = the fractional part of m.

Mode

The mode of a set of data may always be calculated. It is defined as the most frequently occurring value. It may be used directly for categorical data but usually is most useful on grouped or categorized data if the source data are continuous.

$$\begin{aligned} \text{mode}(x_i) &= \max(freq_j) \text{ or} \\ &= \max(freq(x_i)) \quad , \end{aligned}$$

for

$i = 1, \ldots, n$ data values

$j = 1, \ldots, m$ data classes.

Table 9.1 Summary of applicability of 1-dimensional central tendency measures.

		Scale of measurement			
		nominal	ordinal	interval	ratio
Measures of central tendency	mean			×	×
	weighted mean			×	×
	median		×	×	×
	mode	×	×	×	×

See Table 9.1 for a summary of the applicability, based on scale of measurement, of the four one-dimensional measures of central tendency introduced in this section, showing scale of measurement versus measures of central tendency.

9.3.2 Dispersion 1D

In this section we introduce eight measures of dispersion for one-dimensional data. These are, in order, sign, range, mean absolute difference, variance and standard deviation, weighted variance, coefficient of variance, variance ratio, and information measure.

Sign

The sign measure of dispersion is a vector of + and − signs, or often +1 and −1 values that indicate whether a given observation is to the left or right of the sample mean of all the observations. This can give an indication of the symmetry or asymmetry of the dataset. Sign is calculated as follows:

$$sign\,(x_i - \overline{x}) \quad ,$$

where

x_i = the data values for $i = 1, \ldots, n$

$\overline{x}$ = the sample mean

sign indicates only either a positive or negative value of the difference, not its magnitude.

Range

Range as a measure of dispersion is considered for two cases: discrete or continuous data. For discrete data, range is calculated as follows:

$$range = \max(x_i) - \min(x_i) + 1 \;, \text{ for } i = 1, \ldots, n \quad .$$

This means that the range gives the number of discrete data points between and including the endpoints.

For continuous data, range is calculated as follows:

$$range = \max(x_i) - \min(x_i) \quad , \text{ for } i = 1, \ldots, n \quad .$$

In this case range gives basic information about the magnitude of the spread of the observed values.

Mean absolute difference

Sample mean absolute deviation (*MAD*) is a measure of the average deviation of the absolute distance of each observation from the sample mean. The absolute value is introduced to avoid the raw deviations from the mean adding up to zero (see derivation below). *MAD* is calculated as follows:

$$MAD = \frac{\sum_{i=1}^{n} |x_i - \overline{x}|}{n} \quad ,$$

where x_i are data values, $\overline{x}$ is the sample mean, $|\cdot|$ is the absolute value of the contents between the two vertical bars, and n is the number of data values.

Here is why we cannot use the average deviations from the mean:

$$\begin{aligned} \frac{\sum_{i=1}^{n} (x_i - \overline{x})}{n} &= \frac{\sum_{i=1}^{n} x_i}{n} - \frac{\sum_{i=1}^{n} \overline{x}}{n} \\ &= \overline{x} - \frac{n \cdot \overline{x}}{n} \\ &= 0 \quad . \end{aligned}$$

Variance and standard deviation

Variance

Sample variance is a measure of the average squared deviation of each measurement from the sample mean. It avoids the average difference terms summing to zero by squaring each term. Note that this algebraic fact was avoided in the *MAD* measure above by using absolute brackets around the difference. Variance is calculated using this formula:

$$s^2 = \frac{\sum_{i=1}^{n} (x_i - \overline{x})^2}{n-1} \quad ,$$

where x_i are data values, $\overline{x}$ is the sample mean, and n is the number of data values. Note that the denominator is $n-1$, not n. The technical reason for this is that using n results in a biased estimator of the true variance value. Now the previous sentence should feel unsatisfactory to the astute reader who notices that the denominator is n in the definition of *MAD* above. We provide a more complete derivation of the biased estimator reasoning for s^2 in Appendix III.

Standard Deviation

Sample standard deviation is the positive square root of the variance and returns the measure of dispersion calculated by the variance to the original units,

$$s = \sqrt{s^2} \quad .$$

Weighted variance

Weighted sample variance is a measure of the average squared deviation from the sample mean. It is used with grouped observations or in cases where it is sensible to give unequal weights to the raw observations. It is calculated using this formula:

$$s_w^2 = \frac{\sum_{i=1}^{n} w_i \cdot (x_i - \overline{x})^2}{\sum_{i=1}^{n} w_i} \quad ,$$

where x_i are data values, w_i are the observation weights or class frequencies, $\overline{x}$ is the sample mean, and n is the number of data values.

Coefficient of variation

The coefficient of variation (CV) standardizes the sample standard deviation by the sample mean to allow for comparisons of dispersion between datasets with different ranges of data values. CV is defined as

$$CV = \frac{s}{\overline{x}} \quad .$$

Variance ratio

The variance ratio V can be used with data of all scales of measurement. V measures the probability of an observation *not* being at the mode and is defined as

$$V = 1 - \frac{\max(freq_j)}{n} \quad .$$

Note how this equation captures the notion of *not* being at the mode, using the idea that $P(\text{not } A) = 1 - P(A)$. Further, $\frac{\max(freq_j)}{n}$ is the probability of a given observation being the mode, since the mode is defined as the observation with the highest frequency, and dividing this value by the total number of observations n, we get the desired probability. A value of V close to 0 means that a lot of the observations were at the mode and thus there is little dispersion in the dataset. Conversely, a value of V near 1 means that no one observed value dominated the measurements, so the data are dispersed over many values.

Information measure

The information measure H can be used with data of all scales of measurement. H equates variability with surprise, information content, or novelty. Roughly speaking, high-probability events contain less information than less-likely events. Further, information about multiple events is additive. These considerations lead to the following definition of an information-based measure of dispersion.

$$\begin{aligned} H &= -\sum_{j=1}^{m} \left(\frac{f_j}{n}\right) \cdot \ln\left(\frac{f_j}{n}\right) \\ &= \ln(n) - \frac{1}{n} \cdot \sum_{j=1}^{m} f_j \cdot \ln(f_j) \quad , \end{aligned}$$

where

f_j = the frequency of observation in class j

n = the total number of observations

ln = the natural logarithm (logarithm base e)

and for $f_j = 0$ take $f_j \cdot \ln(f_j) = 0 \cdot \ln(0) = 0$.

A proof of the equivalence of the two forms of H is given in Appendix IV; also included is an explanation of the result, $0 \cdot \ln(0) = 0$. Note that the $\frac{f_j}{n}$ terms are the probabilities of class j, that is, the frequency f_j of observations of class j divided by the total number of observations n. Consequently, you may have come across the information measure expressed as

$$H = -\sum_{j=1}^{m} p_j \cdot \ln(p_j) \quad ,$$

where p_j is the probability of class j.

Finally we note that H is maximum when all the probabilities are equal; that is, for m classes $p_j = \frac{1}{m}$ and this $\max(H) = \ln(m)$, as shown here:

$$\begin{aligned} H &= -\sum_{j=1}^{m} p_j \cdot \ln(p_j) \\ &= -\sum_{j=1}^{m} \frac{1}{m} \cdot \ln\left(\frac{1}{m}\right) \\ &= -\frac{1}{m} \cdot \sum_{j=1}^{m} \ln\left(\frac{1}{m}\right) \\ &= -\frac{1}{m} \cdot \sum_{j=1}^{m} (\ln(1) - \ln(m)) \\ &= -\frac{1}{m} \cdot \sum_{j=1}^{m} (0 - \ln(m)) \\ &= \frac{1}{m} \cdot \sum_{j=1}^{m} \ln(m) \\ &= \frac{m}{m} \cdot \ln(m) \\ &= \ln(m) \quad . \end{aligned}$$

In Table 9.2 we summarize the applicability, based on scale of measurement, of the eight one-dimensional measures of dispersion introduced in this section, showing scale of measurement versus measures of dispersion.

Table 9.2 Summary of applicability of 1-dimensional dispersion measures.

		Scale of measurement			
		nominal	ordinal	interval	ratio
Measures of dispersion	sign			×	×
	range		×	×	×
	mean absolute difference			×	×
	variance			×	×
	standard deviation			×	×
	weighted variance			×	×
	coefficient of variation			×	×
	variance ratio	×	×	×	×
	information measure	×	×	×	×

9.3.3 Central Tendency 2D

We now move on to the first two two-dimensional measures of central tendency—sample spatial mean and sample weighted spatial mean—both of which translate straightforwardly from the one-dimensional case. Two-dimensional data refers here to paired observations such as the x and y coordinates of a point in two-dimensional Euclidean space. It is often useful to use a spatial analogy for interpretation and analyis even if the source data are not spatial.

Spatial mean

The spatial mean is the average in each variable of the paired observations

$$(\overline{x}_c\ ,\ \overline{y}_c) = \left(\frac{\sum_{i=1}^{n} x_i}{n}\ ,\ \frac{\sum_{i=1}^{n} y_i}{n} \right)\ ,$$

where $(x_i, y_i),\ i = 1, \ldots, n$ are paired observations.

Weighted spatial mean

The weighted spatial mean is the weighted mean of each variable in a paired dataset:

$$(\overline{x}_{wc}\ ,\ \overline{y}_{wc}) = \left(\frac{\sum_{i=1}^{n} (w_i \cdot x_i)}{\sum_{i=1}^{n} w_i}\ ,\ \frac{\sum_{i=1}^{n} (w_i \cdot y_i)}{\sum_{i=1}^{n} w_i} \right)\ ,$$

where $(x_i, y_i),\ i = 1, \ldots, n$ are paired observations
and w_i are the weights associated with each of the ith paired observations.

Spatial median

There is no canonical ordering in two dimensions as opposed to the one-dimensional case where the natural (number line) ordering allows us to unambiguously sort a set of values and choose the middle value (or the average of the two middle values for an even number of values) as the median. Thus, we must determine another way to define the "middle value" in two dimensions and above. One common way to do this is the spatial median, which is defined as the value that minimizes the Euclidean distance between itself and all the other values in the dataset.

$$(\overline{x}_{\text{med}}, \overline{y}_{\text{med}}) \quad \text{satisfies}$$

$$\min \left(\sum_{i=1}^{n} \sqrt{(x_i - \overline{x}_{\text{med}})^2 + (y_i - \overline{y}_{\text{med}})^2} \right)\ .$$

To solve, let the current estimate for the spatial median be given by

$$u_t = \overline{x}_{\text{med}} \text{ and } v_t = \overline{y}_{\text{med}}$$

and let the initial estimate (starting value for the iteration) for the spatial median be the spatial mean,

$$u_0 = \overline{x}_c \text{ and } v_0 = \overline{y}_c \quad .$$

Then solve iteratively for $t = 1, 2, \ldots$ until a stopping condition is reached:

$$u_t = \frac{\sum_{i=1}^{n} \left(\frac{x_i}{\sqrt{(x_i - u_{t-1})^2 + (y_i - v_{t-1})^2}} \right)}{\sum_{i=1}^{n} \left(\frac{1}{\sqrt{(x_i - u_{t-1})^2 + (y_i - v_{t-1})^2}} \right)}$$

$$v_t = \frac{\sum_{i=1}^{n} \left(\frac{y_i}{\sqrt{(x_i - u_{t-1})^2 + (y_i - v_{t-1})^2}} \right)}{\sum_{i=1}^{n} \left(\frac{1}{\sqrt{(x_i - u_{t-1})^2 + (y_i - v_{t-1})^2}} \right)} \quad .$$

Note that the best we can do in solving for the spatial median is the iterative solution given above, since a closed form or analytic solution is not possible, as is shown in the derivation below. The stopping condition is that the distance between the current estimated median value and the previous step's estimate of the median is below some threshold. This threshold may be determined by the context of the dataset being analyzed. We now provide a short derivation of the above results.

For the two-dimensional case in general to find

$$\min(f(x\ ,\ y)) \quad ,$$

we solve the following system of equations:

$$\frac{\partial}{\partial x} f(x\ ,\ y) = 0$$

$$\frac{\partial}{\partial y} f(x\ ,\ y) = 0 \quad .$$

In our case let

$$\overline{x}_{\text{med}} = u_t \ ,\ \overline{y}_{\text{med}} = v_t \quad .$$

First, solving for u_t to satisfy

$$\min\left(\sum_{i=1}^{n}\sqrt{(x_i-u_t)^2+(y_i-v_t)^2}\right) \quad ,$$

we solve

$$\frac{\partial}{\partial u_t}\left(\sum_{i=1}^{n}\sqrt{(x_i-u_t)^2+(y_i-v_t)^2}\right)=0 \quad .$$

Since we are differentiating with respect to u_t here, let $c=(y_i-\overline{v}_t)^2$. Further, we represent the radical (square root symbol) in its fractional form, as this is convenient for differentiation. Thus we have

$$\frac{\partial}{\partial u_t}\left(\sum_{i=1}^{n}\left((x_i-u_t)^2+c\right)^{\frac{1}{2}}\right)=0$$

$$\sum_{i=1}^{n}\frac{\partial}{\partial u_t}\left((x_i-u_t)^2+c\right)^{\frac{1}{2}}=0$$

$$\sum_{i=1}^{n}\frac{1}{2}\left((x_i-u_t)^2+c\right)^{-\frac{1}{2}}\cdot(2)\cdot(x_i-u_t)\cdot(-1)=0$$

$$\sum_{i=1}^{n}\frac{(x_i-u_t)\cdot(-1)}{\left((x_i-u_t)^2+c\right)^{\frac{1}{2}}}=0$$

$$\sum_{i=1}^{n}\frac{(-x_i)}{\left((x_i-u_t)^2+c\right)^{\frac{1}{2}}}+\sum_{i=1}^{n}\frac{u_t}{\left((x_i-u_t)^2+c\right)^{\frac{1}{2}}}=0$$

$$\sum_{i=1}^{n}\frac{x_i}{\left((x_i-u_t)^2+c\right)^{\frac{1}{2}}}=\sum_{i=1}^{n}\frac{u_t}{\left((x_i-u_t)^2+c\right)^{\frac{1}{2}}}$$

$$\sum_{i=1}^{n}\frac{x_i}{\left((x_i-u_t)^2+c\right)^{\frac{1}{2}}}=u_t\cdot\sum_{i=1}^{n}\frac{1}{\left((x_i-u_t)^2+c\right)^{\frac{1}{2}}},$$

giving

$$u_t=\frac{\displaystyle\sum_{i=1}^{n}\frac{x_i}{\left((x_i-u_t)^2+c\right)^{\frac{1}{2}}}}{\displaystyle\sum_{i=1}^{n}\frac{1}{\left((x_i-u_t)^2+c\right)^{\frac{1}{2}}}} \quad .$$

Substituting back for c, reintroducing the radical notation for the square root, and adding extra brackets to ensure proper interpretation of the summations,

$$u_t = \frac{\sum_{i=1}^{n}\left(\frac{x_i}{\sqrt{(x_i - u_t)^2 + (y_i - v_t)^2}}\right)}{\sum_{i=1}^{n}\left(\frac{1}{\sqrt{(x_i - u_t)^2 + (y_i - v_t)^2}}\right)} \quad .$$

A similar derivation yields

$$v_t = \frac{\sum_{i=1}^{n}\left(\frac{y_i}{\sqrt{(x_i - u_t)^2 + (y_i - v_t)^2}}\right)}{\sum_{i=1}^{n}\left(\frac{1}{\sqrt{(x_i - u_t)^2 + (y_i - v_t)^2}}\right)} \quad .$$

Thus we have an expression of the form

$$(u_t\ ,\ v_t) = (f\,(u_t\ ,\ v_t)\ ,\ f\,(u_t\ ,\ v_t)) \quad .$$

So we solve this by fixed-point iteration (starting at some u_0, v_0) and iteratively generate

$$(u_t\ ,\ v_t) = (f\,(u_{t-1}\ ,\ v_{t-1})\ ,\ f\,(u_{t-1}\ ,\ v_{t-1})) \quad .$$

Note that we usually let

$$u_0 = \overline{x}_c \text{ and } v_0 = \overline{y}_c \quad .$$

In summary, the iterative formulas to estimate the spatial median are

$$u_t = \frac{\sum_{i=1}^{n}\left(\frac{x_i}{\sqrt{(x_i - u_{t-1})^2 + (y_i - v_{t-1})^2}}\right)}{\sum_{i=1}^{n}\left(\frac{1}{\sqrt{(x_i - u_{t-1})^2 + (y_i - v_{t-1})^2}}\right)}$$

and

$$v_t = \frac{\sum_{i=1}^{n} \left(\frac{y_i}{\sqrt{(x_i - u_{t-1})^2 + (y_i - v_{t-1})^2}} \right)}{\sum_{i=1}^{n} \left(\frac{1}{\sqrt{(x_i - u_{t-1})^2 + (y_i - v_{t-1})^2}} \right)} .$$

9.3.4 Dispersion 2D

We now discuss two two-dimensional measures of dispersion: standard distance and the standard deviational ellipse. They are used for isotropic and anisotropic data, respectively. The term *isotropic* refers to data with no clear directional trend, while the term *anisotropic* refers to data with evidence of a directional trend.

Standard distance

If the dataset appears to be isotropic, a reasonable measure of dispersion in two dimensions is standard distance (SD):

$$SD = \sqrt{\frac{\sum_{i=1}^{n} (x_i - \overline{x})^2 + \sum_{i=1}^{n} (y_i - \overline{y})^2}{n}} .$$

As may be observed, this is a variety of pooled variance in two dimensions. The SD measure is often plotted on the original dataset as a circle of radius *SD* centred on the spatial mean of the dataset. Multiples of SD may be displayed as concentric circles centred on the spatial mean.

Standard deviational ellipse (SDE)

For anisotropic data we would like to capture the directionality of the dispersion. We do this via the standard deviational ellipse (SDE). Basically, we are fitting an ellipse centred on the spatial mean and rotated about its origin so that the semimajor axis aligns with the direction of maximum variance (see Figure 9.4). The SDE is calculated as follows. First, the coordinates are transformed to an origin at the mean centre,

$$x'_i = x_i - \overline{x}_c$$

$$y'_i = y_i - \overline{y}_c \quad .$$

Figure 9.4 An example of an SDE plot: Selected settlements in Southern Ontario.

The angle of rotation θ of the SDE is calculated via

$$\tan\theta = \frac{\left(\sum_{i=1}^{n} x_i'^2 - \sum_{i=1}^{n} y_i'^2\right) + \sqrt{\left(\sum_{i=1}^{n} x_i'^2 - \sum_{i=1}^{n} y_i'^2\right)^2 + 4\cdot\left(\sum_{i=1}^{n} x_i'\cdot y_i'\right)^2}}{2\cdot\sum_{i=1}^{n} x_i'\cdot y_i'}$$

$$\theta = \arctan(\tan\theta) \quad .$$

The lengths of the semimajor and semiminor axes of the SDE are given by

$$\delta_x = \sqrt{\frac{\sum_{i=1}^{n}\left(x_i'\cdot\cos\theta - y_i'\cdot\sin\theta\right)^2}{n}}$$

$$\delta_y = \sqrt{\frac{\sum_{i=1}^{n}\left(x_i'\cdot\sin\theta + y_i'\cdot\cos\theta\right)^2}{n}} \quad .$$

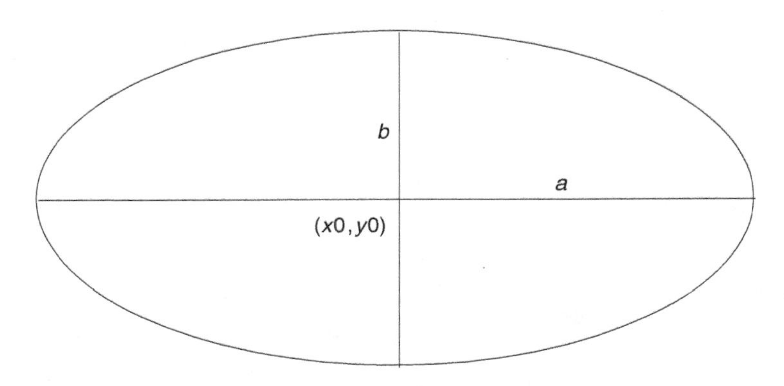

Figure 9.5 An ellipse.

Now we provide a derivation of the above results. First, coordinates are transformed to an origin at the mean centre,

$$x_i' = x_i - \overline{x}_c$$
$$y_i' = y_i - \overline{y}_c \quad .$$

Next, recall the ellipse. Its position is determined by a centre, its shape by the lengths of its semimajor (a) and semiminor (b) axes (see Figure 9.5). The equation of the ellipse is given below, where x_0, y_0 is the origin and a and b are, respectively, the lengths of the semimajor and semiminor axes.

$$\frac{(x - x_0)^2}{a^2} + \frac{(y - y_0)^2}{b^2} = 1 \quad .$$

Next, we equate the standard deviation of the x-coordinates with the major semi-axis of the ellipse and the standard deviation of the y-coordinates with the minor semi-axis of the ellipse:

$$\delta_x' = \sqrt{\frac{\sum_{i=1}^{n} (x_i')^2}{n}}$$

$$\delta_y' = \sqrt{\frac{\sum_{i=1}^{n} (y_i')^2}{n}} \quad .$$

Note that x_i' and y_i' are defined as above, so if we substitute back their definitions we recover the standard form of standard deviation. We want to capture the direction of the anisotropy, so we express these deviations in more general terms that include a rotation θ. Further, we would like the semimajor axis of the ellipse to be in the direction of the maximum deviation. To do this we first express δ_x and δ_y in a more general form, including a rotation θ:

$$\delta_x = \sqrt{\frac{\sum_{i=1}^{n} (x'_i \cdot \cos\theta - y'_i \cdot \sin\theta)^2}{n}}$$

$$\delta_y = \sqrt{\frac{\sum_{i=1}^{n} (x'_i \cdot \sin\theta + y'_i \cdot \cos\theta)^2}{n}} \quad .$$

Observe that if $\theta = 0$ (no rotation), it follows that $\cos 0 = 1$ and $\sin 0 = 0$, so $\delta_x = \delta'_x$ and $\delta_y = \delta'_y$.

Also note that the form of introducing the rotation into the expression is the standard computer graphics rotation transformation (see Appendix V for a simple derivation of this result):

$$\begin{bmatrix} u \\ v \end{bmatrix} = \begin{bmatrix} \cos\theta & -\sin\theta \\ \sin\theta & \cos\theta \end{bmatrix} \cdot \begin{bmatrix} x \\ y \end{bmatrix} \quad .$$

Then to find the θ that gives the maximum value for δ_x—that is, to equate the direction of maximum dispersion to the semimajor axis of the ellipse—we take the derivative of δ_x with respect to θ, set this equal to 0, and solve for θ:

$$0 = \frac{d\delta_x}{d\theta} = \frac{d}{d\theta}\left(\left(\frac{\sum_{i=1}^{n} (x'_i \cdot \cos\theta - y'_i \cdot \sin\theta)^2}{n}\right)^{\frac{1}{2}}\right) \quad .$$

In the derivation that follows, we make temporary simplifications to notation. Let

$$x = x'_i \; , \quad y = y'_i \; , \quad \sum = \sum_{i=1}^{n} \quad .$$

And we use the common practice of concatenating terms to indicate multiplication. This gives

$$0 = \frac{d}{d\theta}\left(\left(\frac{\sum (x\cos\theta - y\sin\theta)^2}{n}\right)^{\frac{1}{2}}\right)$$

$$= \frac{d}{d\theta}\left(\left(\frac{\sum \left(x^2\cos^2\theta - 2xy\cos\theta\sin\theta + y^2\sin^2\theta\right)}{n}\right)^{\frac{1}{2}}\right)$$

$$= \left(\frac{1}{2}\right)\left(\frac{1}{n^{\frac{1}{2}}}\right)\frac{\frac{d}{d\theta}\left(\sum \left(x^2\cos^2\theta - 2xy\cos\theta\sin\theta + y^2\sin^2\theta\right)\right)}{\left(\sum \left(x^2\cos^2\theta - 2xy\cos\theta\sin\theta + y^2\sin^2\theta\right)\right)^{\frac{1}{2}}}.$$

Since the left side of the equation is 0 and the denominator is always positive (except for the degenerate case of no variance in the x values), we can just consider the numerator in solving for θ. We also keep the (1/2) term, since it conveniently cancels out in a few lines:

$$0 = \left(\frac{1}{2}\right)\frac{d}{d\theta}\left(\sum \left(x^2\cos^2\theta - 2xy\cos\theta\sin\theta + y^2\sin^2\theta\right)\right)$$

$$= \left(\frac{1}{2}\right)\frac{d}{d\theta}\left(\cos^2\theta\sum x^2 - 2\cos\theta\sin\theta\sum xy + \sin^2\theta\sum y^2\right)$$

$$= \left(\frac{1}{2}\right)\left(2\cos\theta\,(-\sin\theta)\sum x^2 - 2\,(\cos\theta\cos\theta - \sin\theta\sin\theta)\sum xy \right.$$
$$\left. + \; 2\sin\theta\cos\theta\sum y^2\right)$$

$$= -\sin\theta\cos\theta\sum x^2 + \left(\sin^2\theta - \cos^2\theta\right)\sum xy + \sin\theta\cos\theta\sum y^2$$

$$= \left(\sum xy\right)\sin^2\theta + \left(\sum y^2 - \sum x^2\right)\sin\theta\cos\theta - \left(\sum xy\right)\cos^2\theta$$

$$= \left(\sum xy\right)\frac{\sin^2\theta}{\cos^2\theta} + \left(\sum y^2 - \sum x^2\right)\frac{\sin\theta\cos\theta}{\cos^2\theta} - \left(\sum xy\right)\frac{\cos^2\theta}{\cos^2\theta}$$

$$= \left(\sum xy\right)\tan^2\theta + \left(\sum y^2 - \sum x^2\right)\tan\theta - \left(\sum xy\right).$$

Next let

$$a = \sum xy \ , \ b = \left(\sum y^2 - \sum x^2\right) \ , \ c = -\left(\sum xy\right) \ ,$$

so we get a quadratic equation in the variable $\tan\theta$:

$$0 = a\tan^2\theta + b\tan\theta + c \ .$$

Thus, we have the familiar solutions

$$\tan\theta = \frac{-b \pm \sqrt{b^2 - 4ac}}{2a} \ .$$

We choose the positive case,

$$\tan\theta = \frac{-\left(\sum y^2 - \sum x^2\right) + \sqrt{\left(\sum y^2 - \sum x^2\right)^2 - 4\left(\sum xy\right)\left(-\sum xy\right)}}{2\left(\sum xy\right)} \ .$$

So finally, by restoring the original notation we have

$$\tan\theta = \frac{\left(\sum_{i=1}^{n} x_i'^2 - \sum_{i=1}^{n} y_i'^2\right) + \sqrt{\left(\sum_{i=1}^{n} x_i'^2 - \sum_{i=1}^{n} y_i'^2\right)^2 + 4 \cdot \left(\sum_{i=1}^{n} x_i' \cdot y_i'\right)^2}}{2 \cdot \sum_{i=1}^{n} x_i' \cdot y_i'}$$

$$\theta = \arctan(\tan\theta) \ .$$

9.4 Measuring Geographic Patterns

In this section, for reasons of space we stick mainly to detailed discussions of point pattern analysis. There will be some illustrations of tools for area patterns, and line patterns are discussed more completely in the sections of the text covering mathematical graph theory applications. We will use for most of the examples a dataset that is a small spatial sample of a much larger dataset of the occurrence of outbreak of mountain pine beetle infestation in British Columbia, Canada. This is a point dataset that records the location, year, and number of trees in the cluster of infestation. Figures 9.6 and 9.7 show the entire sample. Note that we do not include any edge corrections in the examples below.

9.4.1 Point Patterns

First-order effects in a point pattern indicate dependency on absolute location (global trends), and density-based measures are used to characterize these effects. These types

Figure 9.6 Sample mountain pine beetle outbreak data, BC, Canada, 2000.

of measures include overall intensity, quadrat counts, and kernel density estimation. Second-order effects in point patterns indicate dependency on relative location (local interaction), and distance-based measures are used to characterize them. Examples of distance-based measures include average nearest-neighbour distance, $G(d)$, $F(d)$, $K(d)$, and $J(d)$. These measures are described in some detail in the following sections.

First-order processes suspected

Overall Intensity

A measure of global average density is given by λ where

$$\lambda = \frac{n}{a} = \frac{\text{number of events in the study region}}{\text{area of the study region}}.$$

Results are sensitive to the definition of the study region.

	OBJECTID	TREES	YR	POINT_X	POINT_Y	UID
1	659	3	2000	931875.8	1030520	1
2	660	3	2000	932417.6	1030839	2
3	662	5	2000	932950.1	1030858	3
4	661	3	2000	932459.8	1030949	4
5	655	5	2000	930381.6	1031352	5
6	658	2	2000	930728.7	1031390	6
7	657	2	2000	930833.2	1031404	7
8	656	3	2000	930937.7	1031435	8
9	654	3	2000	930256.7	1031465	9
10	663	4	2000	932661.1	1031623	10
11	664	5	2000	932488.4	1031907	11
12	653	4	2000	931209.2	1033346	12
13	646	5	2000	932718.5	1033415	13
14	647	4	2000	932589.9	1033540	14
15	652	7	2000	932192.3	1033609	15
16	651	10	2000	932229.2	1033720	16
17	650	9	2000	932242.2	1033882	17
18	648	4	2000	932520.6	1033922	18
19	649	5	2000	932516.3	1034178	19

Figure 9.7 Sample mountain pine beetle outbreak data, BC, Canada, 2000.

Using our mountain pine beetle data, the area of the study region is calculated as the area of the minimum bounding rectangle, as follows:

$$\begin{aligned} a &= (\max(x) - \min(x)) \cdot (\max(y) - \min(y)) \\ &= (932950.1 - 930256.7) \cdot (1034178 - 1030520) \\ &= 9851669 \quad . \end{aligned}$$

Thus,

$$\begin{aligned} \lambda = \frac{n}{a} &= \frac{19}{9851669} \\ &= 1.928607\text{e}{-06} \quad . \end{aligned}$$

This amounts to about two outbreaks per square kilometre.

Quadrat Counts

There are two types of quadrat counts. The first is a regular tiling of quadrats, which comprises a complete enumeration of the study area. The second is a random location or sampling of the quadrats. With both approaches we create a frequency distribution of the number of events per quadrat, which we can then compare to the expected frequency, assuming complete spatial randomness (CSR). CSR is usually modelled with a homogeneous (constant density parameter λ) Poisson distribution. For a clustered point pattern we should observe a few quadrats with many events. For a dispersed point pattern there should be many quadrats with few events.

Kernel Density Estimation

In kernel density estimation (KDE) we interpolate a density surface by creating a regular lattice of sampling points and then sampling a subset of the study area around each of the lattice points to determine local intensity. The simplest variation of this method is referred to as the naive method and is calculated as follows.

Naive method The estimate of intensity at point $\underline{p}$ is

$$\hat{\lambda}_p = \frac{n}{\pi \cdot r^2} = \frac{\text{number of events within a circle of radius } r \text{ centred at } \underline{p}}{\text{area of circle of radius } r} .$$

Note that the choice of r, the kernel bandwidth, affects the estimated density surface in the following way:

$$\begin{aligned} &\text{as } r \rightarrow \text{study region size} \\ &\hat{\lambda}_p \rightarrow \text{overall intensity, } \lambda \\ \text{conversely,}& \\ &\text{as } r \rightarrow 0 \\ &\hat{\lambda}_p \rightarrow \begin{cases} 0 & \text{no point} \\ \infty & \text{one point} \end{cases} . \end{aligned}$$

We attempt to set the bandwidth based on an objective, study-specific rationale such as some physical property of the point process being modelled. More commonly than the naive method, a quadratic kernel is used in KDE. In this case nearby points are weighted more heavily than those farther away in the density estimates. Figure 9.8 shows the KDE surface for an expanded sample of the mountain pine beetle data (the small sample used elsewhere in this chapter can be seen in the upper middle of this larger dataset).

Second-order processes suspected

Nearest-neighbour Analysis

Empirical The average nearest-neighbour distance $\overline{NND}$ is a global measure of clustering and is calculated as follows:

$$\overline{NND} = \frac{\sum_{i=1}^{n} NND_i}{n} .$$

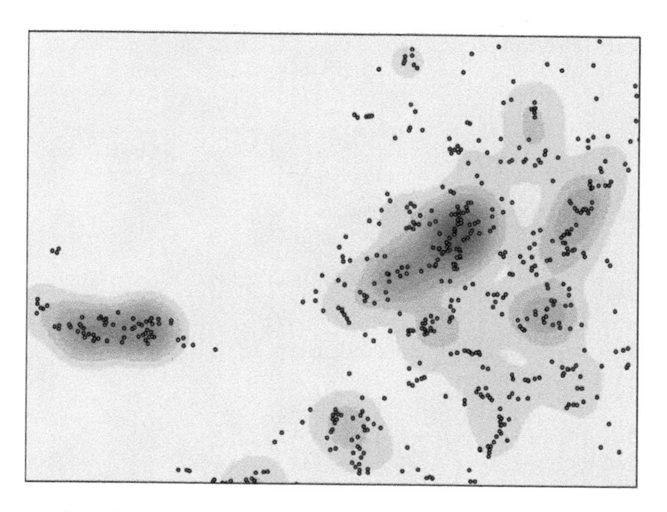

Figure 9.8 KDE surface for sample mountain pine beetle outbreak data, BC, Canada, 2000.

We describe the characteristics of this measure and its test of significance. Then we present a worked example with a spatial sample of the mountain pine beetle data.

Random If a point pattern of global density λ is assumed to come from a random point process, the expected value of average nearest-neighbour distance is calculated as follows (a derivation of this result is provided shortly):

$$\overline{NND}_{\mathrm{R}} = \frac{1}{2\sqrt{\lambda}} \quad .$$

Maximally dispersed If a point pattern of global density λ is maximally dispersed, the expected value of average nearest-neighbour distance is calculated as follows (a derivation of this result is also provided shortly):

$$\overline{NND}_{\mathrm{D}} \doteq \frac{1.07453}{\sqrt{\lambda}} \quad .$$

Standardized neighbour index A standardized measure of average nearest-neighbour distance is calculated as follows:

$$R = \frac{\overline{NND}}{\overline{NND}_{\mathrm{R}}} \quad .$$

If the point pattern is random,

$$\overline{NND} = \overline{NND}_{\text{R}}\ ,\ \text{so}\quad R = \frac{\overline{NND}_{\text{R}}}{\overline{NND}_{\text{R}}} = 1\quad .$$

If the point pattern is maximally clustered,

$$\overline{NND}_{\text{C}} = 0,\ \text{thus}\ R = 0\quad .$$

If the point pattern is maximally dispersed, then

$$R = \frac{\overline{NND}_{\text{D}}}{\overline{NND}_{\text{R}}} \doteq \frac{\dfrac{1.07453}{\sqrt{\lambda}}}{\dfrac{1}{2\sqrt{\lambda}}} \doteq 2.149\quad .$$

Proof of* R $\doteq$ *2.149 for a point pattern maximally dispersed Let $\lambda = density$. The geometric argument is guided by the following figure (see Figure 9.9), the known formulas for the areas of hexagons and triangles, and the fact that for a maximally dispersed point pattern the points are arranged as centres in a hexagonal lattice. First, recall the following two area definitions:

$$A_{\text{HEX}} = \frac{3}{2} \cdot \sqrt{3} \cdot s^2$$

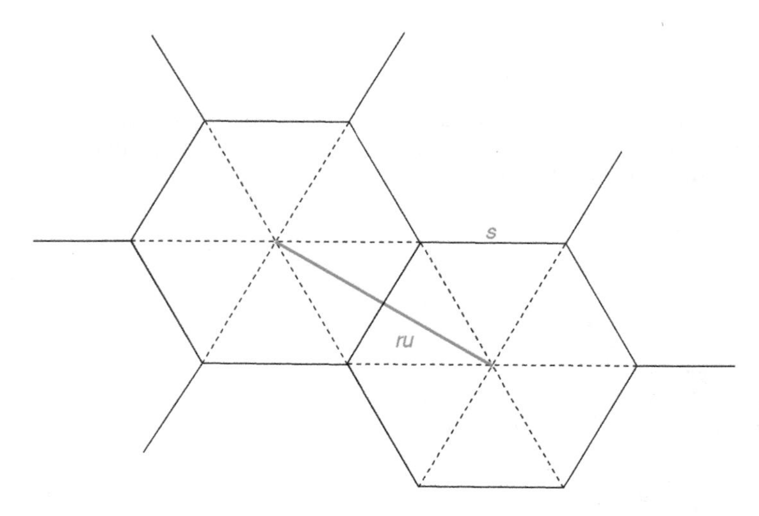

Figure 9.9 Geometry of the proof of the maximum value of R.

and

$$\begin{aligned} A_{\triangle} &= \frac{1}{2} \cdot base \cdot height \\ &= \frac{1}{2} \cdot s \cdot \frac{ru}{2} \\ &= \frac{s \cdot ru}{4} \quad . \end{aligned}$$

Using these two area definitions we can create the following equality:

$$\begin{aligned} A_{\text{HEX}} &= 6 \cdot A_{\triangle} \\ \frac{3}{2} \cdot \sqrt{3} \cdot s^2 &= \frac{6 \cdot s \cdot ru}{4} \\ \sqrt{3} \cdot s &= ru \\ ru &= \sqrt{3} \cdot s = \overline{NND_{\text{D}}}. \end{aligned}$$

Now we express density in terms of hexagonal area since there is one point at the centre of each hexagon in our maximally dispersed point configuration.

$$\begin{aligned} \lambda &= \frac{1}{A_{\text{HEX}}} \\ &= \frac{1}{\frac{3}{2} \cdot \sqrt{3} \cdot s^2} \\ &= \frac{1}{\frac{3^{\frac{3}{2}} \cdot s^2}{2}} \\ &= \frac{2}{3^{\frac{3}{2}} \cdot s^2} \quad . \end{aligned}$$

So we can take the square root of both sides to give

$$\sqrt{\lambda} = \frac{\sqrt{2}}{3^{\frac{3}{4}} \cdot s} \quad ,$$

which we then further manipulate for use:

$$\frac{1}{2 \cdot \sqrt{\lambda}} = \frac{3^{\frac{3}{4}} \cdot s}{2^{\frac{3}{2}}} \quad .$$

Now recall from above that

$$R_{\max} = \frac{\overline{NND}_{\mathrm{D}}}{\overline{NND}_{\mathrm{R}}} \quad .$$

So, substituting the results we derived above, we get

$$\begin{aligned} R_{\max} &= \frac{\sqrt{3} \cdot s}{\dfrac{1}{2 \cdot \sqrt{\lambda}}} \\ &= \frac{\sqrt{3} \cdot s}{\dfrac{3^{\frac{3}{4}} \cdot s}{2^{\frac{3}{2}}}} \\ &= \frac{2^{\frac{3}{2}}}{3^{\frac{1}{4}}} \\ &\doteq 2.149 \quad . \end{aligned}$$

We can perform statistical significance tests of the nearest-neighbour distance by assuming a null hypothesis of complete spatial randomness. The expected value of $\overline{NND}_{\mathrm{R}}$ (used above) and its variance are given here:

$$\overline{NND}_{\mathrm{R}} = \frac{1}{2 \cdot \sqrt{\lambda}} \quad ,$$

where $\lambda = \textit{density}$

$$\operatorname{var}\left[NND_{\mathrm{R}}\right] = \frac{4 - \pi}{4 \cdot \lambda \cdot \pi} \quad .$$

Then we can calculate the following test statistic and compare it to a standard normal distribution:

$$z_n = \frac{\overline{NND} - \overline{NND}_{\mathrm{R}}}{\sqrt{\dfrac{\operatorname{var}\left[NND_{\mathrm{R}}\right]}{n}}} \quad .$$

As an illustration of the type of reasoning used in the spatial analysis literature, we now derive the expected value of the nearest-neighbour distance and variance as given above. First note that

$$\lambda = \frac{\text{total points}}{\text{total area}} = \frac{\text{number of points}}{\text{unit area}} \quad .$$

Thus,

$$\lambda \cdot \text{unit area} = \text{number of points} \quad .$$

Now recall that the Poisson distribution models a random point process,

$$Pr\,(X = x) = \frac{\lambda^{x} \cdot e^{-\lambda}}{x!} \quad , \quad x = 0, 1, 2, \ldots \quad .$$

Here the random variable x is the number of points per unit area. Using the results above we can also consider the following:

$$Pr\,(X = x) = \frac{(\lambda \cdot a)^{x} \cdot e^{-\lambda \cdot a}}{x!} \quad , \quad x = 0, 1, 2, \ldots \quad ,$$

where now the random variable x is the number of points and a is the unit area. If we consider a random circle of radius r, then $a = \pi \cdot r^2$, and the probability that it contains x points is

$$\frac{\left(\lambda \cdot \pi \cdot r^2\right)^{x} \cdot e^{-\lambda \cdot \pi \cdot r^2}}{x!} \quad , \quad x = 0, 1, 2, \ldots \quad .$$

Next, if instead we consider a circle centred on a randomly chosen point, the probability that it contains x points (other than itself) is again given by

$$\frac{\left(\lambda \cdot \pi \cdot r^2\right)^{x} \cdot e^{-\lambda \cdot \pi \cdot r^2}}{x!} \quad , \quad x = 0, 1, 2, \ldots \quad .$$

The probability that this circle contains no other points (other than the one in the centre) (i.e., $X = 0$) is

$$\frac{\left(\lambda \cdot \pi \cdot r^2\right)^{0} \cdot e^{-\lambda \cdot \pi \cdot r^2}}{0!} = e^{-\lambda \cdot \pi \cdot r^2} \quad .$$

In other words, this is the probability that no other points are within distance r of the point at the centre of the circle. Thus the probability of at least one point at a distance $\leq r$ is

$$1 - e^{-\lambda \cdot \pi \cdot r^2} \quad .$$

We can consider this a function of r with parameter λ and *noting especially that this makes* r *the nearest-neighbour distance.* Further, we note that this expression is also the equation of the exponential cumulative distribution function (CDF):

$$f(r) = Pr(R \leq r) = 1 - e^{-\lambda \cdot \pi \cdot r^2} \quad .$$

The probability density function (pdf) of a CDF is obtained by taking the derivative of the CDF:

$$\text{pdf} = \frac{d}{dr}(f(r)) = 2 \cdot \lambda \cdot \pi \cdot r \cdot e^{-\lambda \cdot \pi \cdot r^2} \quad .$$

The expected value of r is calculated (see Appendix I, section 4) as

$$\begin{aligned} \mathrm{E}[r] &= \int_0^\infty (r) \cdot \left(2 \cdot \lambda \cdot \pi \cdot r \cdot e^{-\lambda \cdot \pi \cdot r^2}\right) dr \\ &= 2 \cdot \lambda \cdot \pi \int_0^\infty r^2 \cdot e^{-\lambda \cdot \pi \cdot r^2} dr \quad . \end{aligned}$$

The integral can now be solved using the following known identity:

$$\int_0^\infty x^{2 \cdot n} \cdot e^{-a \cdot x^2} dx = \frac{1 \cdot 3 \cdot 5 \cdot \ \cdots \ \cdot (2 \cdot n - 1)}{2^{n+1} \cdot a^n} \cdot \sqrt{\frac{\pi}{a}} \quad .$$

In our case $n = 1$, $a = \lambda \cdot \pi$, $x = r$, so we get

$$\begin{aligned} \mathrm{E}[r] &= 2 \cdot \lambda \cdot \pi \cdot \left(\frac{1}{4 \cdot \lambda \cdot \pi} \cdot \sqrt{\frac{\pi}{\lambda \cdot \pi}}\right) \\ &= \frac{1}{2 \cdot \sqrt{\lambda}} \quad . \end{aligned}$$

Thus,

$$\overline{NND}_{\mathrm{R}} = \frac{1}{2 \cdot \sqrt{\lambda}} \quad .$$

Now we conclude with a derivation of the variance of the nearest-neighbour distance r. First note that if $\pi \cdot r^2$ has an exponential distribution e^λ, then $2 \cdot \lambda \cdot \pi \cdot r^2$ has a chi-squared distribution with two degrees of freedom, $\chi^2_{(2)}$. A property of this distribution is that the expected value is 2. Next, using the above results and the properties of mathematical expectation and variance (see Appendix I), we derive the variance for r:

$$\begin{aligned}
\mathrm{E}\left[2 \cdot \lambda \cdot \pi \cdot r^2\right] &= 2 \cdot \lambda \cdot \pi \cdot \mathrm{E}\left[r^2\right] \\
2 &= 2 \cdot \lambda \cdot \pi \cdot \mathrm{E}\left[r^2\right] \\
2 &= 2 \cdot \lambda \cdot \pi \cdot \left((\mathrm{E}[r])^2 + \mathrm{var}[r]\right) \\
1 &= \lambda \cdot \pi \cdot \left((\mathrm{E}[r])^2 + \mathrm{var}[r]\right) \\
\mathrm{var}[r] + (\mathrm{E}[r])^2 &= \frac{1}{\lambda \cdot \pi} \\
\mathrm{var}[r] &= \frac{1}{\lambda \cdot \pi} - (\mathrm{E}[r])^2 \\
&= \frac{1}{\lambda \cdot \pi} - \left(\frac{1}{2 \cdot \sqrt{\lambda}}\right)^2 \\
&= \frac{1}{\lambda \cdot \pi} - \frac{1}{4 \cdot \lambda} \\
&= \frac{4 - \pi}{4 \cdot \lambda \cdot \pi} \quad .
\end{aligned}$$

Thus,

$$\mathrm{var}\left[NND_{\mathrm{R}}\right] = \frac{4 - \pi}{4 \cdot \lambda \cdot \pi} \quad .$$

Figure 9.10 shows the interim calculations for $\overline{NND}$ for the small sample of mountain pine beetle data. Using the information in this table, we get

$$\begin{aligned}
\overline{NND} &= \frac{\sum_{i=1}^{n} NND_i}{n} \\
&= 261.5823 \quad .
\end{aligned}$$

Now using the λ calculated above, we get

$$\begin{aligned}
\overline{NND}_{\mathrm{R}} &= \frac{1}{2 \cdot \sqrt{\lambda}} \\
&\doteq \frac{1}{2 \cdot \sqrt{1.928607\mathrm{e}{-06}}} \\
&\doteq 360.0378 \quad .
\end{aligned}$$

Event	Nearest neighbour	Distance
1	2	628.4277
2	4	117.5226
3	4	498.5858
4	2	117.5226
5	9	168.6257
6	7	105.3994
7	6	105.3994
8	7	109.0668
9	5	168.6257
10	11	331.8519
11	10	331.8519
12	15	1017.7945
13	14	179.8274
14	13	179.8274
15	16	117.1036
16	15	117.1036
17	16	161.8650
18	19	256.8313
19	18	256.8313

Figure 9.10 $\overline{NND}$ calculation intermediate results.

So,

$$R = \frac{\overline{NND}}{\overline{NND}_{\mathrm{R}}} \doteq \frac{261.5823}{360.0378} \doteq 0.7265412 \quad .$$

Further, since

$$\mathrm{var}\left[NND_{\mathrm{R}}\right] = \frac{4-\pi}{4 \cdot \lambda \cdot \pi} \doteq \frac{4-\pi}{4 \cdot 1.928607\mathrm{e}{-06} \cdot \pi} \doteq 35419.28 \quad ,$$

we get

$$z_n = \frac{\overline{NND} - \overline{NND}_{\mathrm{R}}}{\sqrt{\frac{\mathrm{var}\left[\overline{NND}_{\mathrm{R}}\right]}{n}}}$$

$$\doteq \frac{261.5823 - 360.0378}{\sqrt{\frac{35419.28}{19}}}$$

$$\doteq -2.280326 \quad .$$

This result indicates that at the 95% significance level the sample pattern is statistically distinct from a random point pattern, and since the z value is less than 0 it suggests evidence of clustering, as the observed $\overline{NND}$ is less than the value expected if the points were randomly distributed in space. Finally, note that to avoid complications arising from considerations of edge effects and sampling assumptions, most analysts today test for deviations from CSR by simulating a population of random point patterns and seeing where the empirical measure falls in this distribution.

The $\overline{NND}$ discards a lot of information about the nature of the clustering and only considers the pairwise nearest-neighbour distances. The following measures were developed to address these issues.

G(d) ***function*** The G function is a CDF of the set of nearest-neighbour distances (i.e., the distance to the nearest neighbour for each point is determined). $G(d)$ is calculated as follows:

$$G(d) = \frac{\left| d_{\min}(\underline{s}_i) < d \right|}{n} \quad ,$$

where

$$d_{\min} = \text{the NND}$$
$$\underline{s}_i = (x_i, y_i)$$
$$n = \text{the number of points}$$
$$i = 1, \ldots, n \quad .$$

Note that G is a function of distance, so for a set of distances we count the number of nearest-neighbour distances (one for each point s_i) that are less then a given d. Further, since G is cumulative it indicates the fraction of all NNDs that are less than d as a function of d. The expected value of $G(d)$ under the assumption of CSR is

$$\mathrm{E}\left[G(d)\right] = 1 - e^{-\lambda \cdot \pi \cdot d^2} \quad .$$

This result follows from the spatial statistical reasoning given above for $\overline{NND}_{\mathrm{R}}$.

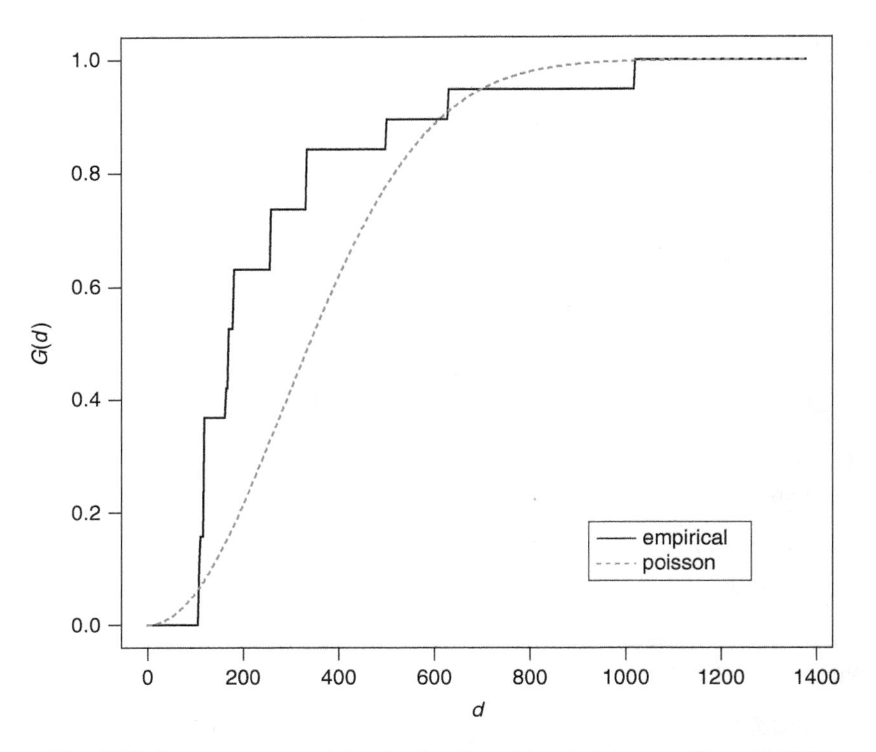

Figure 9.11 $G(d)$ for sample mountain pine beetle outbreak data, BC, Canada, 2000.

Figure 9.11 is a plot of $G(d)$ for the sample dataset. The curve of $G(d)$ for a random point pattern is also shown. The plot shows that there are more shorter NNDs than expected, indicating a clustered pattern.

F(d) *function* $F(d)$ is also a CDF but is generated from the NNDs between a second set of m randomly generated points, $\underline{p}_j$, and the original set of points. It is calculated as follows:

$$F(d) = \frac{\left|d_{\min}(\underline{p}_j, \underline{s}_i) < d\right|}{m},$$

where

$$\underline{p}_j = (x_j, y_j) \; j = 1, \ldots, m$$

$$\underline{s}_i = \text{the original set of points}$$

$$d_{\min}(\underline{p}_i, \underline{s}) = \text{the NND from the random point set to the original point dataset.}$$

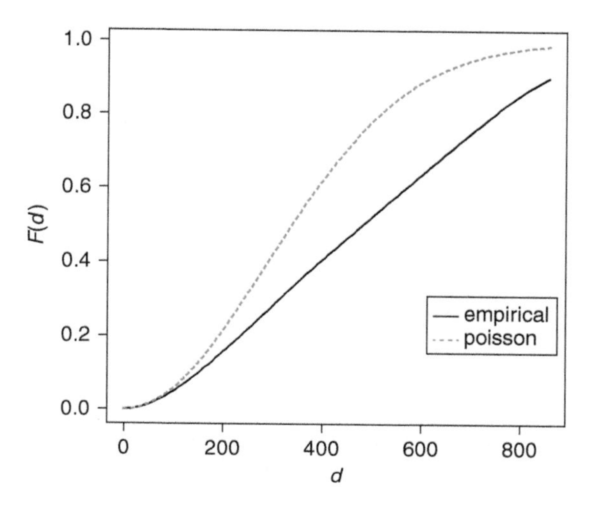

Figure 9.12 $F(d)$ for sample mountain pine beetle outbreak data, BC, Canada, 2000.

The F function in a sense samples the whole study area via NNDs while the G function is tied only to existing point locations in the original dataset. The expected value of $F(d)$ under the assumption of CSR is the same as for $G(d)$,

$$\mathrm{E}\left[F(d)\right] = 1 - e^{-\lambda \cdot \pi \cdot d^2} \quad .$$

Figure 9.12 shows $F(d)$ plotted for the sample data. On the same plot, $F(d)$ for a random point pattern is displayed. Since the observed curve is below that expected for a random pattern, it suggests that there are fewer longer NNDs than expected and so is some evidence for a clustered point pattern.

K(d) ***function*** The $F(d)$ and $G(d)$ functions use only the NNDs to characterize clustering. The $K(d)$ function is based on all the distances between points in a pattern $\underline{s}_i$. For each $\underline{s}_i$, a series of concentric circles with a set of radii d and centred on $\underline{s}_i$ is constructed. The number of other points $\underline{s}_j$ within each circle is counted, and the average for each radius distance is calculated. Finally, this average value is divided by the overall study area point density λ. In summary, the calculation formula is

$$K(d) = \frac{\sum_{i=1}^{n} \left| \underline{s}_j \in C(\underline{s}_i, d) \right|}{n \cdot \lambda} \quad , \; j \neq i \quad ,$$

where

$$\begin{aligned} C(\underline{s}_i, d) &= \text{a circle of radius } d \text{ centred on } \underline{s}_i \\ \underline{s}_i &= (x_i, y_i) \\ n &= \text{the number of points} \\ i &= 1, \ldots, n \quad . \end{aligned}$$

The expected value of $K(d)$ under the assumption of CSR is

$$\mathrm{E}\left[K(d)\right] = \pi \cdot d^2 \quad .$$

The expected value is derived via the following reasoning:

$$\begin{aligned} \lambda &= \frac{\text{number of events}}{\text{unit area}} \\ \text{number of events} &= \lambda \cdot \text{unit area} \\ &= \lambda \cdot \pi \cdot d^2 \quad . \end{aligned}$$

Thus,

$$\begin{aligned} \mathrm{E}\left[K(d)\right] &= \frac{\mathrm{E}\left[\text{number of events}\right]}{\lambda} \\ &= \frac{\lambda \cdot \pi \cdot d^2}{\lambda} \\ &= \pi \cdot d^2 \quad . \end{aligned}$$

J(d) ***function*** The J-function is a ratio of the previously defined $G(d)$ and $F(d)$ functions as an index of spatial association and has become widely used. The value of the function for distance d is given by

$$J(d) = \frac{1 - G(d)}{1 - F(d)} \quad .$$

Figure 9.13 shows a plot of $J(d)$ for the sample mountain pine beetle outbreak data. Note that the curve is above the CSR line for distances less than about 100 m, indicating dispersion below this distance. Empirical data below the CSR curve indicates clustered data.

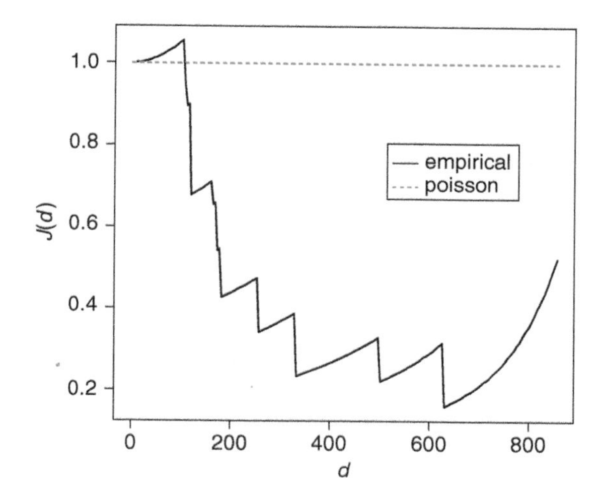

Figure 9.13 $J(d)$ for sample mountain pine beetle outbreak data, BC, Canada, 2000.

9.4.2 Spatial Autocorrelation

Not surprisingly, spatial autocorrelation (SA) is a translation of non-spatial correlation measures into a spatial context. It may be thought of as a formal way to analyze Tobler's assertion that things closer together in geographic space tend to be more similar in characteristics, and things farther away from each other in geographic space tend to be dissimilar. It is an important concept to formalize because most classical aspatial statistical inference is based on assumptions of independence of the observations, and it is well recognized and perhaps defines geographic understanding that geographic spatial data are often not independent. Thus classical statistical analysis is problematic in this case. In fact, it is the spatial pattern of dependence that is often of interest.

In the following sections, we discuss in turn three global measures of SA: join counts, Moran's I, and Geary's C; Moran scatter plots; and four local indicators of spatial association (LISAs): local Moran's I, local Geary's C, Getis's G, and Getis's G^*.

Join counts

Join counts are historically the first (mid-20th century) and conceptually the simplest SA measures. They rely on the identification of neighbourhood structure and count neighbouring zones with similar or dissimilar attributes. Imagine a street block with a collection of residential and retail buildings. We define one building as near another one if it is beside or directly across the street from the other building. This defines the neighbourhood structure (formally a rook's neighbourhood). Note that this is a formal sense of the set of near buildings, not the common usage of the word *neighbourhood*. Figure 9.14 shows this spatial structure explicitly with lines, referred to as joins, shown to indicate the near relationships as defined above. In this figure we have not indicated

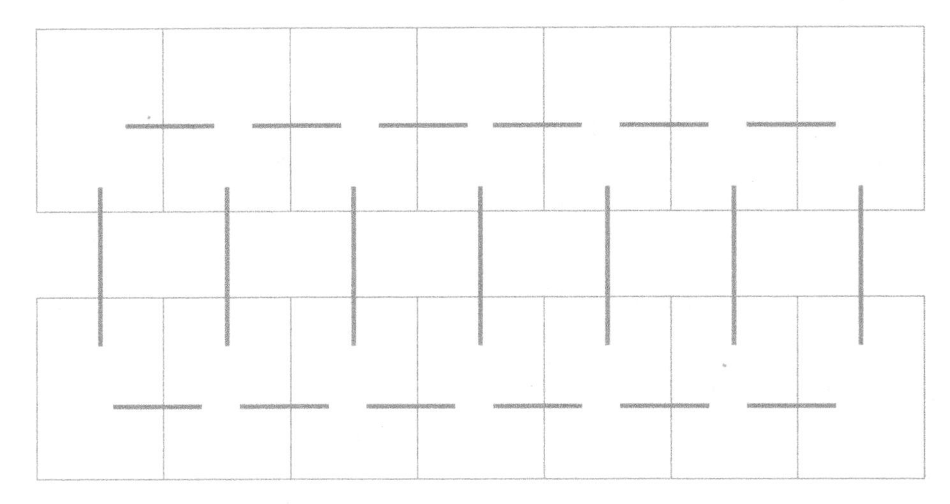

Figure 9.14 A street block example of join spatial structure.

building type. So now consider the building types. We will shade the residential buildings black and leave the retail buildings white. We set the building types arbitrarily in Figure 9.15 and note that we can objectively measure an aspect of the spatial structure of the data by counting the number of joins between buildings of the same type, J_{BB} and J_{WW}, and joins between buildings of different types, J_{BW}. If there are many more joins of type J_{BB} and J_{WW} than J_{BW} we can infer a positive spatial autocorrelation.

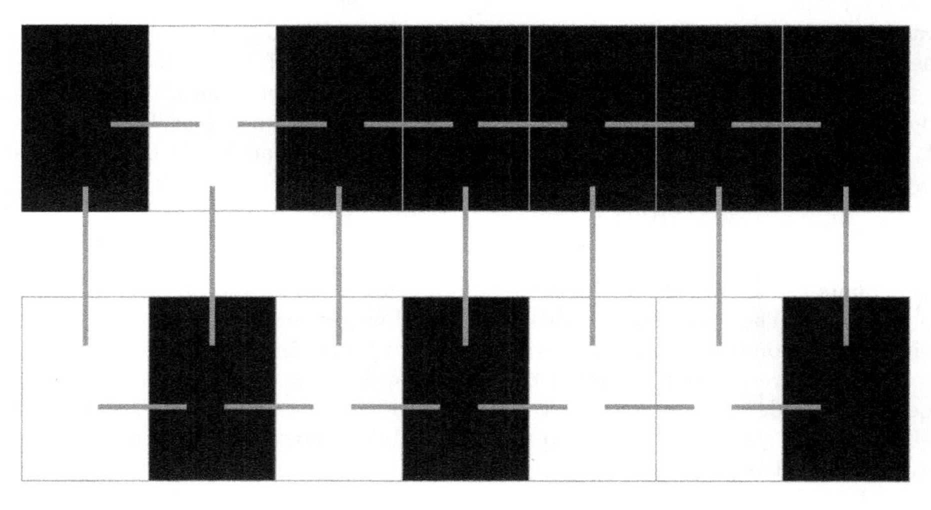

Figure 9.15 A street block example of join counts.

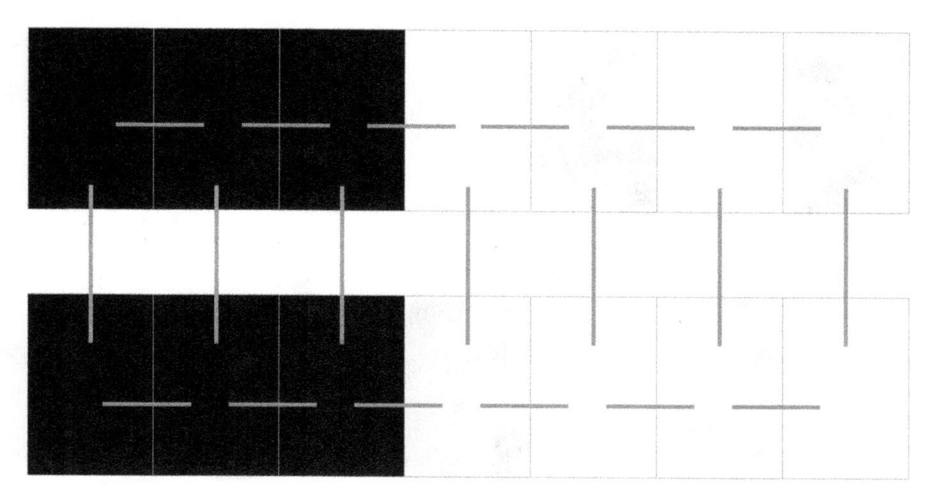

Figure 9.16 A street block example of join counts, positive spatial autocorrelation.

If there are many more J_{BW} than the other two types, we can infer negative spatial auto correlation. In this case we have intermediate results,

$$J_{BB} = 6 \quad , \quad J_{WW} = 1 \quad \text{and} \quad J_{BW} = 12 \quad .$$

There are 7 same-colour joins and 12 different-colour joins. It turns out that these kinds of intermediate results suggest that there is no statistically discernible spatial autocorrelation and that the empirical pattern could have come from a random arrangement of feature attributes. The careful reader might have noticed that elsewhere we have stated that almost all geographic data feature spatial autocorrelation, but note that we said "discernible" autocorrelation. For better or worse, this assumption of complete spatial randomness is our current best analogue of the classical statistical null hypothesis. Now consider another scenario of the distribution of building types along this street block, as provided by Figure 9.16. In this case the join count results are

$$J_{BB} = 7 \quad , \quad J_{WW} = 10 \quad , \text{and} \quad J_{BW} = 2 \quad ,$$

suggesting positive spatial autocorrelation, which seems confirmed by the clustering evident by visual inspection of the figure—perhaps reflecting a traditional landuse planning notion of separation of uses.

Another extreme is provided by the final scenario in Figure 9.17. In this case the join count results are

$$J_{BB} = 0 \quad , \quad J_{WW} = 0 \quad , \text{and} \quad J_{BW} = 19 \quad ,$$

suggesting negative spatial autocorrelation—perhaps an example of an enforced mixed-use type of landuse planning.

Theoretical results have been obtained that allow us, given certain assumptions, to estimate the expected value and variance of the various join counts and thus derive a

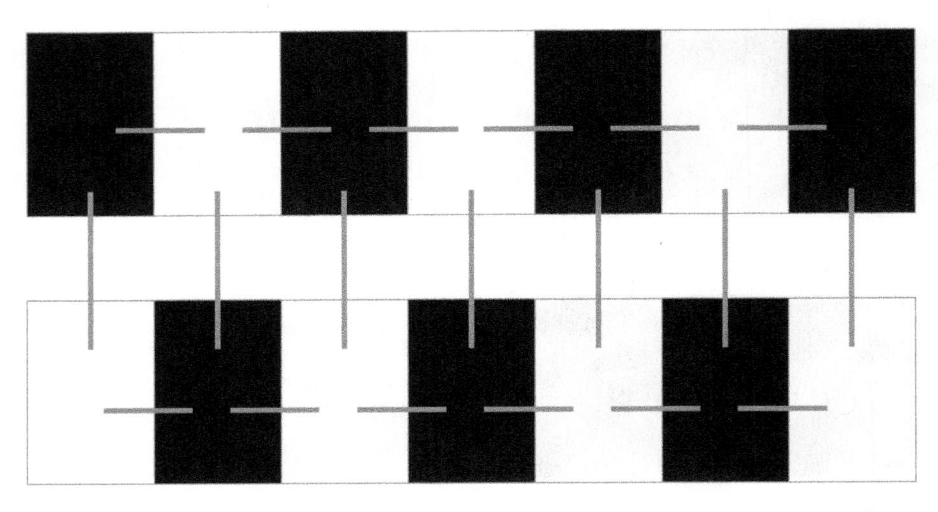

Figure 9.17 A street block example of join counts, negative spatial autocorrelation.

z-score of statistical significance. In the following we derive these results for the case of sampling with replacement. This means that we have some knowledge *a priori* of the probability of all of the attribute values of the features.

Derivation of $\mathrm{E}[J_{\mathrm{BB}}]$

We will show that the expected number of black-black joins, $\mathrm{E}[J_{\mathrm{BB}}]$, is given by

$$\mathrm{E}[J_{\mathrm{BB}}] = k \cdot {p_{\mathrm{B}}}^2 \quad ,$$

$$\text{where, } \; k = \text{ the total number of joins}$$

$$p_{\mathrm{B}} = \text{ the probability of getting a black zone.}$$

First we define a "counting" function as follows:

$$\text{Let } \; Q_i = \begin{cases} 1 & \text{if } J_i \text{ is a BB join} \\ 0 & \text{otherwise} \end{cases} \quad .$$

$$\text{Further, let } \; Q = \sum_{i=1}^{k} Q_i, \; \text{ which gives the number of } J_{\mathrm{BB}} \text{ across all } k \text{ joins.}$$

Next, recall the definition

$$\mathrm{E}[X] = \sum_{x \in S} x \cdot p(x) \quad .$$

Here x is the value of the random variable taken over its range S, and $p(x)$ is the probability of getting the value x. Thus,

$$\begin{aligned} \mathrm{E}[Q_i] &= \sum_S Q_i \cdot p(Q_i) \\ &= 1 \cdot {p_\mathrm{B}}^2 \quad + \quad 0 \cdot (1 - {p_\mathrm{B}}^2) \\ &= {p_\mathrm{B}}^2 \quad . \end{aligned}$$

To summarize the above result, in our case Q_i has only two possible values, 0 or 1, and the probability that $Q_i = 1$ is ${p_\mathrm{B}}^2$, since it requires both zones participating in the join to be black, and each each of these possibilities is independent, so we multiply their probabilities, which are each p_B. Further, since there are only two possible values for the colour of the zone and they are mutually exclusive, this means $\mathrm{not}(p_\mathrm{B}) = 1 - p_\mathrm{B}$. Now, using the property that mathematical expectation E[ ] is a linear operator, we have

$$\begin{aligned} \mathrm{E}[Q] &= \mathrm{E}\left[\sum_{i=1}^{k} Q_i\right] \\ &= \sum_{i=1}^{k} \mathrm{E}[Q_i] \\ &= \sum_{i=1}^{k} {p_\mathrm{B}}^2 \\ &= k \cdot {p_\mathrm{B}}^2 \quad . \end{aligned}$$

Thus,

$$\mathrm{E}[J_\mathrm{BB}] = k \cdot {p_\mathrm{B}}^2 \quad .$$

Derivation of $\mathrm{var}[J_\mathrm{BB}]$

We will show that the variance of black-black joins, $\mathrm{var}[J_\mathrm{BB}]$, is given by

$$\mathrm{var}[J_\mathrm{BB}] = k \cdot {p_\mathrm{B}}^2 + 2 \cdot m \cdot {p_\mathrm{B}}^3 - (k + 2 \cdot m) \cdot {p_\mathrm{B}}^4$$

$$\begin{aligned} \text{where, } k &= \text{the total number of joins,} \\ m &= \frac{1}{2} \cdot \sum_{i=1}^{k} k_i \cdot (k_i - 1) \\ k_i &= \text{the number of joins for the } i\text{th area} \\ p_\mathrm{B} &= \text{the probability of getting a black zone.} \end{aligned}$$

Recall that mathematical variance var[ ] may be expressed in terms of mathematical expectation, E[ ], as follows:

$$\begin{aligned}\mathrm{var}[X] &= \mathrm{E}[(X - \mathrm{E}[X])^2] \\ &= \mathrm{E}[X^2] - \{\mathrm{E}[X]\}^2 \quad .\end{aligned}$$

Thus,

$$\begin{aligned}\mathrm{var}[J_{\mathrm{BB}}] &= \mathrm{E}[(J_{\mathrm{BB}})^2] - \{\mathrm{E}[J_{\mathrm{BB}}]\}^2 \\ &= \mathrm{E}[(J_{\mathrm{BB}})^2] - \{k \cdot {p_{\mathrm{B}}}^2\}^2 \\ &= \mathrm{E}[(J_{\mathrm{BB}})^2] - k^2 \cdot {p_{\mathrm{B}}}^4 \quad .\end{aligned}$$

So we must now deal with the first term. Recall that above we defined

$$J_{\mathrm{BB}} = Q = \sum_{i=1}^{k} Q_i \quad .$$

Using again this substitution, we get

$$\begin{aligned}\mathrm{E}[(J_{\mathrm{BB}})^2] &= \mathrm{E}[Q^2] \\ &= \mathrm{E}\left[\left\{\sum_{i=1}^{k} Q_i\right\}^2\right] \quad ,\end{aligned}$$

and expanding terms, we get

$$= \mathrm{E}\left[\sum_{i=1}^{k} Q_i^2 + \sum\sum_{i \neq j} Q_i \cdot Q_j\right] \quad .$$

This leads us to consider three cases as follows:

1.	2.	3.
2 joins share 2 areas in common	2 joins share 1 area in common	2 joins share no areas in common
○ J_i/J_j ○	○ J_i ○ J_j ○	○ J_i ○ ○ J_j ○

For Case 1, this gives an expected value of ${p_{\mathrm{B}}}^2$ over the k joins; i.e., $k \cdot {p_{\mathrm{B}}}^2$.

For Case 2, we must have three areas coloured black, and this happens with probability ${p_B}^3$. Further, for each area with k_i joins we have $k_i - 1$ choices for the second join. Thus, summed over all the areas, we get

$${p_B}^3 \cdot \sum_{i=1}^{n} k_i \cdot (k_i - 1) = 2 \cdot m \cdot {p_B}^3 \quad ,$$

where m is defined as above.

For Case 3, we must have four areas coloured black, and this happens with probability ${p_B}^4$. Further, we have k choices for the first edge and $k - 1$ choices for the second edge, giving ${p_B}^4 \cdot k \cdot (k - 1)$. However, we do not want cases where the second edge is adjacent to the first edge, but we just calculated that circumstance in Case 2 above, so we can subtract these from our first value, giving

$$(k \cdot (k - 1) - 2 \cdot m) \cdot {p_B}^4 \quad .$$

So finally putting this all together, we have

$$\begin{aligned}
\text{var}[J_{BB}] &= \text{Case 1} + \text{Case 2} + \text{Case 3} - \{\text{E}[J_{BB}]\}^2 \\
&= k \cdot {p_B}^2 + 2 \cdot m \cdot {p_B}^3 + (k \cdot (k - 1) - 2 \cdot m) \cdot {p_B}^4 - \{k \cdot {p_B}^2\}^2 \\
&= k \cdot {p_B}^2 + 2 \cdot m \cdot {p_B}^3 + (k^2 - k - 2 \cdot m - k^2) \cdot {p_B}^4 \\
&= k \cdot {p_B}^2 + 2 \cdot m \cdot {p_B}^3 - (k + 2 \cdot m) \cdot {p_B}^4 \quad .
\end{aligned}$$

Moran's I

Join counts can work for all levels of data, including nominal as shown above. If we have interval or ratio data, we have additional measures of SA available. The first we discuss is Moran's I, the formula for which is

$$I = \frac{n}{\sum_{i=1}^{n} (x_i - \overline{x})^2} \cdot \frac{\sum_{i=1}^{n} \sum_{j=1}^{n} w_{i,j} \cdot (x_i - \overline{x}) \cdot (x_j - \overline{x})}{\sum_{i=1}^{n} \sum_{j=1}^{n} w_{i,j} (x_i - \overline{x})^2} \quad ,$$

where

x_i = the value of the observation in zone i

x_j = the value of the observation in zone j

$\overline{x}$ = the sample mean value of all the observations

$w_{i,j}$ = a weighting matrix, usually binary, that indicates adjacency relationships between zones (the local neighbourhood structure).

A vector notation version of Moran's I is useful for direct translation into numerical computing environments.

$$I = \frac{n}{(\underline{x} - \underline{\overline{x}})^T \cdot (\underline{x} - \underline{\overline{x}})} \cdot \frac{(\underline{x} - \underline{\overline{x}})^T \cdot \mathbf{W} \cdot (\underline{x} - \underline{\overline{x}})}{\underline{1}^T \cdot \mathbf{W} \cdot \underline{1}} ,$$

where

$$\underline{1} = \begin{bmatrix} 1 \\ 1 \\ \vdots \\ 1 \end{bmatrix}, \quad \underline{x} = \begin{bmatrix} x_1 \\ x_2 \\ \vdots \\ x_n \end{bmatrix}, \quad \underline{\overline{x}} = \begin{bmatrix} \overline{x} \\ \overline{x} \\ \vdots \\ \overline{x} \end{bmatrix}, \text{ and } \quad \mathbf{W} = \begin{bmatrix} w_{1,1} & w_{1,2} & \cdots & w_{1,n} \\ w_{2,1} & w_{2,2} & \cdots & w_{2,n} \\ \vdots & \vdots & \ddots & \vdots \\ w_{n,1} & w_{n,2} & \cdots & w_{n,n} \end{bmatrix} .$$

The values of I range between -1 and 1. A value of -1 indicating strong negative SA that is observations close together are dissimilar in value. A value of 1 indicating strong positive SA that is observations close together are similar in value. Values of I close to 0 suggest CSR.

We now give a brief description of calculating Moran's I for the TREES variable on a sample of the mountain pine beetle data, this time with sample size equal to 189. To keep the notion of adjacency simple, we create a Voronoi diagram of the sample outbreak points, and the binary weights matrix records Voronoi cells that share a common edge. Figure 9.18 shows this dataset; the shading refers to the TREES variable, which indicates the number of trees affected at the outbreak location, and the labels are ID numbers for the outbreak locations. The value of n is 189. The mean of the TREES variable is 5.31746, so the mean vector is this value repeated 189 times. The ones vector consists of 189 ones. These vector representations look as sketched below.

$$\underline{1} = \left.\begin{bmatrix} 1 \\ 1 \\ \vdots \\ 1 \end{bmatrix}\right\} \text{189 rows}, \quad \underline{x} = \left.\begin{bmatrix} 3 \\ 3 \\ \vdots \\ 8 \end{bmatrix}\right\} \begin{matrix}\text{189 rows—the} \\ \text{TREES values}\end{matrix} ,$$

$$\underline{\overline{x}} = \left.\begin{bmatrix} 5.31746 \\ 5.31746 \\ \vdots \\ 5.31746 \end{bmatrix}\right\} \text{189 rows, and} \quad \mathbf{W} = \left.\begin{bmatrix} 0 & 0 & \cdots & 0 \\ 0 & 0 & \cdots & 0 \\ \vdots & \vdots & \ddots & \vdots \\ 0 & 0 & \cdots & 0 \end{bmatrix}\right\} \begin{matrix}\text{189 rows by} \\ \text{189 columns}\end{matrix} .$$

Note that it so happens in this case that there are no adjacent cells in the sketched locations, so to better illustrate this part of the calculation we provide a contiguous portion from the middle of the binary weights matrix $\mathbf{W}$ in Figure 9.19. Note also that surrounding the matrix partition, ID numbers are used to index the rows and columns to the Voronoi cell labels. The matrix can be compared to Figure 9.18 to see how the adjacencies of the cells map into $\mathbf{W}$. For instance, look at the bottom row for some

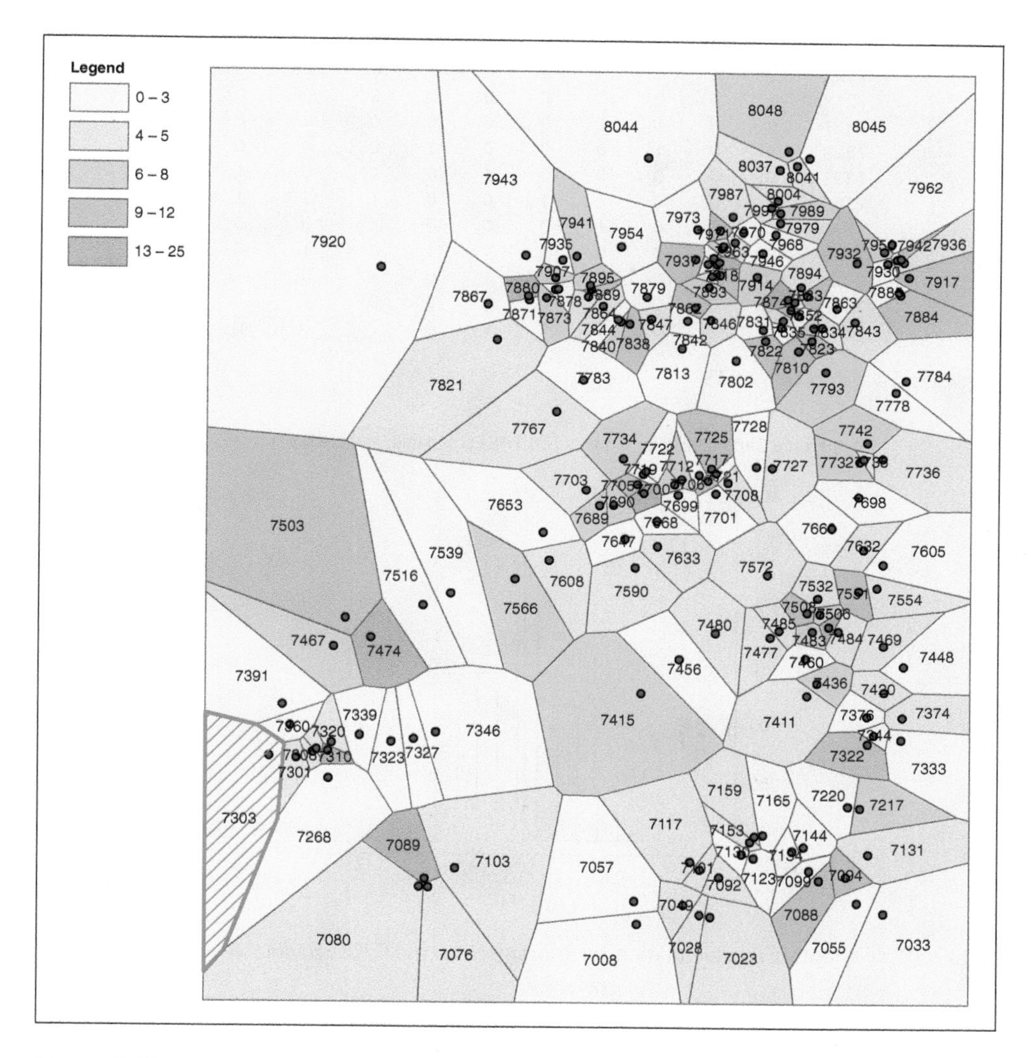

Figure 9.18 Voronoi diagram for sample mountain pine beetle outbreak data, TREES variable and feature IDs, BC, Canada, 2000.

of the adjacencies of cell 7303. Calculating Moran's I using software and the vector formulation of the inputs gives

$$I = -0.007254153 \quad ,$$

which suggests a random pattern for the TREES variable. We can use Monte Carlo simulation to test the significance of this result by permuting the TREES variable over the outbreak locations, calculating I for each of the permutations, and then comparing where our empirical I falls in this distribution. If we do this for 1000 permutations we

	7322	7217	7327	7323	7391	7360	7339	7320	7311	7308	7303
7322	0	1	0	0	0	0	0	0	0	0	0
7217	1	0	0	0	0	0	0	0	0	0	0
7327	0	0	0	1	0	0	0	0	0	0	0
7323	0	0	1	0	0	0	1	0	0	0	0
7391	0	0	0	0	0	1	0	1	0	0	1
7360	0	0	0	0	1	0	0	1	1	1	1
7339	0	0	0	1	0	0	0	1	0	0	0
7320	0	0	0	0	1	1	1	0	1	0	0
7311	0	0	0	0	0	1	0	1	0	1	0
7308	0	0	0	0	0	1	0	0	1	0	0
7303	0	0	0	0	1	1	0	0	0	0	0

Figure 9.19 A portion of $\mathbf{W}$.

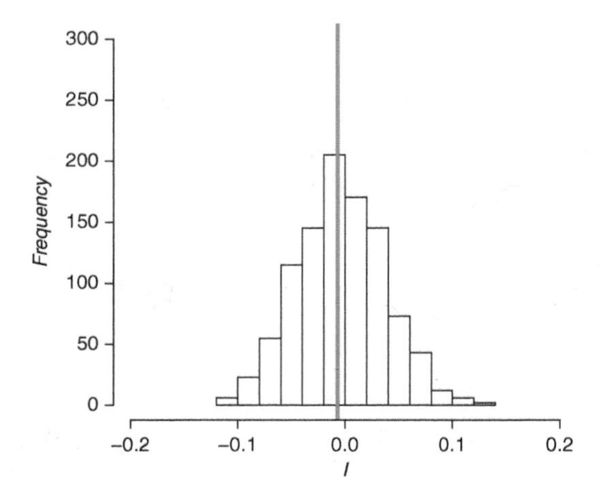

Figure 9.20 Monte Carlo simulation for Moran's I on the TREES variable.

get the results shown in Figure 9.20, which shows evidence that indeed the TREES variable is typical of a random distribution. Note that the vertical line indicates our empirical I value.

Geary's C

Another measure of SA for interval or ratio data is Geary's C. It often provides similar results to Moran's I—though not similar values (see below)—but not always. If both measures are available, it is best to apply both to look for potential anomalies. Geary's C is defined as follows:

$$C = \frac{n-1}{\sum_{i=1}^{n}(x_i - \overline{x})^2} \cdot \frac{\sum_{i=1}^{n}\sum_{j=1}^{n} w_{i,j} \cdot (x_i - x_j)^2}{2 \cdot \sum_{i=1}^{n}\sum_{j=1}^{n} w_{i,j}},$$

where

x_i = the value of the observation in zone i

x_j = the value of the observation in zone j

$\overline{x}$ = the sample mean value of all the observations

$w_{i,j}$ = a weighting matrix, usually binary, that indicates adjacency relationships between zones.

A vector notation version of Geary's C:

$$C = \frac{1}{2} \cdot \frac{n-1}{(\underline{x} - \underline{\overline{x}})^T \cdot (\underline{x} - \underline{\overline{x}})} \cdot \frac{\underline{1}^T \cdot (\mathbf{W} \times (\underline{x} \cdot \underline{1}^T - \underline{1} \cdot \underline{x}^T) \wedge 2) \cdot \underline{1}}{\underline{1}^T \cdot \mathbf{W} \cdot \underline{1}},$$

where

$$\underline{1} = \begin{bmatrix} 1 \\ 1 \\ \vdots \\ 1 \end{bmatrix}, \quad \underline{x} = \begin{bmatrix} x_1 \\ x_2 \\ \vdots \\ x_n \end{bmatrix}, \quad \underline{\overline{x}} = \begin{bmatrix} \overline{x} \\ \overline{x} \\ \vdots \\ \overline{x} \end{bmatrix}, \quad \mathbf{W} = \begin{bmatrix} w_{1,1} & w_{1,2} & \cdots & w_{1,n} \\ w_{2,1} & w_{2,2} & \cdots & w_{2,n} \\ \vdots & \vdots & \ddots & \vdots \\ w_{n,1} & w_{n,2} & \cdots & w_{n,n} \end{bmatrix},$$

and

$\times$ = element by element array multiplication

$\wedge$ = element by element array exponentiation.

The values of c are interpreted as

$$0 < C < 1 \text{ indicates positive SA}$$
$$C = 1 \text{ indicates no SA}$$
$$C > 1 \text{ indicates negative SA.}$$

Moran scatter plot

Recall from above the vector form of Moran's global measure of SA,

$$I = \frac{n}{(\underline{x} - \underline{\overline{x}})^T \cdot (\underline{x} - \underline{\overline{x}})} \cdot \frac{(\underline{x} - \underline{\overline{x}})^T \cdot \mathbf{W} \cdot (\underline{x} - \underline{\overline{x}})}{\underline{1}^T \cdot \mathbf{W} \cdot \underline{1}}.$$

We now make a simple notational simplification by letting $\underline{y} = \underline{x} - \underline{\overline{x}}$, which gives

$$I = \frac{n}{\underline{y}^T \cdot \underline{y}} \cdot \frac{\underline{y}^T \cdot \mathbf{W} \cdot \underline{y}}{\underline{1}^T \cdot \mathbf{W} \cdot \underline{1}}.$$

If the spatial weights matrix **W** is row standardized so the elements of each row sum to 1,

$$\underline{1}^T \cdot \mathbf{W} \cdot \underline{1} = n$$

and

$$I = \frac{\underline{y}^T \cdot \mathbf{W} \cdot \underline{y}}{\underline{y}^T \cdot \underline{y}} \quad .$$

Thus, since the $\underline{y}$ are deviations from the mean, I is formally equivalent to a regression coefficient of $\mathbf{W} \cdot \underline{y}$ on $\underline{y}$. Recall simple linear regression,

$$\hat{y} = b_1 \cdot x + b_0 \quad .$$

The slope of the best-fit line that the equation for $\hat{y}$ models is

$$b_1 = \frac{\sum_{i=1}^{n} (x_i - \overline{x}) \cdot (y_i - \overline{y})}{\sum_{i=1}^{n} (x_i - \overline{x})^2} \quad ,$$

or

$$b_1 = \frac{(\underline{x} - \underline{\overline{x}})^T \cdot (\underline{y} - \underline{\overline{y}})}{(\underline{x} - \underline{\overline{x}})^T \cdot (\underline{x} - \underline{\overline{x}})} \quad .$$

Comparing this to the previously presented form for I shows the structural equivalence,

$$I = \frac{\underline{y}^T \cdot (\mathbf{W} \cdot \underline{y})}{\underline{y}^T \cdot \underline{y}} \quad .$$

To emphasize, here $\underline{y} = (\underline{x} - \underline{\overline{x}})$, and we are predicting the spatially lagged $\underline{y}$, which is $\mathbf{W} \cdot \underline{y}$. The Moran scatter plot is a plot of each of the centred observations $y = x - \overline{x}$ versus the spatially lagged centred observations $\mathbf{W} \cdot \underline{y}$. Figure 9.21 gives some guidance to the interpretation of the spatially lagged scatter plot. The Moran scatter plot can be augmented by a best-fit line with I as its slope, as described above.

Figure 9.22 gives an example of a Moran scatter plot for the TREES variable with the same sample of mountain pine beetle outbreak data as above. Some of the data points are labelled with ID numbers, so you can see especially the high values

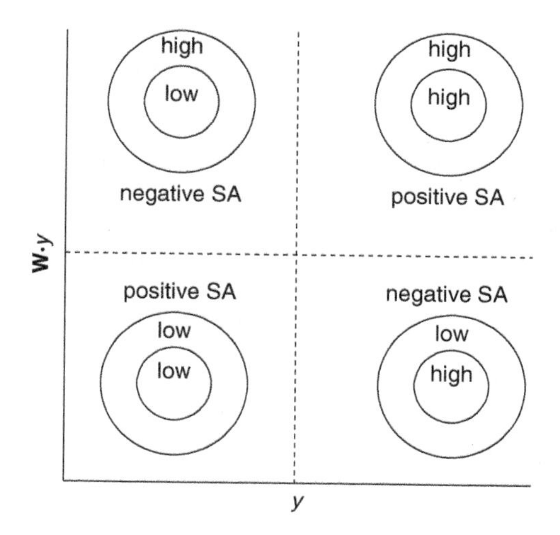

Figure 9.21 The interpretation of a Moran scatter plot.

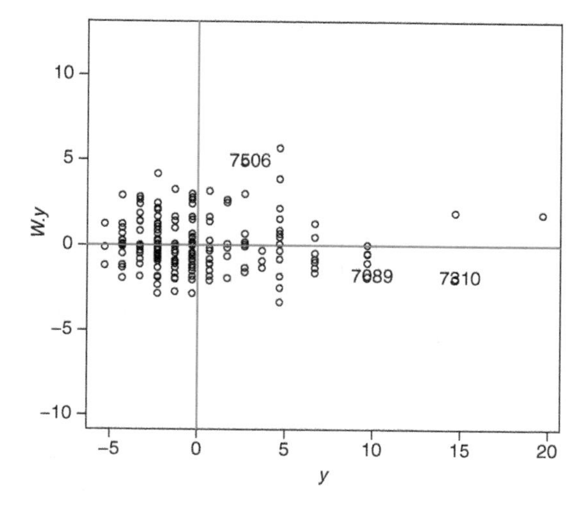

Figure 9.22 Moran scatter plot for TREES variable.

surrounded by low values (e.g., 7089 and 7310) and data point 7506, which is a hot spot of a high value surrounded by a high values. Compare this also back to the plot of the TREES variable, in Figure 9.18.

In the following section we describe local indicators of spatial association (LISAs) and provide a plot of local Moran's I values for the TREES variable, in which you can note the location of the significant local values and compare these to those noted on the Moran scatter plot.

Local indicators of spatial association (LISAs)

LISAs are used to find and plot areas of local non-stationarity. They spatialize the spatial dependency measures discussed in the previous sections. Note that there are two commonly used views of spatial association. First is the neighbourhood view most often encountered in geography, which uses a spatial weights matrix, and measurements are compared to their values when spatially lagged. The second view, which is more prevalent in geostatistics, is the distance view, where spatial association is examined via a continuous function of distance, the variogram. The spatial interpolation technique of Kriging is based on this approach and is discussed in Chapter 11. We briefly describe four local measures of spatial association, local Moran's I, local Geary's C, Getis's G, and Getis's G^*.

Local Moran's I

For each zone i in a study area, local Moran's I measure of spatial association is calculated as follows:

$$I_i = z_i \cdot \sum_{\substack{j=1 \\ j \neq i}}^{n} w_{i,j} \cdot z_j \quad ,$$

where

$$z_j = \frac{(x_i - \overline{x})}{s}$$

$$\overline{x} = \text{the sample mean of } x$$

$$s = \text{the sample standard deviation of } x$$

$$w_{i,j} = \text{the } i, jth \text{ element of the row normalized adjacency matrix } \mathbf{W}.$$

Figure 9.23 gives an example of a local Moran's I for the TREES variable, in this case indicating only statistically significant results. Note again data points 7089, 7310, and 7506, and compare this also back to the plot of the TREES variable in Figure 9.18 and the Moran scatter plot in Figure 9.22. We now list the form and some of the moments for other useful LISAs.

Local Geary's C

$$C_i = \sum_{j=1}^{n} w_{i,j} \cdot (y_i - y_j)^2 \quad .$$

Getis's G

$$G_i(d) = \frac{\sum_{j=1}^{n} w_{i,j}(d) \cdot x_i}{\sum_{j=1}^{n} x_i} \quad , \quad j \neq i \quad .$$

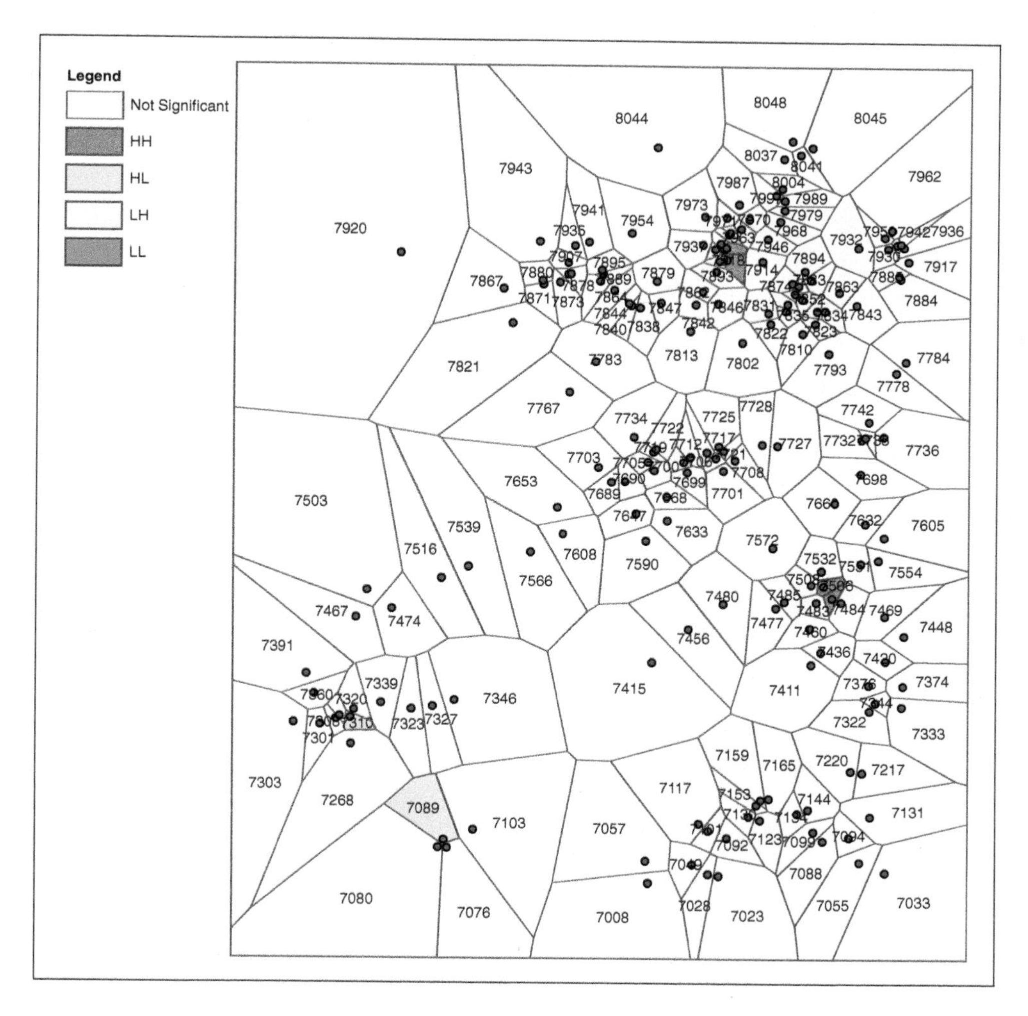

Figure 9.23 Local Moran's I, significant values, for sample mountain pine beetle outbreak data, TREES variable and feature IDs, BC, Canada, 2000.

The expected value and variance of $G_i(d)$, assuming CSR, are, respectively

$$\mathrm{E}\left[G_i(d)\right] = \frac{W_i}{(n-1)}$$

and

$$\operatorname{var}\left[G_i(d)\right] = \frac{W_i \cdot (n-1-W_i) \cdot Y_{i2}}{(n-1)^2 \cdot (n-2) \cdot Y_{i1}^2} \quad ,$$

where

$$j \neq i$$

$$W_i = \sum_{j=1}^{n} w_{i,j}(d)$$

$$Y_{i1} = \frac{\sum_{j=1}^{n} x_i}{(n-1)}$$

$$Y_{i2} = \frac{\sum_{j=1}^{n} x_i^2}{(n-1)} - Y_{i1}^2 \quad .$$

Getis's G^*

$$G_i^*(d) = \frac{\sum_{j=1}^{n} w_{i,j}(d) \cdot x_i}{\sum_{j=1}^{n} x_i} \quad .$$

The expected value and variance of $G_i^*(d)$, assuming CSR, are, respectively,

$$\mathrm{E}\left[G_i^*(d)\right] = \frac{W_i*}{n}$$

and

$$\mathrm{var}\left[G_i^*(d)\right] = \frac{W_i^* \cdot (n - W_i^*) \cdot Y_{i2}^*}{n^2 \cdot (n-1) \cdot (Y_{i1}^*)^2} \quad ,$$

where

$$W_i^* = \sum_{j=1}^{n} w_{i,j}(d)$$

$$Y_{i1}^* = \frac{\sum_{j=1}^{n} x_i}{n}$$

$$Y_{i2}^* = \frac{\sum_{i=1}^{n}\sum_{j=1}^{n} (x_i \cdot x_j)^2}{n} - (Y_{i1}^*)^2 \quad .$$

9.4.3 Landscape Metrics

It is often useful to quantify the composition and shape of features across a landscape. This might be, for example, to plan for or design a conservation network or to study spatial patterns of landuse. So we end this chapter with a brief description of some of the tools contained in a well-known suite of code designed for this purpose.

There has existed for some time an open source software package of metrics called FRAGSTATS for analyzing landscape structure. We describe here a few representative measures. Note that there are versions for raster and vector data, and the results are reported at three levels of aggregation: by individual patch, by class type of patch, and for a whole landscape. Please note that elsewhere in this text we have generally used the term *zone* to refer to an identifible partition of a landscape. However, in this section we defer to landscape ecology terminology and refer to such a partition as *patch*. We give the individual patch vector versions for the first two measures and the whole landscape version of the edge density measure.

Landscape similarity index

$$LSIM = P_i = \frac{\sum_{j=1}^{n} a_{i,j}}{A} \cdot 100 \quad ,$$

where

$$\begin{aligned} i &= 1, \ldots, m \text{ are the indices to the } m \text{ class types of the patches} \\ j &= 1, \ldots, n \text{ are the indices to the } n \text{ patches} \\ a_{i,j} &= \text{the area in } m^2 \text{ of patch } i, j \\ A &= \text{the total landscape area} \\ P_i &= \text{the proportion of the landscape occupied by patch class type } i. \end{aligned}$$

Shape index

$$SHAPE_{i,j} = \frac{p_{i,j}}{2 \cdot \sqrt{\pi \cdot a_{i,j}}} \quad ,$$

where

$$\begin{aligned} i &= 1, \ldots, m \text{ are the indices to the } m \text{ class types of the patches} \\ j &= 1, \ldots, n \text{ are the indices to the } n \text{ patches} \\ p_{i,j} &= \text{the perimeter in } m \text{ of patch } i, j \\ a_{i,j} &= \text{the area in } m^2 \text{ of patch } i, j. \end{aligned}$$

Note that $SHAPE_{i,j} = 1$ when the patch is perfectly disc shaped and is ≥ 1 without bound as the shape becomes more irregular.

Edge density

$$ED = \frac{E}{A} \cdot 10{,}000 \quad ,$$

where

$E =$ the of the length of all the edges in the landscape in m
$A =$ the of the length of all the edges in the landscape in m^2.

Note *ED* returns units of m/ha.

Problems

1. When would it be appropriate to use a standard deviational ellipse as a measure for two-dimensional dispersion?
2. Derive the formula for v_t for the spatial median.
3. Another measure for analyzing point patterns under the assumption of second order processes is $L(d)$, defined as

$$L(d) = \sqrt{\frac{K(d)}{\pi}} - d \quad .$$

   $L(d)$ is a scaled version of $K(d)$ so that $\mathrm{E}[L(d)] = 0$ with values greater than 0, indicating a clustered point pattern, and values less than 0, indicating a dispersed point pattern. Calculate $L(d)$ for the sample mountain pine data given in Figure 9.7 and interpret your results.
4. Calculate $\mathrm{E}[J_{\mathrm{BB}}]$ for the join count example in Figure 9.15, assuming $p_{\mathrm{B}} = 0.5$. Compare your result to the empirical result given in the figure.
5. Nearest-neighbour distance can be generalized to spatial lagged neighbours; that is, second, third, or fourth nearest neighbours, etc. Describe some circumstances where it might be useful to consider lagged neighbourhoods in a spatial analysis. The nearest-neighbour distance is asymmetric; what are the implications of this as a model of an interaction process?

Chapter 10

Geographic Relationships

10.1 Introduction

In this chapter we describe and explain further ideas and aspects of spatial relationships as they pertain to geographic data. First we cover basic measures of geographic relationships and forms. Then we introduce some more-advanced concepts.

10.2 Basic Measures of Geographic Relationships and Forms

In this section we discuss distance, direction, adjacency, interaction, neighbourhood, and area. The first two aspects of characterizing geographic space are mostly geometric, although variations and generalizations are mentioned. Adjacency and neighbourhood are mostly topological concepts, whereas interaction is related to both geometry and topology. Area is generally geometric in nature. Note that these concepts are inherently overlapping and thus are not mutually exclusive.

10.2.1 Distance

The Euclidean distance between two points, $\underline{a} = (a_1, a_2)$ and $\underline{b} = (b_1, b_2)$, in a flat two-dimensional space is given by

$$
\begin{aligned}
d(\underline{a}, \underline{b}) &= \sqrt{(a_1 - b_1)^2 + (a_2 - b_2)^2} \\
&= \left[(a_1 - b_1)^2 + (a_2 - b_2)^2\right]^{\frac{1}{2}} \quad .
\end{aligned}
$$

This form generalizes easily to p-dimensions by summing over all the dimensions:

$$
d(\underline{a}, \underline{b}) = \left[\sum_{i=1}^{p} (a_i - b_i)^2\right]^{\frac{1}{2}} \quad .
$$

Even more-general types of distances can be defined in a systematic way from this basic form as follows:

$$d(\underline{a}, \underline{b}) = \left[\sum_{i=1}^{p} |a_i - b_i|^m \right]^{\frac{1}{m}} \quad ,$$

where $|\cdot|$ is the standard absolute value. In other words,

$$|a| = a \text{ if } a \geq 0$$
$$|a| = -a \text{ if } a < 0 \quad ,$$
$$(\text{recalling } -(-a) = a) \quad .$$

Note that if $m = 1$, we call this the Manhattan or city-block metric, and if $m = 2$, we recover the Euclidean metric in p dimensions. The general form given above is referred to as the Minkowski distance.

At global scales we need a different notion of distance between points because we need to take into account the curvature of the earth. At first approximation we can consider the shape of the earth as a sphere and ignore its slightly lumpy, slightly ellipsoid shape. For a sphere of unit radius we have the spherical version of the cosine law for relating the lengths of sides (a, b, c) and angles (A, B, C) of a triangle—in this case a spherical triangle (see Figure 10.1). Note that on a unit sphere, the lengths of the sides are equal to the angle in radians subtended by the endpoints to the centre of the sphere. This is the spherical law of sines:

$$\cos(c) = \cos(a) \cdot \cos(b) + \sin(a) \cdot \sin(b) \cdot \cos(C) \quad .$$

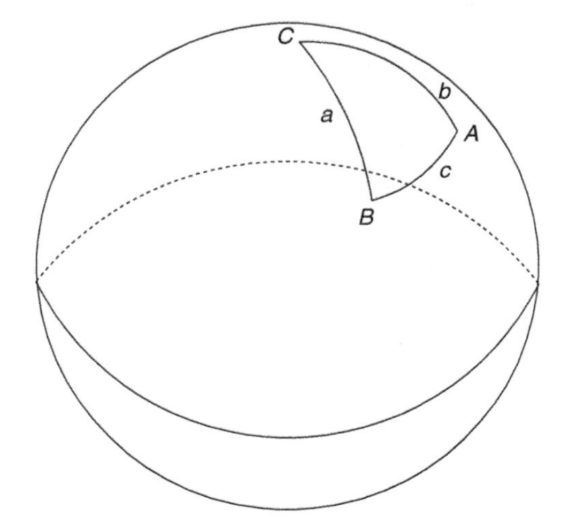

Figure 10.1 A spherical triangle.

If we take the point at angle C of the spherical triangle to be at the north pole of the sphere, the distance c (the distance between two points on the surface of the unit sphere) is given by the above equation. However, for small distances this equation is numerically unstable, so we apply some algebra to get a numerically stable equivalent, using the Haversine function, which is defined as

$$\text{haversin}(\theta) = \sin^2\left(\frac{\theta}{2}\right) = \frac{(1 - \cos(\theta))}{2} \quad .$$

Some slight manipulation gives the identity

$$\cos(\theta) = 1 - 2 \cdot \text{haversin}(\theta) \quad .$$

Recall also the identity

$$\cos(a - b) = \cos(a) \cdot \cos(b) + \sin(a) \cdot \sin(b) \quad .$$

Combining these identities with the spherical law of cosines gives

$$\begin{aligned}
\cos(c) &= \cos(a) \cdot \cos(b) + \sin(a) \cdot \sin(b) \cdot \cos(C) \\
1 - 2 \cdot \text{haversin}(c) &= \cos(a) \cdot \cos(b) + \sin(a) \cdot \sin(b) \cdot (1 - 2 \cdot \text{haversin}(C)) \\
1 - 2 \cdot \text{haversin}(c) &= \cos(a) \cdot \cos(b) + \sin(a) \cdot \sin(b) - \sin(a) \cdot \sin(b) \cdot 2 \cdot \text{haversin}(C) \\
1 - 2 \cdot \text{haversin}(c) &= \cos(a - b) - \sin(a) \cdot \sin(b) \cdot 2 \cdot \text{haversin}(C) \\
1 - 2 \cdot \text{haversin}(c) &= 1 - 2 \cdot \text{haversin}(a - b) - \sin(a) \cdot \sin(b) \cdot 2 \cdot \text{haversin}(C) \\
\text{haversin}(c) &= \text{haversin}(a - b) + \sin(a) \cdot \sin(b) \cdot \text{haversin}(C) \\
\sin^2\left(\frac{c}{2}\right) &= \sin^2\left(\frac{a - b}{2}\right) + \sin(a) \cdot \sin(b) \cdot \sin^2\left(\frac{C}{2}\right) \\
\sin\left(\frac{c}{2}\right) &= \sqrt{\sin^2\left(\frac{a - b}{2}\right) + \sin(a) \cdot \sin(b) \cdot \sin^2\left(\frac{C}{2}\right)} \\
\frac{c}{2} &= \arcsin\left(\sqrt{\sin^2\left(\frac{a - b}{2}\right) + \sin(a) \cdot \sin(b) \cdot \sin^2\left(\frac{C}{2}\right)}\right) \\
c &= 2 \cdot \arcsin\left(\sqrt{\sin^2\left(\frac{a - b}{2}\right) + \sin(a) \cdot \sin(b) \cdot \sin^2\left(\frac{C}{2}\right)}\right) \quad .
\end{aligned}$$

Now notice that for geographic coordinates, our a and b have latitude components $\pi/2-\phi_1$ and $\pi/2-\phi_2$, and C is the difference in longitude $\lambda_2-\lambda_1$. Recall also the identity $\sin(\pi/2-\phi)=\cos(\phi)$. Further, if we consider the earth as the sphere of interest rather than the unit sphere we multiply the resultant length by the radius of the earth R. If we do this, we get

$$\begin{aligned} d_H &= R\cdot 2 \\ &\cdot \arcsin\left(\sqrt{\sin^2\left(\frac{\frac{\pi}{2}-\phi_1-\left(\frac{\pi}{2}-\phi_2\right)}{2}\right)+\sin\left(\frac{\pi}{2}-\phi_1\right)\cdot\sin\left(\frac{\pi}{2}-\phi_2\right)\cdot\sin^2\left(\frac{\lambda_2-\lambda_1}{2}\right)}\right) \\ &= R\cdot 2\cdot\arcsin\left(\sqrt{\sin^2\left(\frac{\phi_2-\phi_1}{2}\right)+\cos(\phi_1)\cdot\cos(\phi_2)\cdot\sin^2\left(\frac{\lambda_2-\lambda_1}{2}\right)}\right) \quad . \end{aligned}$$

For very exacting calculations of geodetic distances, numerical methods exist to get better approximations that take into account the ellipsoid shape of the earth, such as the Vincenty formula, which can give sub-millimetre accuracy.

Distance can also be defined in other ways. A common example is distance along a network; that is, a distance measure constrained to a one-dimensional structure embedded in two-dimensional space. We might find it useful to define distance via, for instance, travel time. We might take into account modes of travel, speed limits, traffic conditions, weather conditions, fuel consumption target, or other details of the travel circumstances. Finally, note that in other parts of this text we have discussed distance between shapes.

10.2.2 Direction

Traditionally, direction occurs in geographic data as a vector quantity of bearing and distance. A whole subfield has recently arisen concerning the statistics of directional data. At the basic level we can define descriptive statistics for the mean and variance of directional data. For more complete coverage of these ideas, including directional distributions, a good source is Mardia and Jupp [78].

Directional mean

Given unit vectors (that is, vectors whose magnitude is one),

$$\underline{x}_1,\ \underline{x}_2,\ \dots,\ \underline{x}_n \quad ,$$

with corresponding angles

$$\theta_i\ ,\ \text{for } i=1\dots n \quad .$$

The Cartesian coordinates of $\underline{x}_i$ are

$$(\cos\theta_i,\ \sin\theta_i)\ ,\ \text{for } i=1\dots n \quad .$$

The Cartesian coordinates of the centre of mass are $(\overline{C},\ \overline{S})$, where

$$\overline{C} = \frac{1}{n} \cdot \sum_{i=1}^{n} \cos\theta_i \quad ,$$

$$\overline{S} = \frac{1}{n} \cdot \sum_{i=1}^{n} \sin\theta_i \quad .$$

The mean resultant length $\overline{R}$ is given by

$$\overline{R} = \left(\overline{C}^{\,2} + \overline{S}^{\,2}\right)^{\frac{1}{2}} \quad .$$

If $\overline{R} = 0$, the directional mean, $\overline{\theta}$, is undefined. When $\overline{R} > 0$, the directional mean is given by

$$\overline{\theta} = \begin{cases} \tan^{-1}\left(\dfrac{\overline{S}}{\overline{C}}\right) & \text{if } \overline{C} \geq 0 \\[2ex] \tan^{-1}\left(\dfrac{\overline{S}}{\overline{C}}\right) + \pi & \text{if } \overline{C} < 0 \end{cases}$$

Please note that for $i = 1 \ldots n$,

$$\overline{\theta} \neq \frac{\theta_1 + \theta_2 + \cdots + \theta_n}{n} \quad .$$

Circular variance

Circular variance (in other words, variance for directional data) is given by

$$V = 1 - \overline{R} \quad ,$$

where $\overline{R}$ is the mean resultant length given above. If the directions are clustered together, $\overline{R} \simeq 1$, and thus $V \simeq 0$. If the directions are dispersed, $\overline{R}$ will approach 0 and V will approach 1.

We now give an example of using directional data, by selecting an arbitrary subset of the mountain pine beetle data and combining it with aspect data from a DEM. Note that the aspect data is given in degrees, with 0 referring to north, and clockwise to 180 degrees referring to south. The original two input datasets are shown in Figure 10.2. The aspect is coloured with grey-scale values—black for no data and moving from dark grey to white for higher angular measures. Figure 10.3 plots the directional input data with an interpolation of a circular histogram. You can see the trend for mostly southward-facing slopes for the beetle outbreak locations, which is consistent with the general observation of the researchers that outbreaks tend to occur more often on south-facing slopes. This observation is reflected in the directional mean calculation given below.

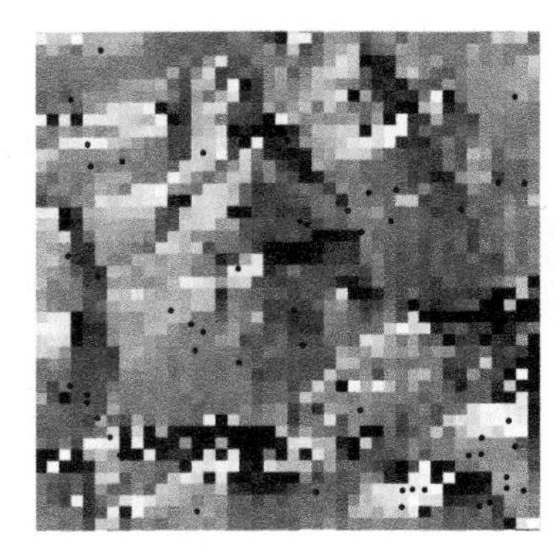

Figure 10.2 Aspect raster and sample mountain pine beetle outbreak locations.

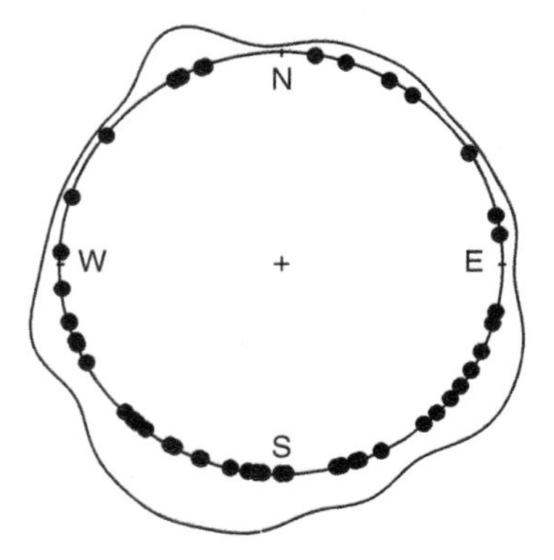

Figure 10.3 Aspect and circular density of aspect of mountain pine beetle outbreak locations.

The following tables summarize the input values and calculations.

θ in degrees:

```
 [1] 225.000000 273.366455 184.763641 120.465546 164.054611 221.987213
 [7] 332.354034 338.702637 332.969147 241.798370  76.701431  29.291363
[13]   8.841815 247.328659 201.501434  17.020525 208.909180 135.000000
[19] 220.683975 307.568604 338.749481 253.610458 217.874985 129.805573
[25] 129.805573 186.340195 165.465546 186.340195 139.513992 188.615646
[31] 193.240524  81.869896 263.290161 106.557068 330.945404  36.528854
[37] 225.000000 210.256439 114.775139 114.775139 178.667786 209.054611
[43] 125.537682 201.801407 153.034286 160.201126 159.102234 201.801407
[49] 201.250504 288.434937 339.775146 103.240517  58.392498 180.000000
[55] 248.498566 254.054611
```

θ in radians:

```
 [1] 3.9269908 4.7711447 3.2247339 2.1025204 2.8632931 3.8744078 5.8006722
 [8] 5.9114762 5.8114079 4.2201777 1.3386925 0.5112307 0.1543188 4.3166994
[15] 3.5168635 0.2970642 3.6461530 2.3561945 3.8516620 5.3680848 5.9122938
[22] 4.4263375 3.8026358 2.2655346 2.2655346 3.2522499 2.8879186 3.2522499
[29] 2.4349785 3.2919640 3.3726834 1.4288992 4.5952802 1.8597717 5.7760869
[36] 0.6375488 3.9269908 3.6696671 2.0032041 2.0032041 3.1183411 3.6486913
[43] 2.1910459 3.5220990 2.6709522 2.7960371 2.7768578 3.5220990 3.5124839
[50] 5.0341393 5.9301950 1.8018869 1.0191413 3.1415927 4.3371182 4.4340894
```

$$n = 56$$

$$\overline{C} = -0.3181003 \ , \ \overline{S} = -0.06214507 \ , \ \overline{R} \doteq 0.324$$

$$\overline{\theta} \doteq 191 \ , \text{ and } V \doteq 0.676 \quad .$$

10.2.3 Adjacency

Adjacency might be considered a nominal or even binary distance. Are you beside something or not? More generally, we might define a threshold distance (with distance defined in any of the above senses) to determine adjacency. Further note that spatially "close" things might not be adjacent. For example, historically to fly from Belfast, Northern Ireland, to Dublin, Ireland, one had to travel via London, England. Measures of adjacency may be lagged: identifying, for example, the nth nearest neighbour in a point pattern analysis. Adjacency is an important idea in network analysis, spatial autocorrelation, and spatial interpolation.

10.2.4 Interaction

Sometimes we are interested in how phenomena in a landscape interact with each other. Often the strength of interaction is inversely related to the distance between things. This may be represented mathematically as

$$w_{i,j} \propto \frac{1}{d^k} \quad ,$$

where $\propto$ means proportional to and can be replaced with an equality to create an equation by multiplying the right side terms by an appropriate constant. Note that if $k = 2$, we get the commonly used gravity models of spatial interaction.

Further, some attribute of the entities being studied can be considered in modelling the strength of the interaction as follows:

$$w_{i,j} \propto \frac{p_i \cdot p_j}{d^k} \quad .$$

For example, p may be the population in two zones. Another case might be p representing the area of the object of study and d being the distance between the centroids of the two areas.

10.2.5 Neighbourhood

A neighbourhood is an attempt to define what is local to or near a given entity or area. It could, for example, be defined by distance, adjacency, attribute, interaction, or proximity polygon (Voronoi polygon). For example, if z_i represents a spatial geographic feature, then a distance-based neighbourhood might be all the z_j's within 10 m of z_i. An adjacency-based neighbourhood would be defined as all the z_j's adjacent to z_i, where adjacency may be defined in any of the ways discussed above. An attribute-based neighbourhood is one consisting of all z_j's sharing a common characteristic with z_i. An interaction-based neighbourhood would be all z_j's interacting beyond a threshold I^* with z_i. Finally, a proximity-based neighbourhood is defined by the area closer to z_i than any other z_j's.

10.2.6 Area

You can calculate the area of a polygon in general position using only the coordinates of its vertices:

$$A_{\text{polygon}} = \sum_{i=1}^{n} (x_{i+1} - x_i) \cdot \left(\frac{(y_{i+1} + y_i)}{2} \right) \quad ,$$

where we take

$$x_{n+1} = x_1$$
$$\text{and}$$
$$y_{n+1} = y_1 \quad .$$

To see where this result comes from, recall that the area of a trapezoid (see Figure 10.4) is given by

$$\begin{aligned} A_{\text{trapezoid}} &= (\text{width}) \cdot (\text{average height}) \\ &= (x_b - x_a) \cdot \left(\frac{(y_b + y_a)}{2} \right) \quad . \end{aligned}$$

Now we can traverse clockwise around the polygon, starting at the bottom left vertex, and calculate the area of the trapezoids created by considering the coordinates of the current vertex and its projection onto the x-axis and its immediate clockwise neighbour and its corresponding projection onto the x-axis. Thus we get a series of areas of trapezoids, each area defined as above (see Figure 10.5). The sum of the areas of this series of trapezoids gives the area of the polygon:

$$\begin{aligned} A_{\text{polygon}} &= \sum_{i=1}^{n} A_{\text{trapezoid}_i} \\ &= \sum_{i=1}^{n} (x_{i+1} - x_i) \cdot \left(\frac{(y_{i+1} + y_i)}{2} \right) \quad , \end{aligned}$$

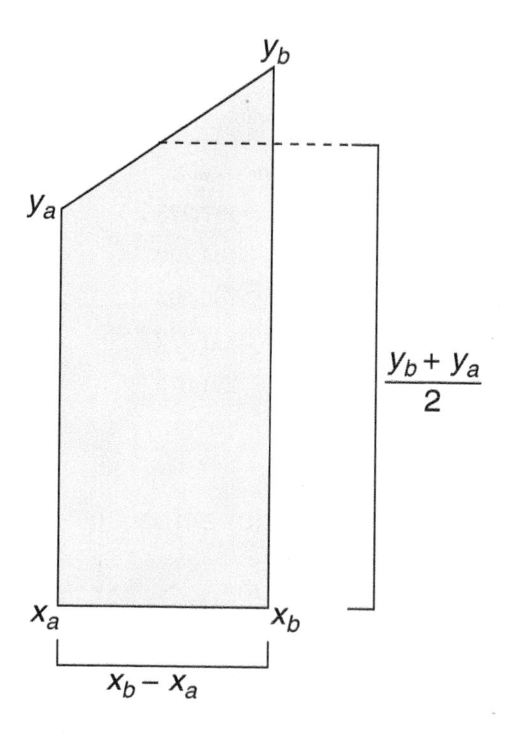

Figure 10.4 Area of a trapezoid.

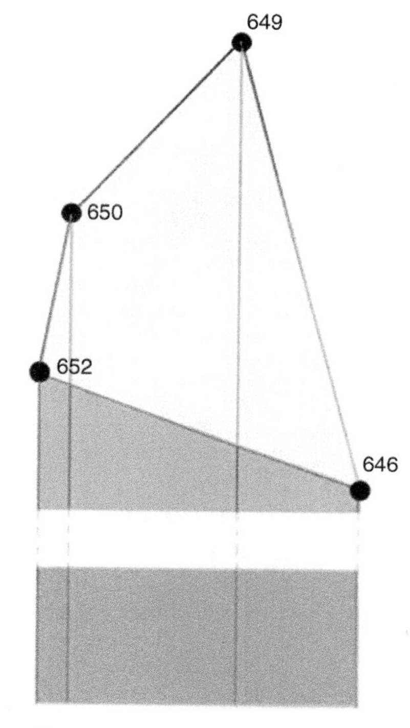

Figure 10.5 Area of a polygon via areas of trapezoids.

with again $x_{n+1} = x_1$ and $y_{n+1} = y_1$. Now note that

$$\text{if } x_{i+1} > x_i \text{ , then } A_{\text{trapezoid}_i} > 0 \quad \text{(a positive area)}$$

or

$$\text{if } x_{i+1} < x_i \text{ , then } A_{\text{trapezoid}_i} < 0 \quad \text{(a negative area)} \quad .$$

The notion of a negative area may seem strange, but *oriented* area is a common concept in advanced geometry and topology. Here it is useful because adding all the areas together gives the area of the polygon as the difference between the positive and negative areas defined by the trapezoids that are defined by the clockwise-ordered vertices of the original polygon. Note that the polygon in Figure 10.5 is the convex hull of a subset of mountain pine beetle data from Chapter 8 and that we have truncated the trapezoids in the figure (the lowest y-coordinate is 3,415, so the actual trapezoids are about five times higher). The coordinates of the convex hull vertices and the calculation of the individual trapezoids are given in the table below. Note also that the coordinates have been transformed by subtracting 932,000 m from the x-coordinate, subtracting 1,030,000 m from the y-coordinate and rounding to the nearest metre. This makes the calculations easier for this example, and the final area value of 211,076 m^2 is only slightly affected by the initial rounding.

trapezoid	id_i	id_{i+1}	x_i	x_{i+1}	y_i	y_{i+1}	$area_{trapezoid_i}$
1	652	650	192	242	3609	3882	187,275
2	650	649	242	516	3882	4178	1,104,220
3	649	646	516	718	4178	3415	766,893
4	646	652	718	192	3415	3609	–1,847,312
							211,076

A 10.3 Advanced Measures of Geographic Relationships and Forms

In this section we examine measures of shape, dimension, and connectivity.

10.3.1 Shape

Shape can be thought of in the normal, informal way and characterized a number of different ways. For instance, polygon shapes can be interrogated for the degree of their compactness or disc-like shape by using the area perimeter ratio (APR). For a circle, the following formula evaluates to 1. For non-circular shapes, the value increases without bound. This captures the geometric property that a circle contains the most area of any shape for a given perimeter. Here is the formula for calculating APR:

$$APR = \frac{\text{perimeter}}{2 \cdot \sqrt{\pi \cdot \text{area}}} \quad .$$

For a circle,

$$\begin{aligned} APR &= \frac{\text{perimeter}_{\text{circle}}}{2 \cdot \sqrt{\pi \cdot \text{area}_{\text{circle}}}} \\ &= \frac{(2 \cdot \pi \cdot \mathrm{r})}{2 \cdot \sqrt{\pi \cdot (\pi \cdot \mathrm{r}^2)}} \\ &= \frac{2 \cdot \pi \cdot \mathrm{r}}{2 \cdot \pi \cdot \mathrm{r}} \\ &= 1 \quad . \end{aligned}$$

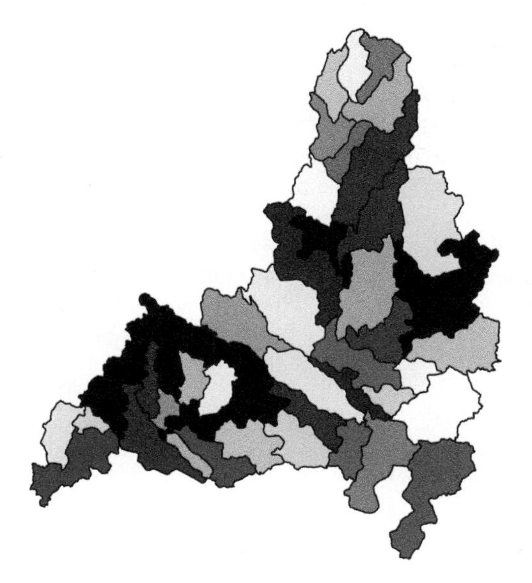

Figure 10.6 APR values for subcatchments in the Grand River Valley.

Figure 10.6 illustrates the area perimeter ratio calculated for subcatchments in the Grand River Valley area of Southern Ontario. Lower values of APR are plotted in a lighter colour, higher values being darker. Note that the more-disk-shaped areas with less-convoluted boundaries have lower APR values.

In a more technical sense, shape can be defined as the property of a geometric figure that is invariant under translation, rotation, and scaling. In Chapter 3 we introduced

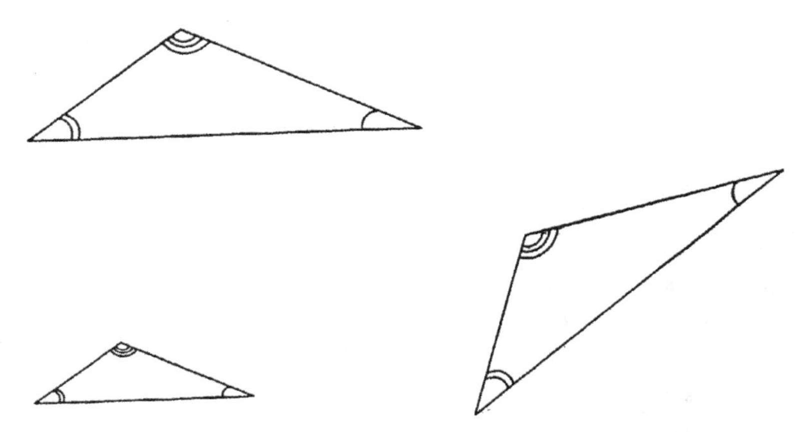

Figure 10.7 Triangles of the same shape.

these ideas. Here we elaborate a little further with some more examples. In Figure 10.7 all three triangles are the same shape. We can consider shapes of k-vertices as points in a $(k-1)$-dimensional space in several ways, and this allows us to consider shape only, without respect to position, orientation, or scale, although we can reintroduce these other characteristics as needed. Below is one formulation due to Kendall [36]. In this case we are transforming the three points that define the triangle vertices into one point in the complex plane.

For Kendall coordinates we first represent the two-dimensional Euclidean coordinates as complex numbers. Complex numbers consist of real (x) and imaginary (y) parts in a combined number, as $z = x \pm i \cdot y$, where $i = \sqrt{-1}$ and $x, y \in \Re$. Here, for $k = 3$,

$$\underline{z}^o = \begin{bmatrix} z_1^o \\ z_2^o \\ z_3^o \end{bmatrix} = \begin{bmatrix} x_1 + i \cdot y_1 \\ x_2 + i \cdot y_2 \\ x_3 + i \cdot y_3 \end{bmatrix} .$$

To remove the location information we pre-multiply $\underline{z}^o$ by the Helmert sub-matrix $\mathbf{H}$ as follows:

$$\underline{z}_H = \mathbf{H} \cdot \underline{z}^o = \begin{bmatrix} \frac{-1}{\sqrt{2}} & \frac{1}{\sqrt{2}} & 0 \\ \frac{-1}{\sqrt{6}} & \frac{-1}{\sqrt{6}} & \frac{2}{\sqrt{6}} \end{bmatrix} \cdot \begin{bmatrix} z_1^o \\ z_2^o \\ z_3^o \end{bmatrix} = \begin{bmatrix} z_1 \\ z_2 \end{bmatrix} .$$

Finally, the following transformation takes our original three points into the complex plane (note that the first two points are the normalized base of the triangle so that only the third point changes for different sets of three vertices):

$$z_1^o \rightarrow \frac{-1}{\sqrt{3}} , \quad z_2^o \rightarrow \frac{1}{\sqrt{3}} , \quad z_3^o \rightarrow u_3^K + i \cdot v_3^K = \frac{z_2}{z_1} .$$

Now consider a trajectory dataset consisting of an ordered set of points with x and y coordinates. If we create groups of three consecutive points, we can consider this a chain of consecutive triangles. We can then transform this trajectory of triangles into a trajectory of points in the complex plane via a series of Kendall transformations, as described above. Figure 10.8 shows how the local shapes of the trajectory map to the shape space. Note that "CW" refers to clockwise motion, whereas "CCW" refers to counterclockwise motion.

The following four figures (10.9, 10.10, 10.11, and 10.12) illustrate, in turn, automatic parsing of clockwise spirals, approximately constant turning movements, zig-zag

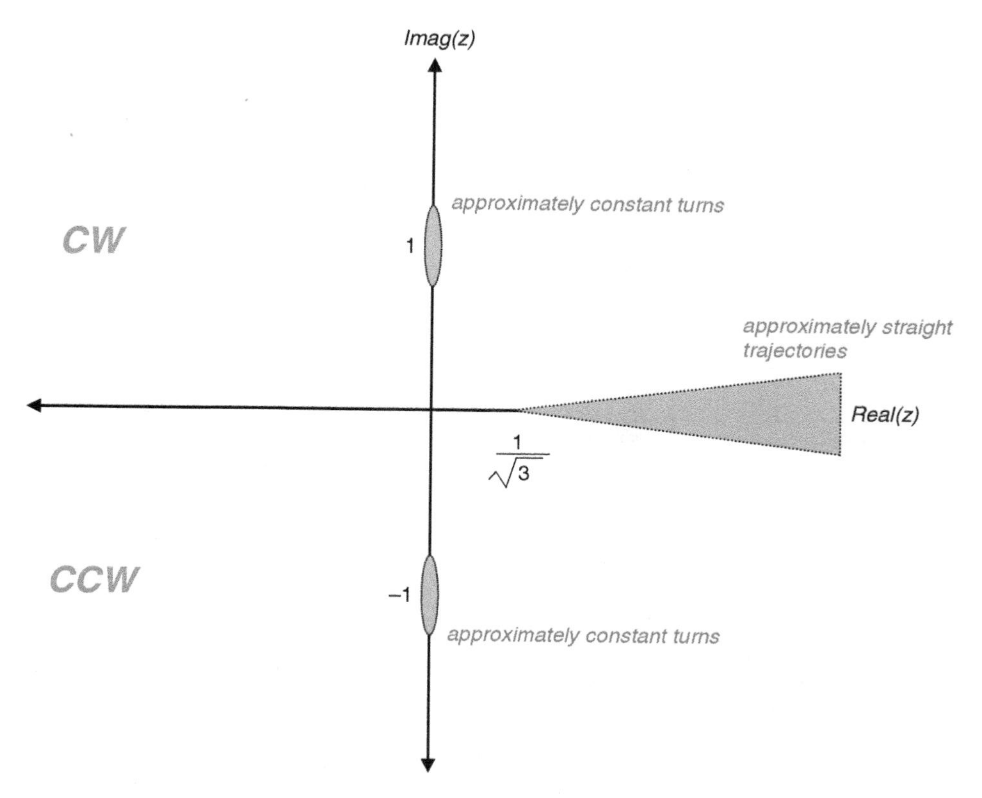

Figure 10.8 Trajectories mapping into shape space.

motions, and approximately straight paths for GPS collar trajectory data for Grizzly bears in Alberta, Canada. They are all classified by the position of the transformed trajectory in the shape space (refer back to Figure 10.8).

10.3.2 Dimension

In this section we discuss standard dimension as well as its generalization in fractal form. For non-fractal shapes we have the following relation between pairs of counts of objects with different lengths of rulers:

$$\frac{N_2}{N_1} = \left(\frac{L_1}{L_2}\right)^D$$

$$N_2 = N_1 \cdot \left(\frac{L_1}{L_2}\right)^D \quad ,$$

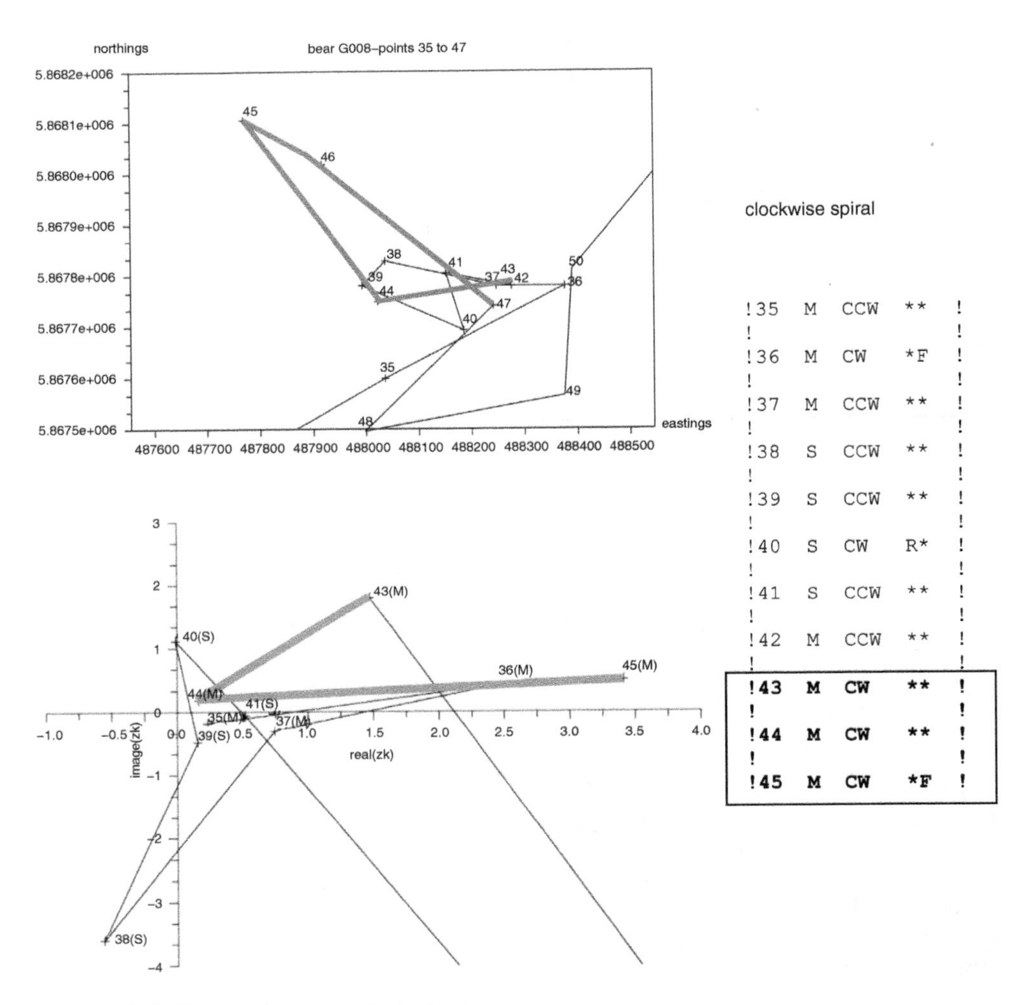

Figure 10.9 Bear trajectory—clockwise turns.

where $N_i \equiv$ the number of elements counted, $L_i \equiv$ the length of the ruler, and $D \equiv$ the dimension. The two examples below (see Figures 10.13 and 10.14) show why this is the case. In one dimension, in the first example we have

$$\begin{aligned} N_2 &= N_1 \cdot \left(\frac{L_1}{L_2}\right)^D \\ 2 &= 1 \cdot \left(\frac{2}{1}\right)^1 \\ &= 2 \quad , \end{aligned}$$

noting that total length measure by either ruler is the same at two units.

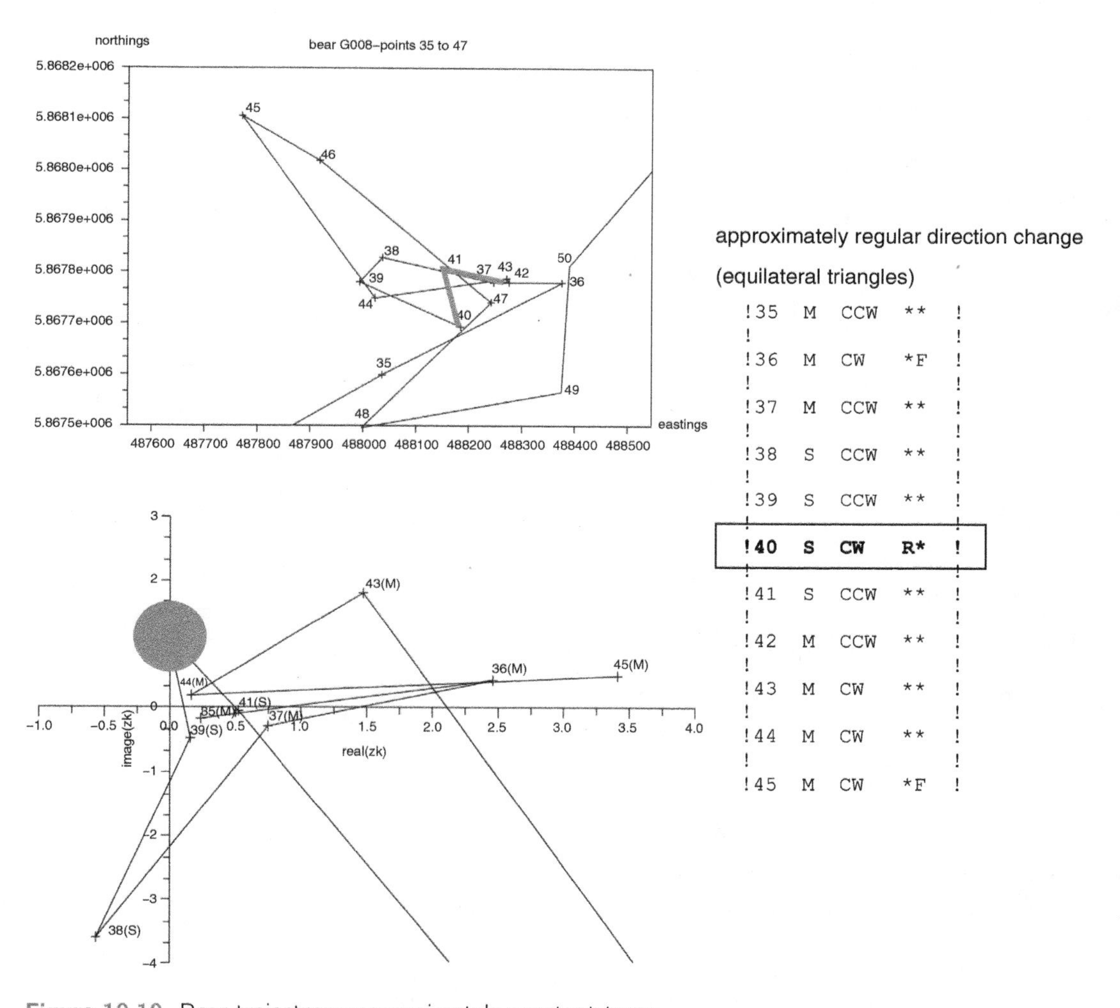

Figure 10.10 Bear trajectory—approximately constant turns.

In two dimensions, in the second example we have

$$N_2 = N_1 \cdot \left(\frac{L_1}{L_2}\right)^D$$

$$9 = 1 \cdot \left(\frac{3}{1}\right)^2$$

$$= 9 \quad .$$

Here again, with either of two ruler lengths the total area measurement remains constant at nine square units.

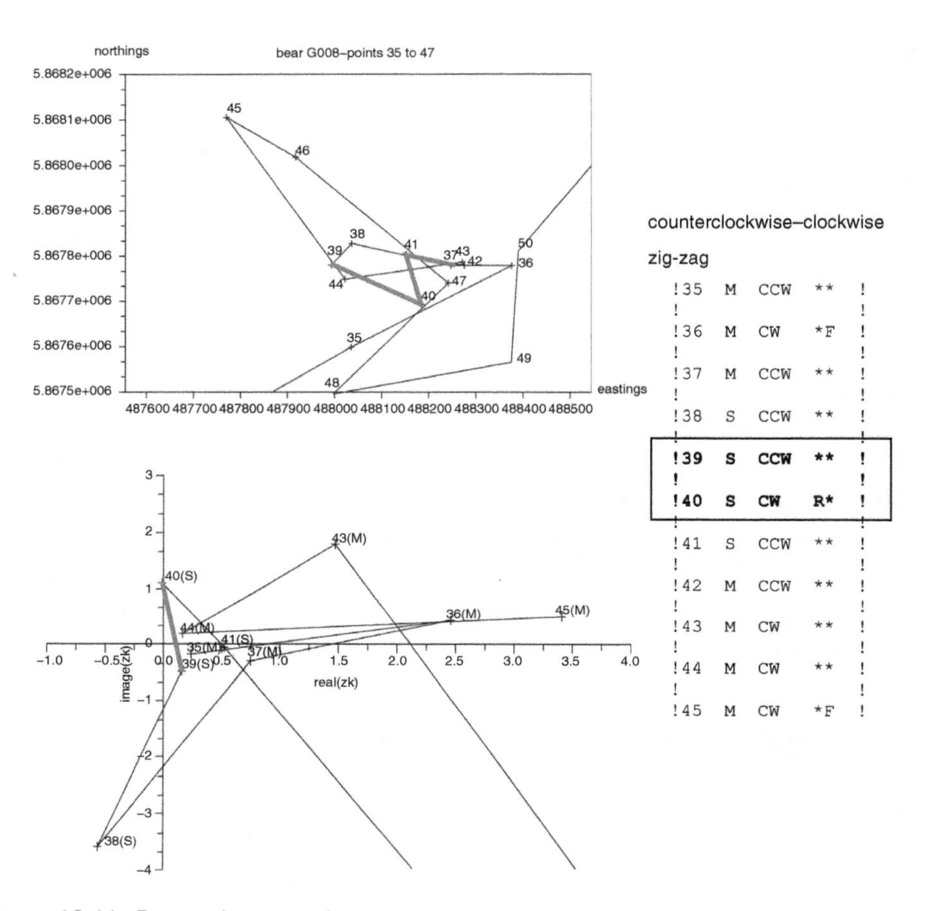

Figure 10.11 Bear trajectory—zig-zag movements.

So anecdotally, at least, this relationship appears to hold. Now notice that we may rearrange the terms in the above expression as follows:

$$\frac{N_2}{N_1} = \left(\frac{L_1}{L_2}\right)^D$$

$$\log\left(\frac{N_2}{N_1}\right) = \log\left(\left(\frac{L_1}{L_2}\right)^D\right)$$

$$\log\left(\frac{N_2}{N_1}\right) = D \cdot \log\left(\frac{L_1}{L_2}\right)$$

$$D = \frac{\log\left(\frac{N_2}{N_1}\right)}{\log\left(\frac{L_1}{L_2}\right)}.$$

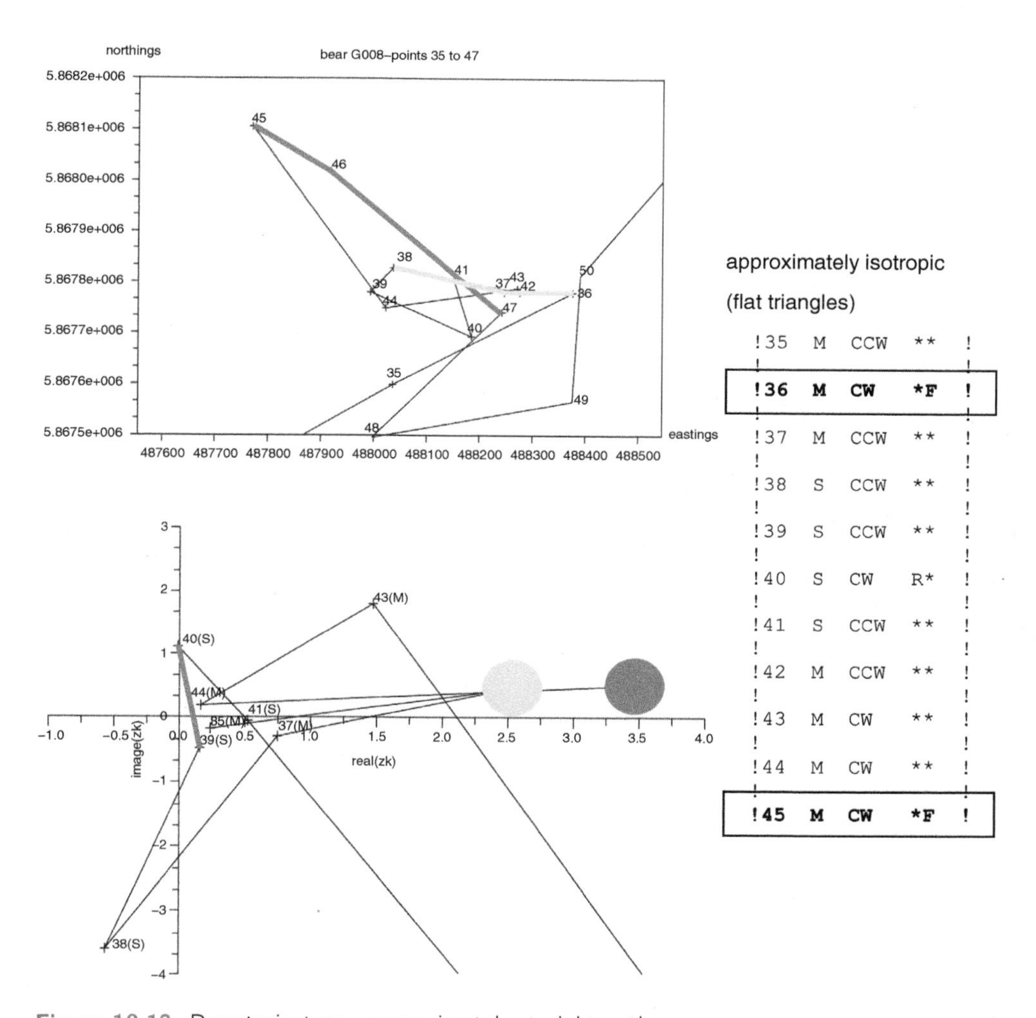

Figure 10.12 Bear trajectory—approximately straight paths.

Thus the dimension D can be expressed as a fraction: the logarithm of the ratio of counts of objects over the logarithm of the ratio of the length of rulers used to count those objects. This suggests that we can define a fractional dimension rather than our traditional notion of integer increments for measuring the dimension of objects.

We now use an example to illustrate how natural geographic features can be characterized by their fractional dimension (referred to as fractal dimension) to try to answer the classic conundrum of how long is a coastline? Figure 10.15 is a Landsat image of a portion of coastline around Victoria, BC. Plotted over the satellite data are two measurements of coastline length, first with a 5 km ruler, next with a 2.5 km ruler. The fractal nature of the coastline is apparent: the length increases as the ruler size decreases. The total length is 15 km with the 5 km and 20 km with the 2.5 km ruler.

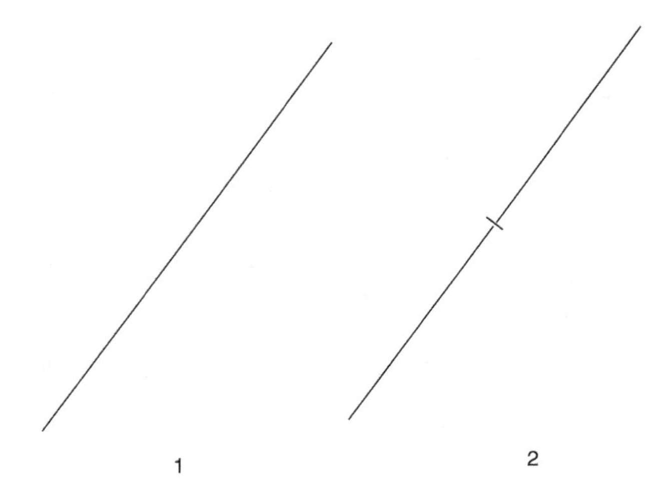

Figure 10.13 Non-fractal shape, one dimension.

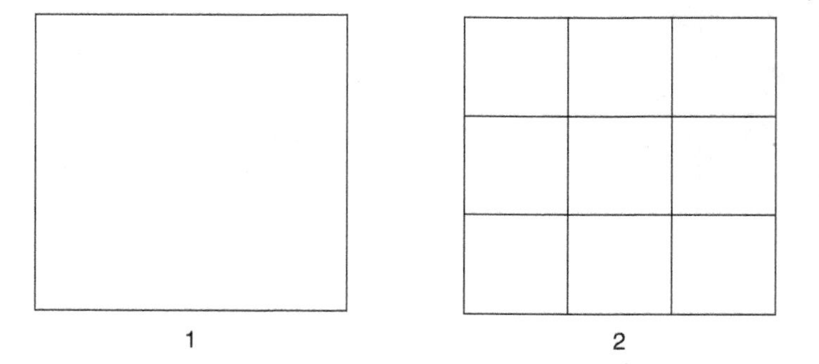

Figure 10.14 Non-fractal shape, two dimensions.

Using the formula given above for estimating the fractal dimension, we get the following result:

$$D = \frac{\log\left(\dfrac{N_2}{N_1}\right)}{\log\left(\dfrac{L_1}{L_2}\right)} = \frac{\log\left(\dfrac{8}{3}\right)}{\log\left(\dfrac{5}{2.5}\right)} \doteq 1.415 \quad .$$

Thus the coastline has an approximate fractal dimension of 1.415. We can get a better estimate by including more ruler lengths and plotting a best-fit line, the slope of which will be an estimate of the fractal dimension. Note that as the rulers get smaller we will eventually hit physical limits to measurement, but long before then we recognize that the shape of any shoreline is constantly changing due to the movement of the land–water interface, so as in any measurement, there is inherent error.

Landsat

Figure 10.15 Measuring the fractal dimension of the coastline around Victoria, BC.

10.3.3 Connectivity Measures for a Connected Planar Graph

We now give two measures, gamma and alpha, of the connectivity of a connected planar graph (please see Appendix II if you are unfamiliar with these ideas from mathematical graph theory). This gives some indication of the number of redundant pathways though a network, or the density of connections between nodes in a network.

Gamma is the ratio of the actual number of edges in a connected planar graph over the maximum number of edges possible for the given number of vertices. That is, how many edges can you add to the graph while respecting the planarity constraint: that is, edges must intersect only at vertices. The equation for calculating gamma in terms of the number of vertices and edges in a connected planar graph is given here:

$$\gamma = \frac{e}{e_{\max}} = \frac{e}{3 \cdot (v-2)} = \frac{e}{3 \cdot v - 6} \quad ,$$

where

$$e = |E|$$

$$v = |V|$$

$$e_{\max} = \max_{\text{for } |V|} |E| \quad .$$

Note that in this case we use the $|\cdot|$ notation to refer to the number of elements in a set. This need not be confusing, given that this usage and its use as absolute value both indicate magnitude—in the latter case the size of a number, in the former the size of a set.

Alpha is the ratio of the number of circuits in a connected planar graph over the maximum number of circuits possible for the given number of vertices. Note that the planar graph with the minimum number of edges for a given number of vertices v, while respecting the connectivity constraint, is the minimum spanning tree, and it has $v-1$ edges. Every edge that is added beyond the minimum spanning tree, while respecting planarity, creates a circuit. The equation for calculating alpha in terms of the number of vertices and edges in a connected planar graph is given here:

$$\alpha = \frac{e - e_{\text{min}}}{e_{\text{max}} - e_{\text{min}}} = \frac{e - (v-1)}{3 \cdot (v-2) - (v-1)} = \frac{e - v + 1}{2 \cdot v - 5} \quad ,$$

where

$$\begin{aligned} e &= |E| \\ v &= |V| \\ e_{\text{min}} &= \min_{\text{for } |V|} |E| \\ e_{\text{max}} &= \max_{\text{for } |V|} |E| \quad . \end{aligned}$$

While the value $e_{\text{min}} = v - 1$ was explained above as a property of the minimum spanning tree of a given number of vertices v, the value $e_{\text{max}} = 3 \cdot (v-2)$ is now justified.

Proof of $e_{\text{max}} = 3 \cdot (v-2)$

For a connected planar graph with $e > 1$ and no multiple edges,

$$e \leq 3 \cdot v - 6 = 3 \cdot (v-2) \quad ,$$

where

$$\begin{aligned} e &= |E| \\ v &= |V| \quad . \end{aligned}$$

Proof

Let the number of edges around each face, f_i, be called the degree of face i, $deg(f_i)$. For a connected planar graph,

$$\sum_{i=1}^{n} deg(f_i) = 2 \cdot e \quad ,$$

since each edge borders two faces, and we consider the surrounding space (in GIS terminology the universe or external polygon) to count as one face. Also note that for an edge with one vertex having only one incident edge, we count each side of the edge in calculating the degree of the face containing this edge. Further, because no multiple edges are allowed,

$$deg(f_i) \geq 3 \quad .$$

Thus over all faces,

$$\sum_{i=1}^{n} deg(f_i) \geq 3 \cdot f \quad ,$$

where

$$f = |F| \quad \text{(the number of faces)} \quad .$$

Combining the previous two results gives

$$2 \cdot e \geq 3 \cdot f \quad \text{or} \quad 3 \cdot f \leq 2 \cdot e \quad .$$

Recalling Euler's equation (from Appendix II),

$$v - e + f = 2 \quad .$$

Rearranging terms, this becomes

$$f = 2 - v + e \quad .$$

Multiplying by 3 gives

$$3 \cdot f = 6 - 3 \cdot v + 3 \cdot e \quad .$$

From above,

$$3 \cdot f \leq 2 \cdot e \quad .$$

Combining the two results gives

$$6 - 3 \cdot v + 3 \cdot e \leq 2 \cdot e \quad .$$

Subtracting $2 \cdot e$ from both sides of the inequality gives

$$6 - 3 \cdot v + e \leq 0 \quad .$$

Subtracting 6 from and adding $3 \cdot v$ to both sides results in

$$e \leq 3 \cdot v - 6 \quad .$$

Factoring out 3 gives, finally,

$$e \leq 3 \cdot (v - 2) \quad .$$

Problems

1. Calculate the distance from Yellowknife, NWT ($62.29°$N, $114.38°$W), to Toronto, ON ($43.40°$N, $79.23°$W), using the Haversine formula.
2. Calculate the area of your home town or home city using the formula given in Section 9.2.6 (assuming a projected coordinate system). Compare your result with a published value. How might you make your calculation more accurate? How accurate is good enough?
3. Calculate an APR and an area value for a contiguous set of about 50 property parcels from downtown Nanaimo. How might this information help planners choose a site for a "contemplative" public garden for the city?
4. What representation issues exist in modelling a single-line street network for calculating the network connectivity measures γ and α? (Hint: think about the assumptions in the derivation of $e_{\max}$).
5. Many spatial analysis methods are sensitive to edge effects; that is, there is a bias due to "missing information" beyond the edge of the study area. How can we define a neighbourhood structure at the edge of the study area to address this issue?

Chapter 11

Geographic Analysis

11.1 Introduction

In this chapter we examine some of the tools of spatial reasoning. First we review some basic ideas that make up the algebra of spatial analysis, Boolean algebra and map algebra. Next we look more broadly at analyzing geographic patterns. We discuss recognizing and classifying geographic patterns, generating new patterns from existing geographic patterns and various interpolation techniques including IDW, membranes and splines, and Kriging. Finally we introduce the concepts of extracting optimal information from geographic patterns.

11.2 The Algebra of Analysis

In this section we introduce Boolean algebra, which most students are familiar with from high school, and map algebra, which is usually equated with raster GIS overlay operations.

11.2.1 Boolean Algebra

Boolean algebra is, broadly speaking, the way we can mathematize logical operations. Often we will use a binary coding system (e.g., $0 =$ FALSE and $1 =$ TRUE). For instance, we may want a query such as

$$\text{LANDUSE} = \text{“Residential” AND ELEVATION} \leq 50 \quad .$$

Boolean algebra gives us the tools to complete this type of query. A good way to look at these operations is to consider truth tables. Below are truth tables for three commonly used logical operators and a fourth useful operator that can be created from combinations of the first three. As you may encounter a few variations, depending on the context, we also provide some alternative notation.

AND

A	B	$A \cdot B$ $A \cap B$
0	0	0
0	1	0
1	0	0
1	1	1

OR

A	B	$A + B$ $A \cup B$
0	0	0
0	1	1
1	0	1
1	1	1

NOT

A	A' $-A$ $\neg A$
0	1
1	0

XOR (exclusive **OR**)

A	B	$A \oplus B$
0	0	0
0	1	1
1	0	1
1	1	0

exclusive **OR** in terms of **AND**, **OR**, and **NOT**

A	B	A'	B'	$A' \cdot B$	$A \cdot B'$	$A' \cdot B + A \cdot B'$
0	0	1	1	0	0	0
0	1	1	0	1	0	1
1	0	0	1	0	1	1
1	1	0	0	0	0	0

11.2.2 Map Algebra

We can group map algebra operations into four classes: local operations, focal operations, zonal operations, and global operations. Operators may be unary (having one argument), binary (having two arguments), or higher. The following subsections describe and give examples of each class of operations.

Local operations

In local operations new cell values depend only on the original value in each particular cell. An example of a unary local operation is thresholding. This classifies a dataset into a raster output based on cell values meeting a threshold value. This is an example of a thresholding operation:

original

1	2	3
2	4	1
2	2	2

$\xrightarrow{\geq 3}$

new

0	0	1
0	1	0
0	0	0

Focal operations

In focal operations the new cell value depends on the values in the neighbourhood of the original cell. As discussed elsewhere in the text, the notion of neighbourhood can be expressed in different ways, depending on context and circumstances. Here is an example of the focal sum operation using the queen's neighbourhood definition:

original

1	2	3
2	4	1
2	2	2

$\xrightarrow{\text{focal sum}}$

new

9	13	10
13	19	14
10	13	9

Note that there are edge-effects issues to consider here, as only the middle cell has a full queen's neighbourhood of values included in the focal sum.

Zonal operations
In zonal operations new cell values are a function of the cell values in the original raster and the values of other cells that are in the same zone. The zones are specified in a second raster. Here is an example of a zonal sum. Note that, as there are two inputs, this is a binary operation:

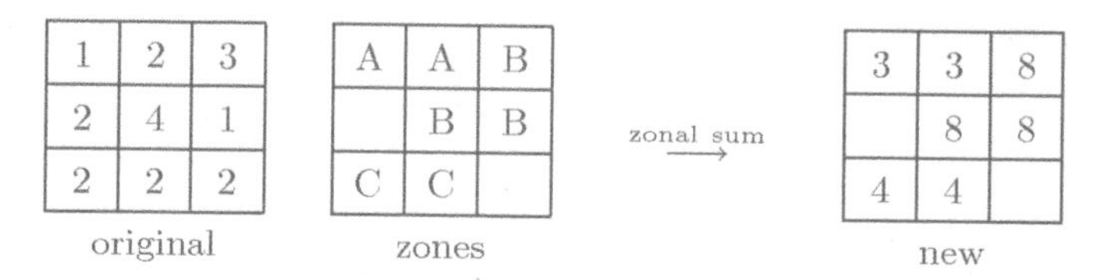

Note that in this example we have missing information in the new raster because the zone designations were not exhaustive in the zones input raster.

Global operations
In global operations new cell values are a function of location and/or the values of all the cells in the original raster. Here is an example of city block distance to nearest source cell (here we have two potential sources):

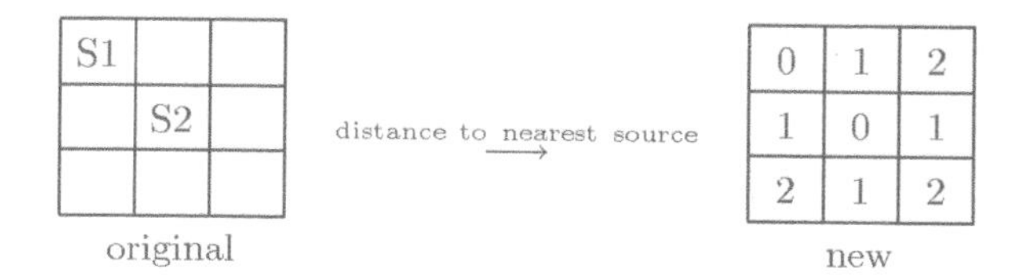

11.3 Analyzing Geographic Patterns

In the following three sections we cover some advanced material on recognizing and classifying geographic patterns, generating new geographic patterns from existing patterns, and extracting optimal information from geographic patterns.

11.3.1 Recognizing and Classifying Geographic Patterns

In this section we discuss some methodologies for extracting geographic patterns from one or more input datasets. In the main we present the techniques for raster data, but in many cases these tools, with appropriate adaptions, can be used with vector data.

Image segmentation—Otsu
We are often interested in separating a dataset into two or more classes. We first present a technique for separating input data into two classes. Figure 11.1 gives a

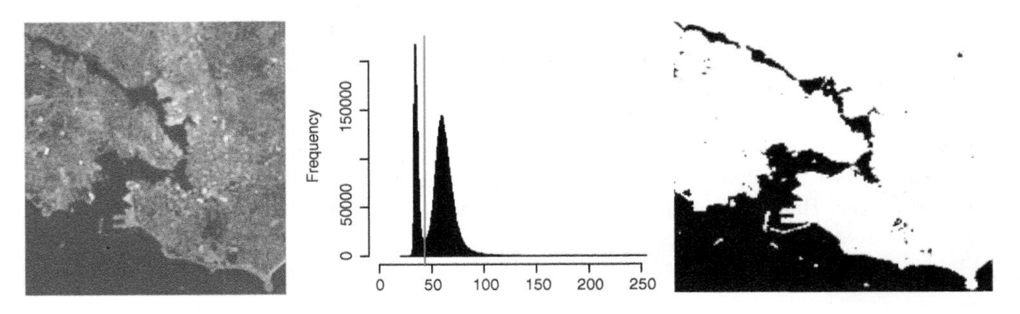

Figure 11.1 A simple example of Otsu image segmentation.

sketch of the idea. Often, class differences are reflected in the distribution of values, and the relative frequencies can suggest a decision boundary, as in this case for Victoria Landsat data segmented into water and not water.

Note that here we assume a discrete set of values in the derivation below, but continuous data may be binned into discrete values to apply this method.

Let cells (pixels) in a given raster dataset (or image) be represented in L discrete values:

$$i \in \{0, 1, 2, \ldots, L-1\} \quad .$$

Let n_i be the number of cells of value i. Then define the probability of a cell being value i:

$$p_i = \frac{n_i}{N} \quad ,$$

where

$$N = \sum_{i=0}^{L-1} n_i \quad .$$

Note that

$$p_i \geq 0 \quad \text{and} \quad \sum_{i=0}^{L-1} p_i = 1 \quad .$$

Suppose we split the cells into two classes, C_0 & C_1 (e.g., background and foreground), by a threshold value S; i.e.,

$$C_0 = \{x, y \mid f(x, y) \in \{0, 1, 2, \ldots, S\}\}$$

$$C_1 = \{x, y \mid f(x, y) \in \{S+1, S+2, \ldots, L-1\}\} \quad .$$

Then,

$$p(C_0) = \sum_{i=0}^{S} p_i$$

and

$$p(C_1) = \sum_{i=S+1}^{L-1} p_i = 1 - p(C_0) \quad .$$

The expected value (mean) of C_0 is

$$\mu_0 = \sum_{i=0}^{S} i \cdot p(i|C_0) \qquad \left(\text{recall, } \mathrm{E}[x] = \sum x \cdot p(x)\right)$$

$$= \frac{\sum_{i=0}^{S} i \cdot p_i}{\sum_{i=0}^{S} p_i}$$

$$\left(\text{by Bayes we have } p(B|A) = p(A|B) \cdot \frac{p(B)}{p(A)} \quad ,\right.$$

$$\text{so in our case this gives } p(i|C_0) = p(C_0|i) \cdot \frac{p(i)}{p(C_0)} \quad ,$$

$$\text{and } \; p(C_0|i) = 1 \qquad (\text{by definition since } 0 \le i \le S) \quad ,$$

$$\text{and } \; p(C_0) = \sum_{i=0}^{S} p_i \quad ,$$

$$\left.\text{and } \; p(i) \equiv p_i\right)$$

$$= \frac{\mu(S)}{\omega(S)} \quad ,$$

where

$$\mu(S) = \sum_{i=0}^{S} i \cdot p_i \qquad (\text{first order cumulative moment up to the } S\text{th level})$$

and

$$\omega(S) = \sum_{i=0}^{S} p_i \qquad (\text{zeroth order cumulative moment up to the } S\text{th level}) \quad .$$

Similarly, the expected value (mean) of C_1 is

$$\begin{aligned}
\mu_1 &= \sum_{i=S+1}^{L-1} i \cdot p(i|C_1) \\
&= \frac{\sum_{i=S+1}^{L-1} i \cdot p_i}{\sum_{i=S+1}^{L-1} p_i} \\
&= \frac{\left(\sum_{i=0}^{L-1} i \cdot p_i - \sum_{i=0}^{S} i \cdot p_i\right)}{\left(1 - \sum_{i=1}^{S} p_i\right)} \\
&= \frac{(\mu_T - \mu(S))}{(1 - \omega(S))}
\end{aligned}$$

where

$$\mu_T = \sum_{i=0}^{L-1} i \cdot p_i$$

Note that $\mu_T = \omega(S) \cdot \mu_0 + (1 - \omega(S)) \cdot \mu_1$, since

$$\begin{aligned}
\mu_T &= \sum_{i=0}^{L-1} i \cdot p_i \\
&= \sum_{i=0}^{S} i \cdot p_i + \sum_{i=S+1}^{L-1} i \cdot p_i \\
&= \frac{\left(\sum_{i=0}^{S} p_i\right) \cdot \left(\sum_{i=0}^{S} i \cdot p_i\right)}{\left(\sum_{i=0}^{S} p_i\right)} + \frac{\left(\sum_{i=S+1}^{L-1} p_i\right) \cdot \left(\sum_{i=S+1}^{L-1} i \cdot p_i\right)}{\left(\sum_{i=S+1}^{L-1} p_i\right)} \\
&= \left(\sum_{i=0}^{S} p_i\right) \cdot \mu_0 + \left(\sum_{i=S+1}^{L-1} p_i\right) \cdot \mu_1 \\
&= \omega(S) \cdot \mu_0 + (1 - \omega(S)) \cdot \mu_1 \quad .
\end{aligned}$$

To determine the threshold value S, we introduce the following:

$$\eta = \frac{\sigma_{\mathrm{B}}^2}{\sigma_{\mathrm{T}}^2} \qquad \left(\frac{\text{between class variance}}{\text{total variance}}\right),$$

where

$$\begin{aligned}
\sigma_{\mathrm{B}}^2 &= (\mu_0 - \mu_{\mathrm{T}})^2 \cdot \omega(S) + (\mu_1 - \mu_{\mathrm{T}})^2 \cdot (1 - \omega(S)) \qquad \left(\mathrm{var}[x] = \sum (x-\mu)^2 \cdot p(x)\right) \\
&= \omega(S) \cdot (\mu_0 - \mu_{\mathrm{T}})^2 + (1 - \omega(S)) \cdot (\mu_1 - \mu_{\mathrm{T}})^2 \\
&= \omega(S) \cdot (1 - \omega(S)) \cdot (\mu_0 - \mu_1)^2 \quad . \ (*)
\end{aligned}$$

(*) Recall,

$$\begin{aligned}
\mu_{\mathrm{T}} &= \omega(S) \cdot \mu_0 + (1 - \omega(S)) \cdot \mu_1 \\
\Rightarrow \quad \mu_0 - \mu_{\mathrm{T}} &= \mu_0 - \omega(S) \cdot \mu_0 - (1 - \omega(S)) \cdot \mu_1 \\
&= (1 - \omega(S)) \cdot \mu_0 - (1 - \omega(S)) \cdot \mu_1 \\
&= (1 - \omega(S)) \cdot (\mu_0 - \mu_1) \\
\Rightarrow \quad (\mu_0 - \mu_{\mathrm{T}})^2 &= (1 - \omega(S))^2 \cdot (\mu_0 - \mu_1)^2 \quad .
\end{aligned}$$

Similarly,

$$\begin{aligned}
\mu_1 - \mu_{\mathrm{T}} &= \mu_1 - \omega(S) \cdot \mu_0 - (1 - \omega(S)) \cdot \mu_1 \\
&= \mu_1 - \omega(S) \cdot \mu_0 - \mu_1 + \omega(S) \cdot \mu_1 \\
&= \omega(S) \cdot (\mu_1 - \mu_0) \\
\Rightarrow \quad (\mu_1 - \mu_{\mathrm{T}})^2 &= (\omega(S))^2 \cdot (\mu_1 - \mu_0)^2 \\
&= (\omega(S))^2 \cdot (\mu_0 - \mu_1)^2 \quad .
\end{aligned}$$

Combining these two results with the expression for σ_{B}^2 above, we get

$$\begin{aligned}
\sigma_{\mathrm{B}}^2 &= \omega(S) \cdot (\mu_0 - \mu_{\mathrm{T}})^2 + (1 - \omega(S)) \cdot (\mu_1 - \mu_{\mathrm{T}})^2 \\
&= \omega(S) \cdot (1 - \omega(S))^2 \cdot (\mu_0 - \mu_1)^2 + (1 - \omega(S)) \cdot (\omega(S))^2 \cdot (\mu_0 - \mu_1)^2 \\
&= \left[\omega(S) \cdot (1 - \omega(S))^2 + (1 - \omega(S)) \cdot (\omega(S))^2\right] \cdot (\mu_0 - \mu_1)^2 \\
&= \left[\omega(S) \cdot (1 - 2 \cdot \omega(S) + \omega(S)^2) + \omega(S)^2 - \omega(S)^3\right] \cdot (\mu_0 - \mu_1)^2 \\
&= \left[\omega(S) - 2 \cdot \omega(S)^2 + \omega(S)^3 + \omega(S)^2 - \omega(S)^3\right] \cdot (\mu_0 - \mu_1)^2 \\
&= \left[\omega(S) - \omega(S)^2\right] \cdot (\mu_0 - \mu_1)^2 \\
&= \omega(S) \cdot (1 - \omega(S)) \cdot (\mu_0 - \mu_1)^2 \quad .
\end{aligned}$$

Further, we have

$$\sigma_{\mathrm{T}}^{2} = \sum_{i=0}^{L-1} (i - \mu_{\mathrm{T}})^{2} \cdot p_i \qquad \text{(again by definition of var}[x]\text{)} \quad ,$$

but since σ_{T}^{2} is constant with respect to S,

$$\max(\eta) = \max(\sigma_{\mathrm{B}}^{2}) \quad ,$$

so the optimal threshold S^{*} can be determined by

$$\sigma_{\mathrm{B}}^{2}(S^{*}) = \max_{0 \leq S \leq L-1} (\sigma_{\mathrm{B}}^{2}(S)) \quad ,$$

which may be solved iteratively through a systematic evaluation of $\sigma_{\mathrm{B}}^{2}(S)$ for the sequence of S values.

Maximum likelihood estimation (MLE)

Maximum likelihood estimation is a general approach to estimate the parameter(s) of a chosen family of probability density functions (pdfs) with a given dataset. The basic idea is to choose the parameter value that would maximize the chance of observing the given data. This is done by optimizing the joint probabilities, assuming the data as given. Let $f(x, \theta)$ be some pdf with parameter θ and given data $\{x_1, x_2, \ldots, x_n\}$, then define the likelihood function as follows:

$$L(\theta \mid x_1, x_2, \ldots, x_n) = f(x_1, \theta) \cdot f(x_2, \theta) \cdot \quad \cdots \quad \cdot f(x_n, \theta) \quad .$$

We then solve for θ in

$$\frac{d\ L(\theta \mid x_1, x_2, \ldots, x_n)}{d\theta} = 0 \quad ,$$

or it is often easier (and equivalent) to solve for θ in

$$\frac{d\ \ln L(\theta \mid x_1, x_2, \ldots, x_n)}{d\theta} = 0 \quad .$$

The introduction of the natural logarithm allows for multiplications to be replaced with sums and for simplification of exponential functions in, for example, the exponential and normal pdfs. This serves to simplify the algebra in solving for θ.

Gaussian maximum likelihood classification

Gaussian maximum likelihood classification uses information about the sample mean, variance, and covariance of a set of exemplar classifications to classify the remaining raster cells in a dataset. We assume the input data are Gaussian (normally) distributed. This is the most common supervised classification method used with RS image data. This technique is a special case of Bayes's classification, as was Otsu's method, described above. Note that we can use more than one input raster for the classification, so we represent the raster cell values as vectors.

Let the classification categories for a dataset be represented by

$$\omega_i \ , \ i = 1, 2, \dots , m \quad ,$$

where m is the total number of categories. We consider the conditional probabilities of determining a category for a raster cell with values $\underline{x}$:

$$p(\omega_i \,|\, \underline{x}) \ , \ i = 1, 2, \dots , m \quad .$$

The above probability gives the likelihood that the correct category is ω_i for a raster cell with values $\underline{x}$.

Unfortunately, $p(\omega_i \,|\, \underline{x})$ are unknown. But with training data for each category, ω_i, we can estimate a pdf for ω_i for the given values $\underline{x}$ (i.e., $p(\underline{x} \,|\, \omega_i)$). This set of functions gives the relative likelihood that a given raster cell lies in each particular category.

We can get the desired $p(\omega_i \,|\, \underline{x})$ via Bayes's theorem,

$$p(\omega_i \,|\, \underline{x}) = \frac{p(\underline{x} \,|\, \omega_i) \cdot p(\omega_i)}{p(\underline{x})} \ , \qquad i = 1, 2, \dots , m \quad ,$$

where $p(\omega_i)$ is the probability of category ω_i occurring in the dataset. These $p(\omega_i)$ may be estimated from prior knowledge, but they can (and often are) assumed to be equally probable.

The classification rule is given by

$$\underline{x} \in \omega_i \quad \text{if} \quad p(\omega_i \,|\, \underline{x}) > p(\omega_j \,|\, \underline{x}) \ , \qquad \text{for all } j \neq i \quad . \tag{11.1}$$

In other words, $\underline{x}$ belongs to category ω_i if $p(\omega_i \,|\, \underline{x})$ is the largest value for all i. Using Bayes's result given above in (11.1) and dividing out $p(\underline{x})$ as a common factor, we get

$$\underline{x} \in \omega_i \quad \text{if} \quad p(\underline{x} \,|\, \omega_i) \cdot p(\omega_i) > p(\underline{x} \,|\, \omega_j) \cdot p(\omega_j) \ , \qquad \text{for all } j \neq i \quad . \tag{11.2}$$

We next let

$$g_i(\underline{x}) = \ln(p(\underline{x} \,|\, \omega_i) \cdot p(\omega_i)) \tag{11.3}$$

$$= \ln(p(\underline{x} \,|\, \omega_i)) + \ln(p(\omega_i)) \quad . \tag{11.4}$$

Thus, (11.2) becomes

$$\underline{x} \in \omega_i \quad \text{if} \quad g_i(\underline{x}) > g_j(\underline{x}) \,, \qquad \text{for all } j \neq i \quad . \tag{11.5}$$

If we have no information about $p(\omega_i)$ we can drop it from (11.4), as it will be the same for all categories if we assume equal prior probabilities.

Until now we have left unused our initial assumption of Gaussian distributed input data. We assume a multivariate normal distribution, which has an explicitly known form and properties. For m categories we have

$$p(\underline{x} \,|\, \omega_i) = (2 \cdot \pi)^{-\frac{m}{2}} \cdot \left| \sum_i \right|^{-\frac{1}{2}} \cdot \exp\left\{ -\frac{1}{2} \cdot (\underline{x} - \underline{m_i})^T \cdot \sum_i^{-1} \cdot (\underline{x} - \underline{m_i}) \right\} ,$$

where $\underline{m_i}$ is the mean vector for category i

$\sum_i$ is the covariance matrix for category i

$\sum_i^{-1}$ is the inverse covariance matrix for category i.

Note that $(2 \cdot \pi)^{-\frac{m}{2}}$ is common to all $p(\underline{x} \,|\, \omega_i)$, so is ignored for the comparison in the decision rule. Substituting the multi-variate normal pdf in (11.4) gives

$$g_i(\underline{x}) = \ln(p(\omega_i)) - \frac{1}{2} \cdot \ln \left| \sum_i \right| - \frac{1}{2} \cdot (\underline{x} - \underline{m_i})^T \cdot \sum_i^{-1} \cdot (\underline{x} - \underline{m_i}) \quad .$$

If $p(\omega_i)$ are *a priori* equally probable and we remove the common $\frac{1}{2}$ factor, we get

$$g_i(\underline{x}) = -\ln \left| \sum_i \right| - (\underline{x} - \underline{m_i})^T \cdot \sum_i^{-1} \cdot (\underline{x} - \underline{m_i}) \quad .$$

Thus $g_i(\underline{x})$ can be calculated using the sample statistics $\underline{m_i}$ and $\sum_i$ of the training datasets for each class ω_i.

K-means classification

Another way of classifying a dataset to identify patterns in geographic data is the K-means algorithm. This algorithm is straightforward to state and implement.

K-means algorithm:

1. Decide on how many classes to be considered (this is the K).
2. Choose a mean vector $\underline{m_i}$ for each class ω_i .

3. Assign each raster cell in the dataset to the class (actually class mean) that it is closest to:

$$\underline{x} \in \omega_i \quad \text{if} \quad d(\underline{x}, \underline{m}_i) < d(\underline{x}, \underline{m}_j) , \qquad \text{for all } j \neq i \quad .$$

4. Recalculate the mean vectors $\underline{m}_i$ based on the new clusters.
5. If any $\underline{m}_i$ changes, repeat steps 2 to 4; otherwise stop.

In general, K-means is sensitive to the starting point (initial mean vectors), the assumed number of clusters, the order in which the raster cells are considered, and the geometric properties of the classes. K-means works well with reasonably compact well-separated clusters.

11.3.2 Generating New Patterns from Existing Geographic Patterns

In this section we cover operations that manipulate an existing geographic pattern to create a new geographic pattern. This is often done because the new pattern offers different or clearer insights into the underlying geographic phenomena represented by the pattern. Table 11.1, adapted from O'Sullivan and Unwin [89], summarizes some of these operations. Those operations marked with an asterisk are covered elsewhere in this textbook. Those operations marked with an open circle are covered in this section. We now look at two families of operations that generate new geographic patterns from existing geographic patterns: mathematical morphology and interpolation.

Mathematical morphology

There is a well-developed theory and practice in image processing referred to as mathematical morphology. We can use some of these ideas to formalize and extend

Table 11.1 Geographic transformations.

		To			
		(*Point, dim-0*)	(*Line, dim-1*)	(*Area, dim-2*)	(*Field, dim-3*)
From	(*Point, dim-0*)	Mean centre*	Network graphs*, MST*	Voronoi polygons*, TIN, Point buffer, SDE*	Interpolation°, KDE*, Distance surfaces
	(*Line, dim-1*)	Intersection*	Shortest distance path*	Line buffer	Distance to nearest line object surface
	(*Area, dim-2*)	Centroid*	Area skeleton	Area buffer°, Polygon overlay*	Pycnophylatic interpolation
	(*Field, dim-3*)	Surface specific points	Edge detection*	Watershed delineation	Raster local operations*

Source: Adapted from O'Sullivan and Unwin [89].

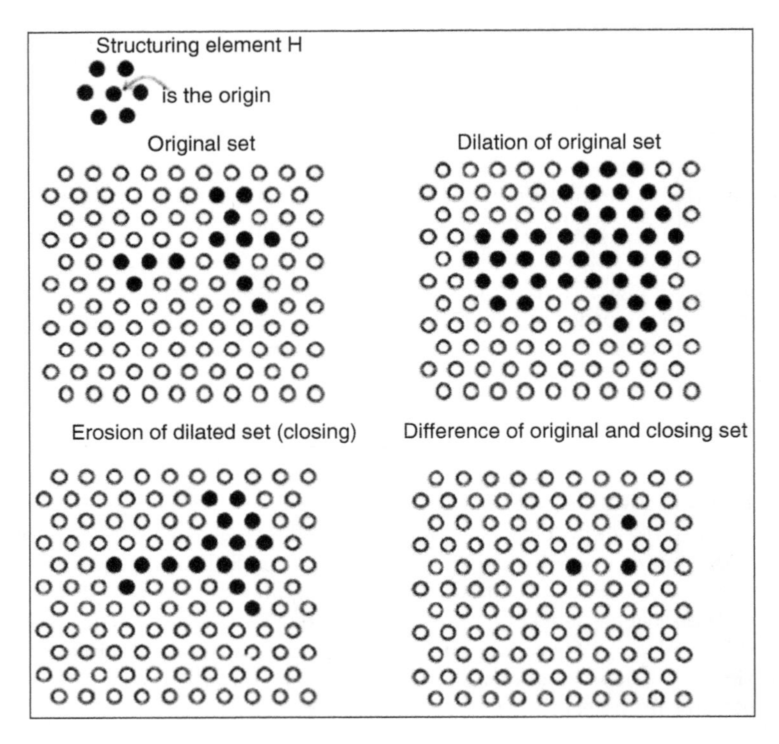

Figure 11.2 Mathematical morphology operators.
Source: From Serra [103], p. 41.

our understanding of some useful pattern-generating operations. Figure 11.2 shows a couple of the basic operations—dilation and erosion—and closing, which is a composite of a dilation followed by an erosion. There is also an opening operation, which is an erosion followed by a dilation. An important thing to note is that a closing operation does not return the original dataset. That is, dilation technically does not admit an inverse. It is this property that will be shown to be of value in the sequel.

We now illustrate the application of these concepts with a couple of examples (see also Roberts et al. [97]). First, Figure 11.3 shows how the mathematical morphology operation of closing can be used to identify potential connectivity between core areas of natural habitat, as well as areas that may be regenerated to create increased natural interior core area. Note that we can usually identify small areas, in absolute terms, that can have a large impact on increasing interior habitat area; for instance, by filling in interstices in existing core areas. In our case we equate the dilation operator with a positive polygon buffer of a certain distance and the erosion operator with a negative polygon buffer of the same distance. A difference operator highlights the new information of potential additional interior habitat and potential areas to

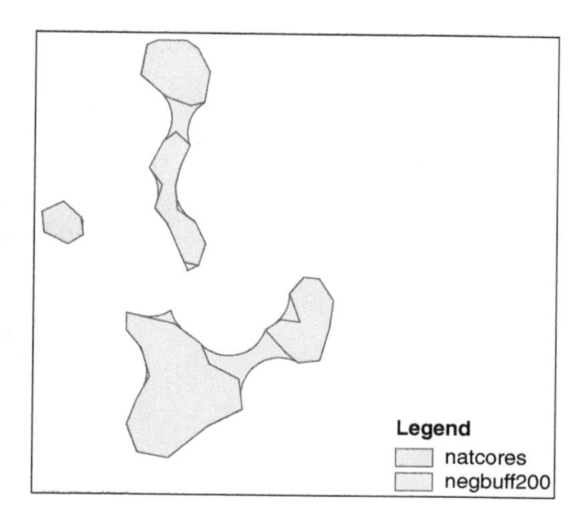

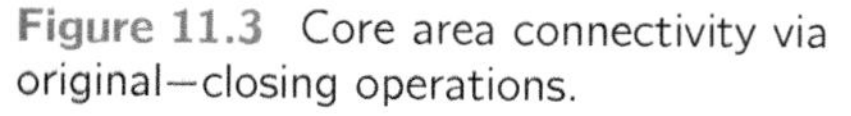
Figure 11.3 Core area connectivity via original—closing operations.

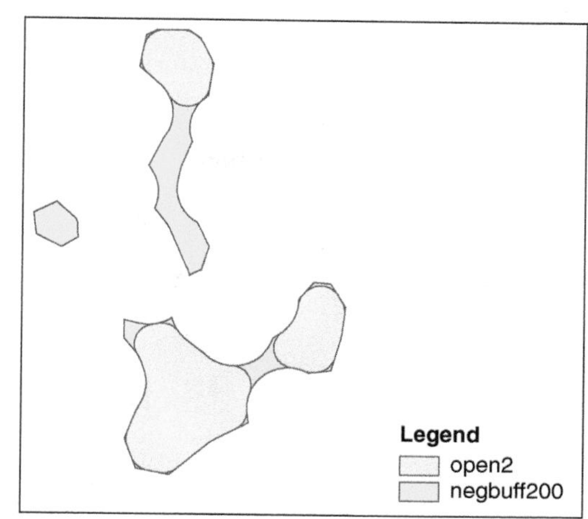

Figure 11.4 Core area compactness via closing and opening operations.

increase connectivity. In summary we found the new patterns by way of the following operations:

$$\begin{aligned}\text{Core area connectivity and new interior} &= \text{closing} - \text{original dataset} \\ &= (\text{dilation} \circ \text{erosion}) - \text{original dataset} \quad ,\end{aligned}$$

where $-$ is set difference and $\circ$ is function composition. In the second example Figure 11.4 shows how the mathematical morphology operation of opening can be used to identify and measure the compactness of potential core areas. In this case the output of the previous example is used as the input for an opening operator. In summary,

$$\begin{aligned}\text{Connected core area compactness} &= \text{opening} \circ \text{closing} \\ &= (\text{dilation} \circ \text{erosion}) \circ (\text{dilation} \circ \text{erosion}) \quad ,\end{aligned}$$

where $\circ$ is function composition. The output polygons of these operations can be measured for area and area perimeter ratio (see Chapter 10) to quantify the extent and compactness of the identified core areas.

Spatial interpolation

In this subsection we describe three selected approaches to spatial interpolation: inverse distance weighting (IDW), stochastic models (membranes and splines), and Kriging. Interpolation allows us to fill in missing data values or generate new values between existing values, using the existing data to help make reasonable choices for these new data values.

Inverse Distance Weighting

As the name implies, in this interpolation technique we estimate an unknown value $\hat{z}_j$ at a known location by incorporating a weighted sum of known values, basing the weights on some function of the inverse distance to these known values. This serves to weight or bias nearer values more than farther values. The methodology is given by the following equations:

$$\hat{z}_j = \sum_{i=1}^{m} w_{i,j} \cdot z_i \quad ,$$

where

$$w_{i,j} \propto \frac{1}{d_{i,j}}$$

$$d_{i,j} \equiv \text{distance from } i\text{th to } j\text{th location} \quad .$$

Often, we have

$$w_{i,j} = \frac{\frac{1}{d_{i,j}}}{\sum_{i=1}^{m} \frac{1}{d_{i,j}}} \quad .$$

We may also have

$$w_{i,j} \propto \frac{1}{d_{i,j}^{k}} \quad .$$

We now show a worked example for IDW using annual rainfall for south-central Saskatchewan in 1980 (see Figure 11.5 and the following table). We assume seven known values and one site with an unknown value; we estimate the latter with IDW and then compare it to the actual value.

Site	Easting	Northing	Rainfall (annual mm, 1980)
CHAPLIN	379437	5595605	222.4
INDIAN HEAD	592195	5595032	347.9
MELFORT	526967	5850097	271.3
MOOSE JAW	457268	5572159	176.3
OUTLOOK	354240	5707520	171.5
REGINA	521321	5583149	229.4
SASKATOON	383819	5784645	219.5

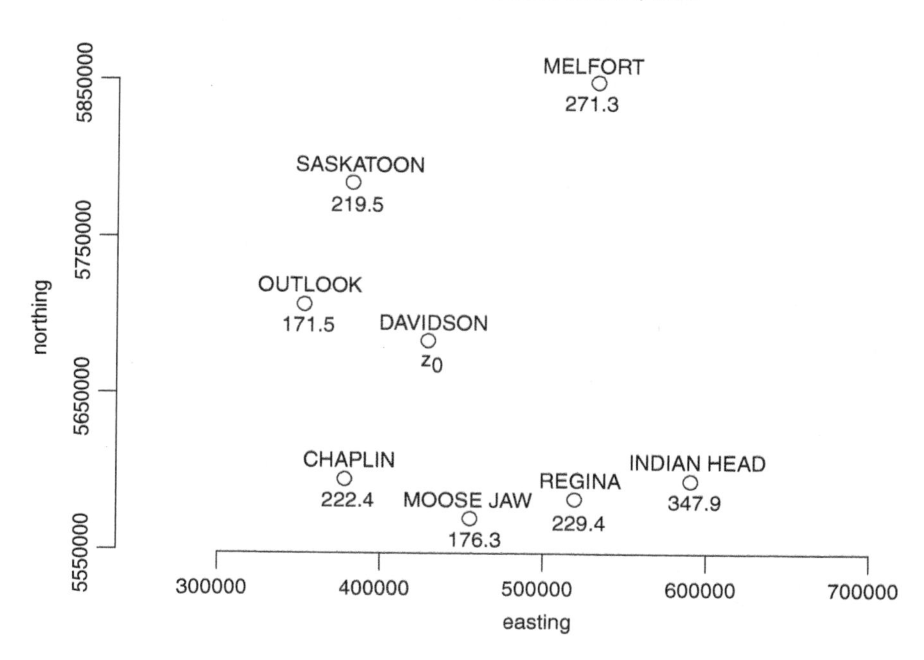

Figure 11.5 Annual rainfall data example.

The site at DAVIDSON, where we wish to estimate a rainfall value, is located at (430284, 5683661). The following table summarizes the IDW calculations.

	Distances to DAVIDSON	Inverse Distances	Normed I.D. (weights)	Weighted values
CHAPLIN	101682.23	9.834560e−06	0.16932322	37.65748
INDIAN HEAD	184581.34	5.417666e−06	0.09327683	32.45101
MELFORT	192479.98	5.195345e−06	0.08944911	24.26754
MOOSE JAW	114720.67	8.716825e−06	0.15007900	26.45893
OUTLOOK	79699.07	1.254720e−05	0.21602715	37.04866
REGINA	135611.20	7.374022e−06	0.12695974	29.12457
SASKATOON	111160.98	8.995962e−06	0.15488495	33.99725
sums		5.808158e−05	1	221.0054

So the interpolated value for annual rainfall in DAVIDSON in 1980 is approximately 221.01 mm. The true value is 199.5 mm.

Stochastic Models

We now describe in a quite general setting some interpolation techniques that rely on an *a priori* model of constraints between adjacent values in an interpolation. Parts of the following discussion are adapted from course notes, which are now contained in Fieguth [38]. We begin with the following model of a measurement:

$$\begin{array}{ccccccc} \underline{y} = & \mathbf{C} \cdot & \underline{x} + & \underline{e} \\ \text{observation} = & \text{physical model} & \times \ \text{true values} & + \ \text{noise} \\ & \text{(deterministic)} & \text{(unknown)} & \text{(error)} \\ (n \times 1) & (n \times m) & (m \times 1) & (n \times 1) \end{array} \quad .$$

Now let $\hat{\underline{x}}$ be the estimate of the true values. Then,

$$\underline{e} = \underline{y} - \mathbf{C} \cdot \hat{\underline{x}} \quad .$$

We want the estimate $\hat{\underline{x}}$ to minimize error (actually squared error),

$$\min \ (\underline{e}^T \cdot \underline{e}) \quad .$$

To find this minimum value, we set

$$\frac{\partial \left(\underline{e}^T \cdot \underline{e} \right)}{\partial \ \hat{\underline{x}}} = 0 \qquad \text{and solve for } \hat{\underline{x}} \quad .$$

Note,

$$\begin{aligned} \underline{e}^T \cdot \underline{e} &= (\underline{y} - \mathbf{C} \cdot \hat{\underline{x}})^T \cdot (\underline{y} - \mathbf{C} \cdot \hat{\underline{x}}) \\ &= \left(\underline{y}^T - (\mathbf{C} \cdot \hat{\underline{x}})^T\right) \cdot (\underline{y} - \mathbf{C} \cdot \hat{\underline{x}}) & (*) \\ &= \left(\underline{y}^T - \hat{\underline{x}}^T \cdot \mathbf{C}^T\right) \cdot (\underline{y} - \mathbf{C} \cdot \hat{\underline{x}}) & (**) \\ &= \underline{y}^T \cdot \underline{y} - \hat{\underline{x}}^T \cdot \mathbf{C}^T \cdot \underline{y} - \underline{y}^T \cdot \mathbf{C} \cdot \hat{\underline{x}} + \hat{\underline{x}}^T \cdot \mathbf{C}^T \cdot \mathbf{C} \cdot \hat{\underline{x}} \\ &= \underline{y}^T \cdot \underline{y} - 2 \cdot \underline{y}^T \cdot \mathbf{C} \cdot \hat{\underline{x}} + \hat{\underline{x}}^T \cdot \mathbf{C}^T \cdot \mathbf{C} \cdot \hat{\underline{x}} & (***) \quad . \end{aligned}$$

$$(*) \qquad (A + B)^T = \left(A^T + B^T\right)$$

$$(**) \qquad (A \cdot B)^T = B^T \cdot A^T$$

$$(***) \qquad -\underline{\hat{x}}^T \cdot \mathbf{C}^T \cdot \underline{y}$$

$$((1 \times m)(m \times n)(n \times 1) = 1 \times 1 \qquad \text{scalar})$$

$$= -\left(\underline{\hat{x}}^T \cdot \mathbf{C}^T \cdot \underline{y}\right)^T \qquad (\text{if } a \text{ scalar, then } a = a^T)$$

$$= -\underline{y}^T \cdot \left(\underline{\hat{x}}^T \cdot \mathbf{C}^T\right)^T \qquad \left((A \cdot B)^T = B^T \cdot A^T\right)$$

$$= -\underline{y}^T \cdot \mathbf{C}^{TT} \cdot \underline{\hat{x}}^{TT} \qquad \left((A \cdot B)^T = B^T \cdot A^T\right)$$

$$= -\underline{y}^T \cdot \mathbf{C} \cdot \underline{\hat{x}} \qquad \left(A^{TT} = A\right)$$

Now back to the main derivation:

$$\frac{\partial\left(\underline{e}^T \cdot \underline{e}\right)}{\partial\,\underline{\hat{x}}} = 0$$

$$\frac{\partial}{\partial\,\underline{\hat{x}}}\left(\underline{y}^T \cdot \underline{y} - 2 \cdot \underline{y}^T \cdot \mathbf{C} \cdot \underline{\hat{x}} + \underline{\hat{x}}^T \cdot \mathbf{C}^T \cdot \mathbf{C} \cdot \underline{\hat{x}}\right) = 0$$

$$0 - 2 \cdot \underline{y}^T \cdot \mathbf{C} + 2 \cdot \underline{\hat{x}}^T \cdot \mathbf{C}^T \cdot \mathbf{C} = 0 \qquad (*)$$

$$\underline{\hat{x}}^T \cdot \mathbf{C}^T \cdot \mathbf{C} = \underline{y}^T \cdot \mathbf{C}$$

$$\left(\underline{\hat{x}}^T \cdot \mathbf{C}^T \cdot \mathbf{C}\right)^T = \left(\underline{y}^T \cdot \mathbf{C}\right)^T$$

$$\left(\mathbf{C}^T \cdot \mathbf{C}\right)^T \cdot \underline{\hat{x}}^{TT} = \mathbf{C}^T \cdot \underline{y}^{TT}$$

$$\mathbf{C}^T \cdot \mathbf{C}^{TT} \cdot \underline{\hat{x}} = \mathbf{C}^T \cdot \underline{y}$$

$$\mathbf{C}^T \cdot \mathbf{C} \cdot \underline{\hat{x}} = \mathbf{C}^T \cdot \underline{y}$$

$$\text{if } \left(\mathbf{C}^T \cdot \mathbf{C}\right)^{-1} \text{ exists}$$

$$\Rightarrow \qquad \underline{\hat{x}} = \left(\mathbf{C}^T \cdot \mathbf{C}\right)^{-1} \cdot \mathbf{C}^T \cdot \underline{y} \quad .$$

This is the solution for the estimator of the true $\underline{x}$ values.

(∗) (differentiation of quadratic forms)

First note that

$$\frac{\partial(x_i)}{\partial x_j} = \begin{cases} 1 & \text{if } i = j \\ 0 & \text{if } i \neq j \end{cases}$$

and

$$\frac{\partial(x_i \cdot x_k)}{\partial x_j} = \begin{cases} 2 \cdot x_j & \text{if } i = k = j \\ x_i & \text{if } k = j \text{ but } i \neq j \\ x_k & \text{if } i = j \text{ but } k \neq j \\ 0 & \text{otherwise} \end{cases} .$$

If we consider

$$\underline{x}^T \cdot \mathbf{A} \cdot \underline{x} = \sum_{i=1}^{n} \sum_{k=1}^{n} a_{i,k} \cdot x_i \cdot x_k \quad ,$$

then

$$\frac{\partial}{\partial x_j} \left(\underline{x}^T \cdot \mathbf{A} \cdot \underline{x} \right) = \frac{\partial}{\partial x_j} \left(\sum_{i=1}^{n} \sum_{k=1}^{n} a_{i,k} \cdot x_i \cdot x_k \right)$$

$$= \frac{\partial}{\partial x_j} \left(a_{j,j} \cdot x_j \cdot x_j + \sum_{i \neq j} a_{i,j} \cdot x_i \cdot x_j + \sum_{k \neq j} a_{j,k} \cdot x_j \cdot x_k + \sum_{i \neq j} \sum_{k \neq j} a_{i,k} \cdot x_i \cdot x_k \right)$$

$$= a_{j,j} \cdot \frac{\partial}{\partial x_j} \left(x_j^2 \right) + \sum_{i \neq j} a_{i,j} \cdot \frac{\partial}{\partial x_j} (x_i \cdot x_j) + \sum_{k \neq j} a_{j,k} \cdot \frac{\partial}{\partial x_j} (x_j \cdot x_k)$$

$$+ \sum_{i \neq j} \sum_{k \neq j} a_{i,k} \cdot \frac{\partial}{\partial x_j} (x_i \cdot x_k)$$

$$= 2 \cdot a_{j,j} \cdot x_j + \sum_{i \neq j} a_{i,j} \cdot x_i + \sum_{k \neq j} a_{j,k} \cdot x_k + 0$$

$$= a_{j,j} \cdot x_j + \sum_{i \neq j} a_{i,j} \cdot x_i + a_{j,j} \cdot x_j + \sum_{k \neq j} a_{j,k} \cdot x_k$$

$$= \sum_{i} a_{i,j} \cdot x_i + \sum_{k} a_{j,k} \cdot x_k$$

$$= \underline{x}^T \cdot \mathbf{A} + \underline{x}^T \cdot \mathbf{A}^T$$

$$= \underline{x}^T \cdot \left(\mathbf{A} + \mathbf{A}^T \right)$$

$$= 2 \cdot \underline{x}^T \cdot \mathbf{A} \qquad \text{(if } \mathbf{A} \text{ is symmetric, then } \mathbf{A} = \mathbf{A}^T\text{)} \quad .$$

For stochastic interpolation problems the physical model matrix $\mathbf{C}$ simply denotes the presence and location of the observed values and the location of data points to be interpolated. For example, if we have five equally spaced points and we know the end and middle point values and wish to interpolate the remaining values, then

$$\mathbf{C} = \begin{bmatrix} 1 & 0 & 0 & 0 & 0 \\ 0 & 0 & 1 & 0 & 0 \\ 0 & 0 & 0 & 0 & 1 \end{bmatrix} .$$

So

$$\mathbf{C}^T = \begin{bmatrix} 1 & 0 & 0 \\ 0 & 0 & 0 \\ 0 & 1 & 0 \\ 0 & 0 & 0 \\ 0 & 0 & 1 \end{bmatrix} ,$$

and

$$\mathbf{C}^T \cdot \mathbf{C} = \begin{bmatrix} 1 & 0 & 0 & 0 & 0 \\ 0 & 0 & 0 & 0 & 0 \\ 0 & 0 & 1 & 0 & 0 \\ 0 & 0 & 0 & 0 & 0 \\ 0 & 0 & 0 & 0 & 1 \end{bmatrix} .$$

Note that for a matrix to be invertible it must be a square matrix. But the numerical calculation of the inverse of a matrix may also be numerically unstable, so a value called the condition number is often calculated to have some confidence in the inverse calculation. In our case, $\det(\mathbf{C}^T \cdot \mathbf{C}) = 0$, so is singular, and there can be no inverse. So some modifications are in order to be able to solve for $\underline{\hat{x}}$. The process of making these adjustments is called regularization. A first attempt at regularization might be to add the $\underline{\hat{x}}$ values to the measurement model matrix to address the singularity. Thus the problem becomes

$$\min \left((\underline{y} - \mathbf{C} \cdot \underline{\hat{x}})^T \cdot (\underline{y} - \mathbf{C} \cdot \underline{\hat{x}}) + \underline{\hat{x}}^T \cdot \underline{\hat{x}} \right) .$$

The solution obtained, following the same general algebraic reasoning as above, is

$$\underline{\hat{x}} = \left(\mathbf{C}^T \cdot \mathbf{C} + \mathbf{I} \right)^{-1} \cdot \mathbf{C}^T \cdot \underline{y} ,$$

where for our example,

$$\mathbf{I} = \begin{bmatrix} 1 & 0 & 0 & 0 & 0 \\ 0 & 1 & 0 & 0 & 0 \\ 0 & 0 & 1 & 0 & 0 \\ 0 & 0 & 0 & 1 & 0 \\ 0 & 0 & 0 & 0 & 1 \end{bmatrix} .$$

Then,

$$(\mathbf{C}^T \cdot \mathbf{C} + \mathbf{I})^{-1} = \begin{bmatrix} 0.5 & 0 & 0 & 0 & 0 \\ 0 & 1 & 0 & 0 & 0 \\ 0 & 0 & 0.5 & 0 & 0 \\ 0 & 0 & 0 & 1 & 0 \\ 0 & 0 & 0 & 0 & 0.5 \end{bmatrix},$$

which gives

$$(\mathbf{C}^T \cdot \mathbf{C} + \mathbf{I})^{-1} \cdot \mathbf{C}^T = \begin{bmatrix} 0.5 & 0 & 0 \\ 0 & 0 & 0 \\ 0 & 0.5 & 0 \\ 0 & 0 & 0 \\ 0 & 0 & 0.5 \end{bmatrix}.$$

If we do this we can solve for $\underline{\hat{x}}$, but all the unmeasured values get set to 0. For instance, if the known measurements are

$$\underline{y} = \begin{bmatrix} 2 \\ 5 \\ 4 \end{bmatrix},$$

then

$$\begin{aligned} \underline{\hat{x}} &= (\mathbf{C}^T \cdot \mathbf{C} + \mathbf{I})^{-1} \cdot \mathbf{C}^T \cdot \underline{y} \\ &= \begin{bmatrix} 0.5 & 0 & 0 \\ 0 & 0 & 0 \\ 0 & 0.5 & 0 \\ 0 & 0 & 0 \\ 0 & 0 & 0.5 \end{bmatrix} \cdot \begin{bmatrix} 2 \\ 5 \\ 4 \end{bmatrix} \\ &= \begin{bmatrix} 1 \\ 0 \\ 2.5 \\ 0 \\ 2 \end{bmatrix}. \end{aligned}$$

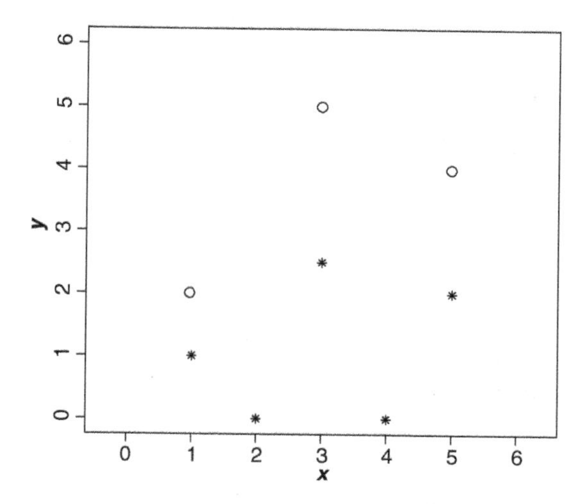

Figure 11.6 First attempt at regularization for stochastic interpolation.

A plot of this results looks like Figure 11.6. If we want to model variation between the points, we need to do more. So we assume a model for the $\underline{\hat{x}}$ values. The simplest model is to assert smoothness by penalizing the difference of adjacent x values. We do this by replacing the identity matrix $\mathbf{I}$ with a new constraints matrix $\mathbf{L}$ structured as shown here (sized for our example, but $(n-1) \times n$ in general):

$$\mathbf{L} = \begin{bmatrix} 1 & -1 & 0 & 0 & 0 \\ 0 & 1 & -1 & 0 & 0 \\ 0 & 0 & 1 & -1 & 0 \\ 0 & 0 & 0 & 1 & -1 \end{bmatrix} .$$

Now the problem is to solve the following:

$$\min\left((\underline{y} - \mathbf{C} \cdot \underline{\hat{x}})^T \cdot (\underline{y} - \mathbf{C} \cdot \underline{\hat{x}}) + (\mathbf{L} \cdot \underline{\hat{x}})^T \cdot (\mathbf{L} \cdot \underline{\hat{x}})\right) ,$$

the solution of which is

$$\underline{\hat{x}} = \left(\mathbf{C}^T \cdot \mathbf{C} + \mathbf{L}^T \cdot \mathbf{L}\right)^{-1} \cdot \mathbf{C}^T \cdot \underline{y} .$$

For our example, this gives the result,

$$\underline{\hat{x}} = \begin{bmatrix} 2.733333 \\ 3.466667 \\ 4.200000 \\ 4.133333 \\ 4.066667 \end{bmatrix} .$$

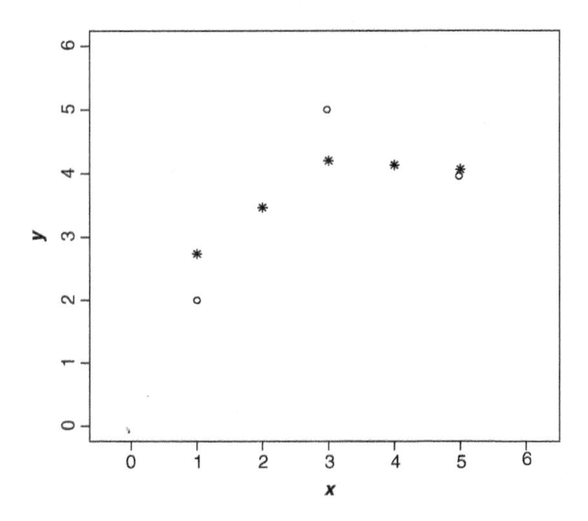

Figure 11.7 Membrane regularization for stochastic interpolation.

A plot of this result looks like Figure 11.7. This solution is known as a membrane constraint and produces a straight-line interpolation between adjacent points. Other constraints are possible. A matrix of the form

$$\mathbf{L} = \begin{bmatrix} 1 & -1 & 0 & 0 & 0 \\ -1 & 2 & -1 & 0 & 0 \\ 0 & -1 & 2 & -1 & 0 \\ 0 & 0 & -1 & 2 & -1 \\ 0 & 0 & 0 & -1 & 1 \end{bmatrix}$$

produces a thin-plate interpolation. We can also add a weighting parameter λ that serves to emphasize the measurements when λ is small and emphasize the smoothing when λ is large. With this parameter, the problem becomes

$$\min \left((\underline{y} - \mathbf{C} \cdot \underline{\hat{x}})^T \cdot (\underline{y} - \mathbf{C} \cdot \underline{\hat{x}}) + \lambda \cdot (\mathbf{L} \cdot \underline{\hat{x}})^T \cdot (\mathbf{L} \cdot \underline{\hat{x}}) \right) \quad ,$$

the solution of which is

$$\underline{\hat{x}} = \left(\mathbf{C}^T \cdot \mathbf{C} + \lambda \cdot \mathbf{L}^T \cdot \mathbf{L} \right)^{-1} \cdot \mathbf{C}^T \cdot \underline{y} \quad .$$

It is interesting to note at this point that the best-fit linear solutions may also be created using the same approaches as above, with some slight modifications to the variables but using a very similar algebraic form—if we make the following substitutions. We replace **C** with the augmented matrix **X**, which consists of a column of ones with row dimension equal to the number of known values and a column of the positions of the known y values. The vector $\underline{y}$ remains the same. Rather than estimating directly the $\hat{\underline{x}}$ values, the output of the expression with these substitutions now estimates the regression coefficients $\underline{b}$ of the best linear-fit line (or surface, in higher dimensions). To demonstrate how this works, let us consider the previous example. Here we have the following inputs:

$$\mathbf{X} = \begin{bmatrix} 1 & 1 \\ 1 & 3 \\ 1 & 5 \end{bmatrix}$$

and

$$\underline{y} = \begin{bmatrix} 2 \\ 5 \\ 4 \end{bmatrix} .$$

As described above, the estimates for $\underline{b}$ are given by

$$\hat{\underline{b}} = (\mathbf{X}^T \cdot \mathbf{X})^{-1} \cdot \mathbf{X}^T \cdot \underline{y} .$$

In our example, the calculations give

$$\hat{\underline{b}} = \begin{bmatrix} 2.166667 \\ 0.500000 \end{bmatrix} .$$

Then, applying the regression coefficients to each of the x values, we get the best-fit line as follows:

$$\hat{\underline{y}} = b_0 + b_1 \cdot \underline{x} .$$

We get (also shown in the plot Figure 11.8)

$$\hat{\underline{y}} = \begin{bmatrix} 2.666667 \\ 3.166667 \\ 3.666667 \\ 4.166667 \\ 4.666667 \end{bmatrix} .$$

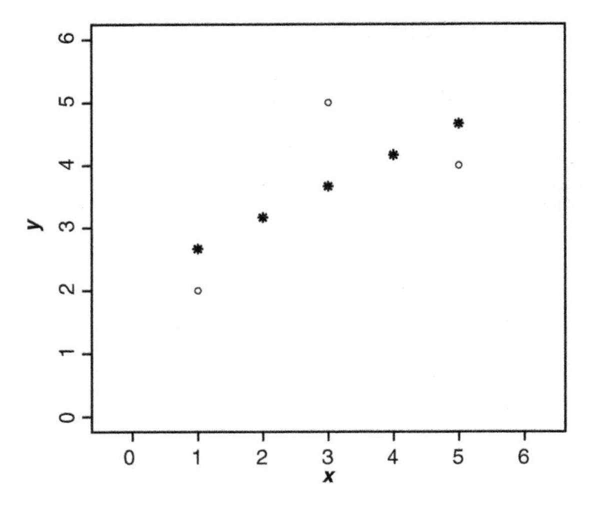

Figure 11.8 Best-fit linear interpolation.

For GIS, we most often want to interpolate in two dimensions. The stochastic methods can be adapted for two dimensions with some minor adjustments. First, the input is transformed from the two-dimensional form to lexicographical order (usually based on a vertical scanning pattern). As a small example, consider a 3×3 raster with known values in the four corners, and we want to interpolate the remaining values.

$$\begin{array}{ccc} 4 & * & 3 \\ * & * & * \\ 7 & * & 2 \end{array} \quad \Rightarrow \quad \begin{array}{c} 4 \\ * \\ 7 \\ * \\ * \\ * \\ 3 \\ * \\ 2 \end{array} .$$

So we have

$$\underline{y} = \begin{bmatrix} 4 \\ 7 \\ 3 \\ 2 \end{bmatrix} .$$

We also have to adjust the constraints matrix $\mathbf{L}$ to account for both the vertical and horizontal smoothing penalties. This results in a banded $\mathbf{L}^T \cdot \mathbf{L}$ matrix as we incorporate the lags introduced via the lexicographical transformation. For a membrane constraint for our example, $\mathbf{L}$ is structured as below (where the six horizontal constraints are listed first, followed by the six vertical constraints),

$$\mathbf{L} = \begin{bmatrix} 1 & 0 & 0 & -1 & 0 & 0 & 0 & 0 & 0 \\ 0 & 1 & 0 & 0 & -1 & 0 & 0 & 0 & 0 \\ 0 & 0 & 1 & 0 & 0 & -1 & 0 & 0 & 0 \\ 0 & 0 & 0 & 1 & 0 & 0 & -1 & 0 & 0 \\ 0 & 0 & 0 & 0 & 1 & 0 & 0 & -1 & 0 \\ 0 & 0 & 0 & 0 & 0 & 1 & 0 & 0 & -1 \\ 1 & -1 & 0 & 0 & 0 & 0 & 0 & 0 & 0 \\ 0 & 1 & -1 & 0 & 0 & 0 & 0 & 0 & 0 \\ 0 & 0 & 0 & 1 & -1 & 0 & 0 & 0 & 0 \\ 0 & 0 & 0 & 0 & 1 & -1 & 0 & 0 & 0 \\ 0 & 0 & 0 & 0 & 0 & 0 & 1 & -1 & 0 \\ 0 & 0 & 0 & 0 & 0 & 0 & 0 & 1 & -1 \end{bmatrix} .$$

Thus we have $\mathbf{L}^T \cdot \mathbf{L}$ as follows:

$$\mathbf{L}^T \cdot \mathbf{L} = \begin{bmatrix} 2 & -1 & 0 & -1 & 0 & 0 & 0 & 0 & 0 \\ -1 & 3 & -1 & 0 & -1 & 0 & 0 & 0 & 0 \\ 0 & -1 & 2 & 0 & 0 & -1 & 0 & 0 & 0 \\ -1 & 0 & 0 & 3 & -1 & 0 & -1 & 0 & 0 \\ 0 & -1 & 0 & -1 & 4 & -1 & 0 & -1 & 0 \\ 0 & 0 & -1 & 0 & -1 & 3 & 0 & 0 & -1 \\ 0 & 0 & 0 & -1 & 0 & 0 & 2 & -1 & 0 \\ 0 & 0 & 0 & 0 & -1 & 0 & -1 & 3 & -1 \\ 0 & 0 & 0 & 0 & 0 & -1 & 0 & -1 & 2 \end{bmatrix} .$$

Again, the solution approach described above can be used, and the vector form of the solution is given by

$$\underline{\hat{x}} = \left(\mathbf{C}^T \cdot \mathbf{C} + \mathbf{L}^T \cdot \mathbf{L}\right)^{-1} \cdot \mathbf{C}^T \cdot \underline{y} \ .$$

When this result is transformed back to two dimensions, the result is as follows:

$$\underline{\hat{x}} = \begin{bmatrix} 4.095238 & 3.857143 & 3.476190 \\ 4.428571 & 4.000000 & 3.571429 \\ 5.190476 & 4.142857 & 3.238095 \end{bmatrix} .$$

Ordinary Kriging

In ordinary Kriging spatial interpolation, we estimate the value of a variable z at point $\underline{s}$ as

$$\widehat{z}_s = \sum_{i=1}^{n} w_i \cdot z_i = \underline{w}^T \cdot \underline{z} \qquad \text{and} \qquad \sum_{i=1}^{n} w_i = 1 \quad ,$$

for an unsampled location $\underline{s}$, where w_i are the unknown weights to be solved for and z_i are the known sample point values. For ordinary Kriging, the following assumptions are in force:

1. There is a constant (unknown) mean μ.
2. The dataset is isotropic.
3. The semi-variogram can be modelled by a simple curve (the semi-variogram will be discussed shortly).
4. The semi-variogram is homogeneous; that is, variation is a function of distance only (2nd order stationarity), $C(z_i, z_j) = C(d_{i,j})$.

Kriging uses the covariance structure of the given data, which is ignored by other interpolation techniques, to estimate the covariance structure across the study area. This covariance structure is used to create the weights used in the interpolation.

We now give a derivation of how to solve this (the unknown weights) in the general case. We want to minimize the expected mean squared error of our estimate $\widehat{z}_0$ of the true value z_0 at location 0,

$$\min \mathrm{E}\left[\{\widehat{z}_0 - z_0\}^2\right] \quad .$$

To solve this problem we need to do some algebra to get the statement into a form such that we can take derivatives and thus find the minimum value. Note that the asterisks refer to asides in the derivation, which will follow the main results.

$$\begin{aligned}
&\mathrm{E}\left[\{\widehat{z}_0 - z_0\}^2\right] \\
&= \mathrm{E}\left[\{\widehat{z}_0 - z_0 + \mu - \mu\}^2\right] \\
&= \mathrm{E}\left[\{(\widehat{z}_0 - \mu) - (z_0 - \mu)\}^2\right] \\
&= \mathrm{E}\left[(\widehat{z}_0 - \mu)^2 - 2 \cdot (\widehat{z}_0 - \mu) \cdot (z_0 - \mu) + (z_0 - \mu)^2\right] \\
&= \mathrm{E}\left[(\widehat{z}_0 - \mu)^2\right] - 2 \cdot \mathrm{E}\Big[(\widehat{z}_0 - \mu) \cdot (z_0 - \mu)\Big] + \mathrm{E}\left[(z_0 - \mu)^2\right] \\
&= \mathrm{E}\left[\left(\sum_{i=1}^{n} w_i \cdot z_i - \mu\right)^2\right] - 2 \cdot \mathrm{E}\left[\left(\sum_{i=1}^{n} w_i \cdot z_i - \mu\right) \cdot (z_0 - \mu)\right] + \sigma^2 \qquad (*)
\end{aligned}$$

$$= \mathrm{E}\left[\left(\sum_{i=1}^{n} w_i \cdot (z_i - \mu)\right)^2\right] - 2 \cdot \mathrm{E}\left[\left(\sum_{i=1}^{n} w_i \cdot (z_i - \mu)\right) \cdot (z_0 - \mu)\right] + \sigma^2 \qquad (**)$$

$$= \sum_{i=1}^{n}\sum_{j=1}^{n} w_i \cdot w_j \cdot \mathrm{cov}\,(z_i, z_j) - 2 \cdot \sum_{i=1}^{n} w_i \cdot \mathrm{E}\left[(z_i - \mu) \cdot (z_0 - \mu)\right] + \sigma^2 \qquad (***)$$

$$= \sum_{i=1}^{n}\sum_{j=1}^{n} w_i \cdot w_j \cdot \mathrm{cov}\,(z_i, z_j) - 2 \cdot \sum_{i=1}^{n} w_i \cdot \mathrm{cov}\,(z_i, z_0) + \sigma^2$$

$$= \underline{w}^T \cdot \mathrm{cov}\,(z_i, z_j) \cdot \underline{w} - 2 \cdot \underline{w}^T \cdot \mathrm{cov}\,(z_i, z_0) + \sigma^2$$

$$= \underline{w}^T \cdot \mathbf{C} \cdot \underline{w} - 2 \cdot \underline{w}^T \cdot \underline{c} + \sigma^2 \quad ,$$

where

$\mathbf{C}$ is the $n \times n$ covariance matrix, $\mathrm{cov}\,(z_i, z_j)$
and
$\underline{c}$ is the $n \times 1$ column vector of covariances, $\mathrm{cov}\,(z_i, z_0)$.

(*) Recall,

$$\mu = \mathrm{E}\,[x] \quad , \quad \sigma^2 = \mathrm{var}\,[x] = \mathrm{E}\left[(x - \mathrm{E}\,[x])^2\right] \quad .$$

(**) Recall,

$$\sum_{i=1}^{n} w_i = 1 \quad ,$$

and thus

$$\sum_{i=1}^{n} w_i \cdot z_i - \mu = \sum_{i=1}^{n} w_i \cdot z_i - \sum_{i=1}^{n} w_i \cdot \mu$$
$$= \sum_{i=1}^{n} w_i \cdot (z_i - \mu) \quad .$$

(***) Recall,

$$\mathrm{var}\,[x] = \mathrm{E}\left[(x - \mathrm{E}\,[x])^2\right]$$

and

$$\mathrm{E}\left[\sum_{i=1}^{n} a_i \cdot x_i\right] = \sum_{i=1}^{n} a_i \cdot \mathrm{E}\left[x_i\right]$$

and

$$\operatorname{var}\left[\sum_{i=1}^{n} a_i \cdot x_i\right] = \sum_{i=1}^{n}\sum_{j=1}^{n} a_i \cdot a_j \cdot \operatorname{cov}\left(x_i, x_j\right) \quad .$$

So to summarize, we can write

$$\mathrm{E}\left[\{\widehat{z}_0 - z_0\}^2\right] = \underline{w}^T \cdot \mathbf{C} \cdot \underline{w} - 2 \cdot \underline{w}^T \cdot \underline{c} + \sigma^2 \quad .$$

To minimize this with the constraint that

$$\sum_{i=1}^{n} w_i = 1 \qquad \text{or} \qquad \underline{w}^T \cdot \underline{1} = 1 \,, \qquad \text{where} \qquad \underline{1} = \begin{bmatrix} 1 \\ 1 \\ \vdots \\ 1 \end{bmatrix} \quad ,$$

we use the method of Lagrange multipliers. To do this we set up a system of two equations,

$$\begin{aligned} f\left(\underline{w}\right) &= \underline{w}^T \cdot \mathbf{C} \cdot \underline{w} - 2 \cdot \underline{w}^T \cdot \underline{c} + \sigma^2 \\ \text{and} \qquad g\left(\underline{w}\right) &= 2 \cdot \left(\underline{w}^T \cdot \underline{1} - 1\right) \quad . \end{aligned}$$

Next we let

$$\begin{aligned} L\left(\underline{w}, \lambda\right) &= f\left(\underline{w}\right) + \lambda \cdot g\left(\underline{w}\right) \\ &= \underline{w}^T \cdot \mathbf{C} \cdot \underline{w} - 2 \cdot \underline{w}^T \cdot \underline{c} + \sigma^2 + \lambda \cdot 2 \cdot \left(\underline{w}^T \cdot \underline{1} - 1\right) \quad , \end{aligned}$$

where λ is the Lagrange multiplier.

We solve this by setting the partial derivatives with respect to $\underline{w}$ and λ to 0, giving

$$\begin{aligned} \frac{\partial L\left(\underline{w}, \lambda\right)}{\partial \underline{w}} &= 2 \cdot \mathbf{C} \cdot \underline{w} - 2 \cdot \underline{c} + 2 \cdot \underline{1} \cdot \lambda = 0 \\ \Rightarrow \qquad \mathbf{C} \cdot \underline{w} + \underline{1} \cdot \lambda &= \underline{c} \end{aligned}$$

and

$$\frac{\partial L(\underline{w}, \lambda)}{\partial \lambda} = 2 \cdot \left(\underline{w}^T \cdot \underline{1} - 1\right) = 0$$
$$\Rightarrow \qquad \underline{w}^T \cdot \underline{1} = 1 \qquad \text{or} \qquad \underline{1}^T \cdot \underline{w} = 1 \quad .$$

This system we can solve simultaneously as

$$\begin{bmatrix} \mathbf{C} & \vdots & \underline{1} \\ \cdots & \cdots & \cdots \\ \underline{1}^T & \vdots & 0 \end{bmatrix} \cdot \begin{bmatrix} \underline{w} \\ \cdots \\ \lambda \end{bmatrix} = \begin{bmatrix} \underline{c} \\ \cdots \\ 1 \end{bmatrix}$$

or

$$\mathbf{C}_+ \cdot \underline{w}_+ = \underline{c}_+ \quad .$$

Finally, by premultiplying both sides by the inverse matrix of $\mathbf{C}_+$, we get

$$\underline{w}_+ = (\mathbf{C}_+)^{-1} \cdot \underline{c}_+ \quad ,$$

which provides the needed weights (ignoring the last tuple, λ, in the vector $\underline{w}_+$).

Note that the Kriging equations can also be rewritten in terms of the semi-variogram directly instead of the covariance. The semi-variogram is defined as

$$\gamma(d_{i,j}) = \frac{1}{2} \cdot \text{var}[(Y_i) - (Y_j)] \quad .$$

Next, using the definition properties of variance (see Appendix I), we can write

$$\begin{aligned} \gamma(d_{i,j}) &= \frac{1}{2} \cdot \text{var}[(Y_i) - (Y_j)] \\ &= \frac{1}{2} \cdot (\text{var}[Y_i] + \text{var}[Y_j] - 2 \cdot \text{cov}[Y_i, Y_j]) \quad . \end{aligned}$$

If $\text{var}[Y_i] = \text{var}[Y_j] = \text{var}[Y] = \sigma^2$ and $\text{cov}[Y_i, Y_j] = \mathbf{C}(d_{i,j})$, then

$$\begin{aligned} \gamma(d_{i,j}) &= \frac{1}{2} \cdot (\sigma^2 + \sigma^2 - 2 \cdot \mathbf{C}(d_{i,j})) \\ &= \sigma^2 - \mathbf{C}(d_{i,j}) \quad , \end{aligned}$$

or

$$\mathbf{C}(d_{i,j}) = \sigma^2 - \gamma\left(d_{i,j}\right) \quad .$$

Rewriting this in matrix notation, we have

$$\mathbf{C} = \sigma^2 \cdot \underline{1} \cdot \underline{1}^T - \Gamma \quad .$$

Similarly,

$$\underline{c} = \sigma^2 \cdot \underline{1} - \underline{\gamma} \quad .$$

Now recall that we can write the system of equations to be solved for the Kriging interpolation weights as

$$\begin{bmatrix} \mathbf{C} & \vdots & \underline{1} \\ \cdots & \cdots & \cdots \\ \underline{1}^T & \vdots & 0 \end{bmatrix} \cdot \begin{bmatrix} \underline{w} \\ \cdots \\ \lambda \end{bmatrix} = \begin{bmatrix} \underline{c} \\ \cdots \\ 1 \end{bmatrix} \quad .$$

Multiplying the block matrix by the block vector on the left-hand side of this expression gives

$$\begin{bmatrix} \mathbf{C} \cdot \underline{w} + \underline{1} \cdot \lambda \\ \cdots \\ \underline{1}^T \cdot \underline{w} \end{bmatrix} = \begin{bmatrix} \underline{c} \\ \cdots \\ 1 \end{bmatrix} \quad .$$

Substituting in for the values of $\mathbf{C}$ and $\underline{c}$ from above gives

$$\begin{bmatrix} \left(\sigma^2 \cdot \underline{1} \cdot \underline{1}^T - \Gamma\right) \cdot \underline{w} + \underline{1} \cdot \lambda \\ \cdots \\ \underline{1}^T \cdot \underline{w} \end{bmatrix} = \begin{bmatrix} \sigma^2 \cdot \underline{1} - \underline{\gamma} \\ \cdots \\ 1 \end{bmatrix} \quad .$$

Distributing $\underline{w}$ in the first term, we get

$$\begin{bmatrix} \sigma^2 \cdot \underline{1} \cdot \underline{1}^T \cdot \underline{w} - \Gamma \cdot \underline{w} + \underline{1} \cdot \lambda \\ \cdots \\ \underline{1}^T \cdot \underline{w} \end{bmatrix} = \begin{bmatrix} \sigma^2 \cdot \underline{1} - \underline{\gamma} \\ \cdots \\ 1 \end{bmatrix} \quad .$$

Recalling by assumption $\underline{1}^T \cdot \underline{w} = 1$ results in

$$\begin{bmatrix} \sigma^2 \cdot \underline{1} - \Gamma \cdot \underline{w} + \underline{1} \cdot \lambda \\ \cdots \\ 1 \end{bmatrix} = \begin{bmatrix} \sigma^2 \cdot \underline{1} - \underline{\gamma} \\ \cdots \\ 1 \end{bmatrix} \quad .$$

Subtracting

$$\begin{bmatrix} \sigma^2 \cdot \underline{1} \\ \cdots\cdots\cdots \\ 1 \end{bmatrix}$$

from both sides leaves

$$\begin{bmatrix} -\Gamma \cdot \underline{w} + \underline{1} \cdot \lambda \\ \cdots\cdots\cdots\cdots\cdots \\ 0 \end{bmatrix} = \begin{bmatrix} -\underline{\gamma} \\ \cdots\cdots \\ 0 \end{bmatrix} .$$

Now, multiplying both sides by -1 gives

$$\begin{bmatrix} \Gamma \cdot \underline{w} - \underline{1} \cdot \lambda \\ \cdots\cdots\cdots\cdots\cdots \\ 0 \end{bmatrix} = \begin{bmatrix} \underline{\gamma} \\ \cdots\cdots \\ 0 \end{bmatrix} .$$

Adding

$$\begin{bmatrix} \underline{0} \\ \cdots\cdots \\ 1 \end{bmatrix}$$

to both sides gives

$$\begin{bmatrix} \Gamma \cdot \underline{w} - \underline{1} \cdot \lambda \\ \cdots\cdots\cdots\cdots\cdots \\ 1 \end{bmatrix} = \begin{bmatrix} \underline{\gamma} \\ \cdots\cdots \\ 1 \end{bmatrix} .$$

Then we substitute $\underline{1}^T \cdot \underline{w}$ for 1 in the first term, giving

$$\begin{bmatrix} \Gamma \cdot \underline{w} - \underline{1} \cdot \lambda \\ \cdots\cdots\cdots\cdots\cdots \\ \underline{1}^T \cdot \underline{w} \end{bmatrix} = \begin{bmatrix} \underline{\gamma} \\ \cdots\cdots \\ 1 \end{bmatrix} .$$

This may be rewritten as

$$\begin{bmatrix} \Gamma & \vdots & \underline{1} \\ \cdots & \cdots & \cdots \\ \underline{1}^T & \vdots & 0 \end{bmatrix} \cdot \begin{bmatrix} \underline{w} \\ \cdots\cdots \\ -\lambda \end{bmatrix} = \begin{bmatrix} \underline{\gamma} \\ \cdots\cdots \\ 1 \end{bmatrix} ,$$

or in more compact notation,

$$\Gamma_+ \cdot \underline{w}_+^* = \underline{\gamma}_+$$
$$\underline{w}_+^* = (\Gamma_+)^{-1} \cdot \underline{\gamma}_+ \quad ,$$

where

$$\mathbf{\Gamma}_+ (d_{i,j}) = \begin{bmatrix} \gamma(d_{1,1}) & \gamma(d_{1,2}) & \cdots & \gamma(d_{1,n}) & 1 \\ \vdots & \vdots & & \vdots & \vdots \\ \gamma(d_{n,1}) & \gamma(d_{n,2}) & \cdots & \gamma(d_{n,n}) & 1 \\ 1 & 1 & \cdots & 1 & 0 \end{bmatrix}, \quad \underline{\gamma}_+ (d_{i,0}) = \begin{bmatrix} \gamma(d_{1,0}) \\ \vdots \\ \gamma(d_{n,0}) \\ 1 \end{bmatrix} \quad ,$$

and where $\gamma(d_{i,j})$ is the semi-variogram (how the data vary with relative distance). Note that the above results are equivalent to the previous calculation using covariance, since the λ in both $\underline{w}_+$ and $\underline{w}_+^*$ is not used as part of the final weights vector.

The semi-variogram is either calculated by using an assumed model, which is often a good approximation, and the choice of model has been shown to be not too critical, or estimated directly from the data, after removing trends. A commonly used model variogram is the spherical model,

$$\hat{\gamma}(d) = \begin{cases} c_0 + c_1 \cdot \left[\frac{3}{2} \cdot \frac{d}{a} - 0.5 \cdot \left(\frac{d}{a}\right)^3\right] & \text{if } 0 \le d \le a \\ c_0 + c_1 & \text{otherwise} \end{cases} .$$

An empirical estimation of the semi-variogram may be calculated as

$$2 \cdot \hat{\gamma}(d) = \frac{1}{n(d)} \cdot \sum_{d_i = d - \frac{\triangle}{2}}^{d + \frac{\triangle}{2}} (z_i - z_j)^2 \quad ,$$

where $n(d)$ is the number of inter-point distances in the distance band $\left[d - \frac{\triangle}{2}, d + \frac{\triangle}{2}\right]$. Using the rainfall dataset given at the beginning of this section, we now show a worked example to estimate the value at z_0. We assume a spherical model for the semi-variogram, with $c_0 = 0$, $c_1 = 3287.151$, which is the variance of the known rainfall data, and $a = 237955$, which is the width of the study area. The input data are given here again for convenience:

Site	Easting	Northing	Rainfall (annual mm, 1980)
CHAPLIN	379437	5595605	222.4
INDIAN HEAD	592195	5595032	347.9
MELFORT	526967	5850097	271.3
MOOSE JAW	457268	5572159	176.3
OUTLOOK	354240	5707520	171.5
REGINA	521321	5583149	229.4
SASKATOON	383819	5784645	219.5

and z_0 is DAVIDSON at location (430284, 5683661). The distance matrix for pairwise distances between the known values is given by

$$\begin{bmatrix} 0.00 & 212758.77 & 294162.0 & 81285.79 & 114716.42 & 142429.71 & 189090.78 \\ 212758.77 & 0.00 & 263273.3 & 136852.00 & 263203.59 & 71863.27 & 281733.29 \\ 294161.99 & 263273.33 & 0.0 & 286544.03 & 223970.57 & 267007.70 & 157401.76 \\ 81285.79 & 136852.00 & 286544.0 & 0.00 & 170109.87 & 64988.98 & 224822.28 \\ 114716.42 & 263203.59 & 223970.6 & 170109.87 & 0.00 & 208288.76 & 82602.56 \\ 142429.71 & 71863.27 & 267007.7 & 64988.98 & 208288.76 & 0.00 & 243941.46 \\ 189090.78 & 281733.29 & 157401.8 & 224822.28 & 82602.56 & 243941.46 & 0.00 \end{bmatrix}.$$

Then, by applying the semi-variogram model and adding the required constants, we get

$$\mathbf{\Gamma}_+ = \begin{bmatrix} 0.000 & 3233.819 & 3287.151 & 1618.827 & 2192.914 & 2598.865 & 3093.460 & 1 \\ 3233.819 & 0.000 & 3287.151 & 2523.095 & 3287.151 & 1443.825 & 3287.151 & 1 \\ 3287.151 & 3287.151 & 0.000 & 3287.151 & 3270.454 & 3287.151 & 2785.861 & 1 \\ 1618.827 & 2523.095 & 3287.151 & 0.000 & 2924.417 & 1313.170 & 3272.408 & 1 \\ 2192.914 & 3287.151 & 3270.454 & 2924.417 & 0.000 & 3213.697 & 1642.877 & 1 \\ 2598.865 & 1443.825 & 3287.151 & 1313.170 & 3213.697 & 0.000 & 3287.151 & 1 \\ 3093.460 & 3287.151 & 2785.861 & 3272.408 & 1642.877 & 3287.151 & 0.000 & 1 \\ 1.000 & 1.000 & 1.000 & 1.000 & 1.000 & 1.000 & 1.000 & 0 \end{bmatrix}$$

Using software, we can calculate the inverse of the above matrix:

$$\Gamma_{+}^{-1} =$$

$$\begin{bmatrix} -4.468791e{-4} & 7.506923e{-6} & 4.855438e{-5} & 2.697389e{-4} & 1.778338e{-4} & -3.105626e{-5} & -2.569862e{-5} & 1.496923e{-1} \\ 7.506923e{-6} & -4.085166e{-4} & 5.620787e{-5} & -3.899160e{-5} & 3.957194e{-5} & 3.114483e{-4} & 3.277324e{-5} & 2.018533e{-1} \\ 4.855438e{-5} & 5.620787e{-5} & -2.494680e{-4} & 2.372649e{-5} & 5.757481e{-8} & 1.496929e{-5} & 1.059524e{-4} & 2.330748e{-1} \\ 2.697389e{-4} & -3.899160e{-5} & 2.372649e{-5} & -6.255866e{-4} & -1.040328e{-5} & 3.671552e{-4} & 1.436091e{-5} & 8.501782e{-2} \\ 1.778338e{-4} & 3.957194e{-5} & 5.757481e{-8} & -1.040328e{-5} & -4.518855e{-4} & -6.927159e{-6} & 2.517526e{-4} & 1.188454e{-1} \\ -3.105626e{-5} & 3.114483e{-4} & 1.496929e{-5} & 3.671552e{-4} & -6.927159e{-6} & -6.748220e{-4} & 1.923263e{-5} & 5.873177e{-2} \\ -2.569862e{-5} & 3.277324e{-5} & 1.059524e{-4} & 1.436091e{-5} & 2.517526e{-4} & 1.923263e{-5} & -3.983732e{-4} & 1.527846e{-1} \\ 1.496923e{-1} & 2.018533e{-1} & 2.330748e{-1} & 8.501782e{-2} & 1.188454e{-1} & 5.873177e{-2} & 1.527846e{-1} & -2.442425e3 \end{bmatrix} .$$

Distances from the unknown value to the known values are calculated as

$$\begin{bmatrix} 101682.23 & 184581.34 & 192479.98 & 114720.67 & 79699.07 & 135611.20 & 111160.98 \end{bmatrix} .$$

Again, using the semi-variogram model these values give

$$\underline{\gamma}_{+} = \begin{bmatrix} 1978.738 \\ 3057.628 \\ 3118.542 \\ 2192.981 \\ 1589.711 \\ 2505.811 \\ 2135.838 \\ 1.000 \end{bmatrix} .$$

Thus,

$$\underline{w}_{+} = \begin{bmatrix} 0.18133505 \\ -0.02926865 \\ 0.03896829 \\ 0.13578701 \\ 0.37107109 \\ 0.14050836 \\ 0.16159885 \\ 46.68894942 \end{bmatrix} .$$

Then using the unaugmented vector $\underline{w}$ with the known values, we get $\widehat{z}_0 = 196.0$. Note that this value is less than 2% from the actual value of 199.5. Further note that this is the estimate for only one location. To get an interpolated surface we need to

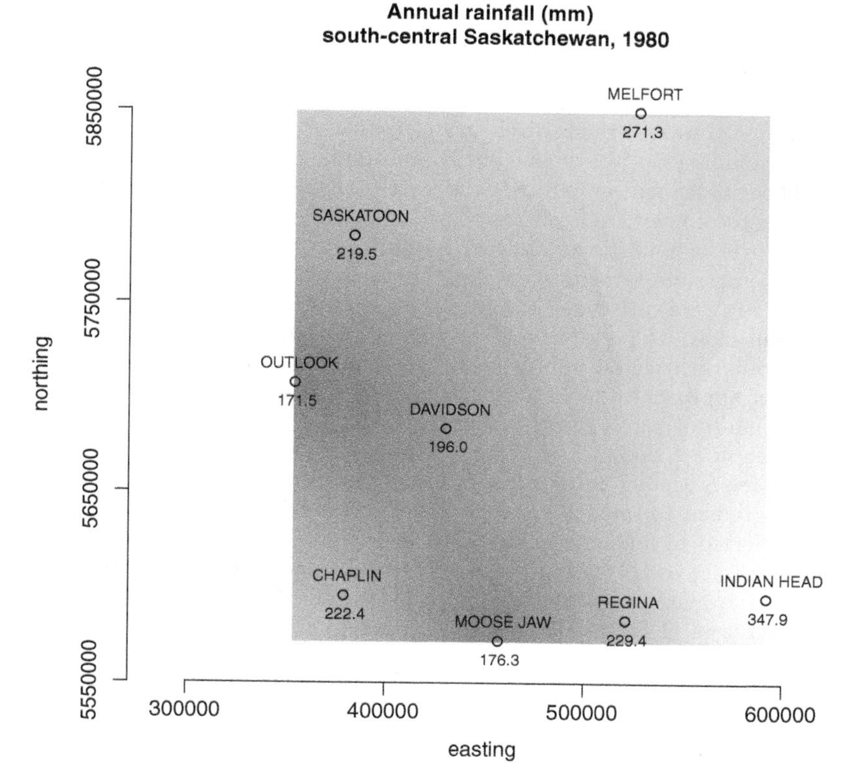

Figure 11.9 Ordinary Kriging interpolation.

calculate estimates for all the unknown values. However, we do not need to redo all the calculations. We only need to recalculate $\underline{\gamma}_+$, $\underline{w}_+$ and $\widehat{z}_i$ for each new location z_i. For the problem above we can calculate these three values over a 238×278 regular grid. Figure 11.9 shows the results of this interpolation, with the original point data included as an overlay. Further details and other variations of Kriging may be found in Bailey and Gatrell [6] and Cressie [31].

11.3.3 Extracting Optimal Information from Geographic Patterns

In this last section of the chapter we discuss multi-objective optimization for landuse and environmental planning decision support. This gives us a chance to introduce the reader to several important and useful tools and ideas for GIS analysis: Pareto optimal solutions, configuration optimization, and genetic algorithms (GA) including specific multi-objective optimization solvers such as NSGA-II.

Combinatorial optimization problems are specified by a discrete set of solutions and cost functions. The set of solutions is often not given explicitly but instead is represented by set of decision values with known ranges. One class of approaches to the solution of combinatorial optimization problems is known as local search strategies. This class of algorithms requires the definition of a neighbourhood around each proposed solution. With this structure in place it is possible to solve some problems iteratively by moving systematically from solutions with known objective function values to nearby, possibly better, solutions.

A GA is one of several local search strategies, including evolutionary algorithms and genetic programming that have come to be known as evolutionary strategies. Here the neighbourhood is defined by chromosome strings that code samples from the solution set. The concept of nearness is abstracted to imply solutions that are reachable in one step by crossover or mutation operators acting on these chromosome strings. Further, a selection operator implements the notion of moving iteratively to better solutions. All evolutionary strategies are types of local search optimizers (see Aarts and Lenstra [1]). GAs have been shown to perform well in a variety of engineering optimization problems, including those with single, multi-modal, and multiple objective functions. In fact, these types of approaches have now been recognized as a new field of study, namely *evolutionary multi-objective optimization* (EMO).

The GA approach does not require an arbitrary convex weighting scheme, and there are generally fewer scaling issues with objective function values, as is the case with hill-climbing or gradient approaches. Evolutionary strategies are members of a class of optimizers referred to as heuristic methods (see Figure 11.10). In a simple GA (SGA), initial solutions are coded as randomly generated binary strings, and better solutions are iteratively evolved using stochastic methods that include selection, crossover, and mutation operators. The algorithm starts by randomly generating a fixed number of initial chromosomes (usually binary strings) that code potential problem solutions. Following the biological genetics terminology, each position on the chromosome is referred to as a locus, and the character value at each locus is referred to as an allele. Further, specific chromosomes are referred to as genotypes, whereas, the specific solutions coded by the chromosomes are called phenotypes.

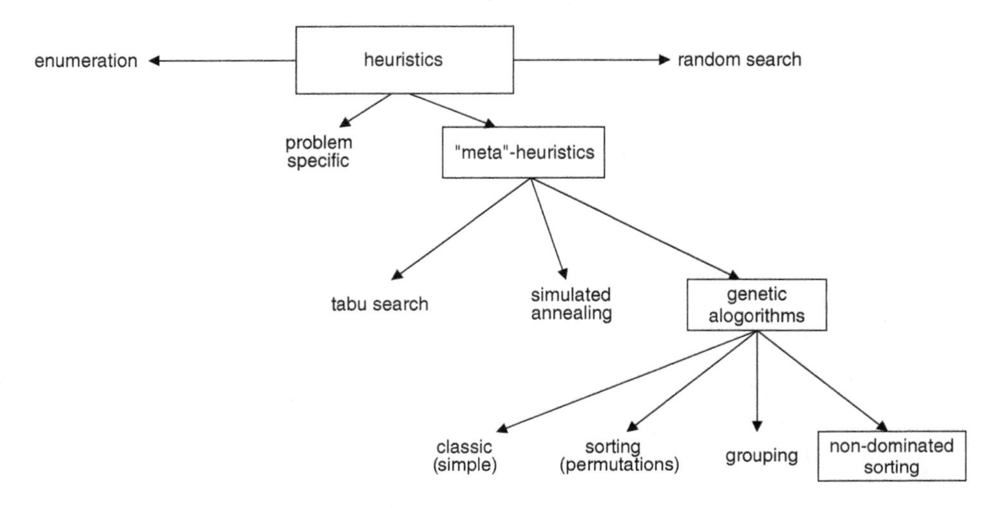

Figure 11.10 Optimization heuristics.

The main part of the algorithm is an iterative process comprising the following steps. All solutions in the current generation are evaluated by fitness functions. The best-performing solutions are selected to be parents. The crossover and mutation operators are applied to these parent chromosomes to produce a pair of offspring solution chromosomes. In other words, combining the best-performing solutions from a population of solutions allows better solutions to evolve. The iterative process is continued until a prescribed maximum number of generations (iterations) have been performed. Note that the maximum number of generations is a user-specified parameter, which most often is dictated by available computer resources.

There are several ways to select the solutions to be processed for the next generation of the GA. A commonly used approach is roulette wheel selection, which works by proportionally selecting population members based on their fitness values. The basic idea of roulette wheel selection is that the values of each population member's fitness are sorted or otherwise ranked, and then a normalized cumulative sum vector is created from these values. Then, by choosing a random number between 0 and 1, and choosing the member whose ranked cumulative fitness is closest to but not greater than the random value, a member is selected as a candidate for the crossover and mutation operations that create new chromosomes. Usually pairs of chromosome are selected as parents for the crossover operator, which is described next.

Single-point crossover is commonly used. Two parent chromosomes are selected as above, with roulette wheel selection, and a random number is generated between 0 and 1 that determines whether the crossover occurs. If this random number is less than a preset crossover rate, crossover occurs. This crossover parameter is usually set quite high (e.g., 0.9). A second random number is then generated between 1 and l, where l is the chromosome length. This defines the point at which each parent chromosome is split. The crossover operation consists of combining the front part of the parent 1 chromosome with the tail of the parent 2 chromosome, and the remaining two pieces are similarly combined, thus creating two new offspring. Other problem-specific chromosome modifying processes exist (e.g., grouping genetic algorithms). The last chromosome operator to be discussed in this subsection is the mutation operator.

The mutation operator randomly changes the value of one allele at a randomly chosen locus of a chromosome. Three random numbers are generated for each of the offspring chromosomes that result from the crossover operation described above. The first random number is a real number between 0 and 1. If this number is less than the preset mutation rate parameter, it triggers a mutation event. The second random number generated is an integer lying between 1 and l. This number determines the position (or locus) of the mutation. The third random number generated is an integer with a value within the range of the chromosome's alphabet. This number determines the mutated allele value at the chosen locus. Note that in some GAs the mutation rate is set dynamically, usually as a response to the premature convergence issues that existed with early implementations of GAs. Typically, a fixed low mutation rate is used (e.g., 0.01). A good overview of GA basics may be found in Mitchell [79].

Recently, approaches to multi-objective optimization problems have been designed around the concept of a trade-off surface or Pareto front (also referred to as an efficient frontier or a non-dominated front). A feasible solution point is Pareto optimal if no other feasible solution scores at least as well in all objective functions, and if the solution is strictly superior in at least one objective function to any other solution. For a maximization problem, this may be stated as follows:

$$\text{Solution } \underline{x}_i \text{ is dominated by } \underline{x}_j \text{ if}$$

$$\mathbf{f}_1(\underline{x}_i) \leq \mathbf{f}_1(\underline{x}_j) \cap \mathbf{f}_2(\underline{x}_i) \leq \mathbf{f}_2(\underline{x}_j) \cap \therefore . \cap \mathbf{f}_k(\underline{x}_i) \leq \mathbf{f}_k(\underline{x}_j)$$

$$\text{and} \quad \mathbf{f}(\underline{x}_i) \neq \mathbf{f}(\underline{x}_j) \quad ,$$

$$\text{where } k \text{ is the number of objective functions.}$$

Further,

$$\text{solution } \underline{x}_i \text{ is non-dominated if}$$

$$\nexists\, \underline{x}_j \text{ in the population of solutions that dominate } \underline{x}_i.$$

The set of all Pareto optimal points form a Pareto front and represent the trade-off surface among the multiple objective functions.

The Pareto front can be thought of as a set of the best trade-off solutions. Further, it is possible to define a hierarchical Pareto ranking by removing, at each generation, the current Pareto front members and then recalculating a new front from the remaining population members (potential solutions). Pareto ranking continues in this manner, removing the solutions belonging to the current nth Pareto front, until all the solutions have been ranked. This methodology is referred to as non-dominated sorting (see Figure 11.11). Pareto ranking is implemented in GAs by replacing the raw fitness values, generated by the objective functions, with dummy fitness values that are assigned based on each solution's membership within the Pareto front ranking. Lower Pareto front numbers are assigned higher dummy fitness values. Note that each solution that is a member of the same Pareto front receives the same dummy fitness value assigned to that front. This basic procedure, including dummy fitness value assignment, is often modified by a introducing a procedure to ensure that solutions are spread along the Pareto fronts. A commonly used algorithm that incorporates many of these ideas is the Non-dominated Sorting Genetic Algorithm-II (NSGA-II) (see Deb et al. [33]).

In a geographic configuration optimization approach, the loci of the chromosome tuples are mapped to all of, or a subset of, a polygon spatial dataset. Further, the length of the chromosome is the number of polygons whose attribute may be changeable in the design process. Thus the loci of the chromosomes are indexed to the index numbers of the changeable sites, and the corresponding alleles take values that characterize the polygonal partitions. The alleles are initialized from a finite alphabet and are evolved by the GA. The values of these alleles represent the various evolved configuration solutions, since each allele codes the feature type of a given partition of the landscape. Figure 11.12 shows a very simple example of these ideas. In this case the objective function counts the number of black-black neighbours on the dual graph of a polygonal partition, with three partitions available for changing attributes (vertices 2, 3, and 8). The functions to be optimized are designed to measure the performance of the configuration against, for example, metrics of ecological sustainability such as those proscribed in Forman's Landscape Ecology [41], [35].

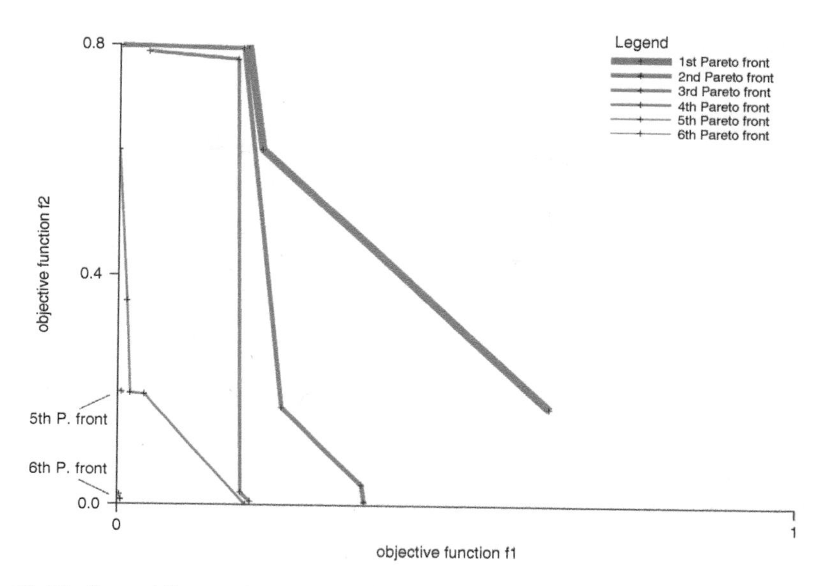

Figure 11.11 Sorted Pareto fronts for two object functions for a maximization problem.

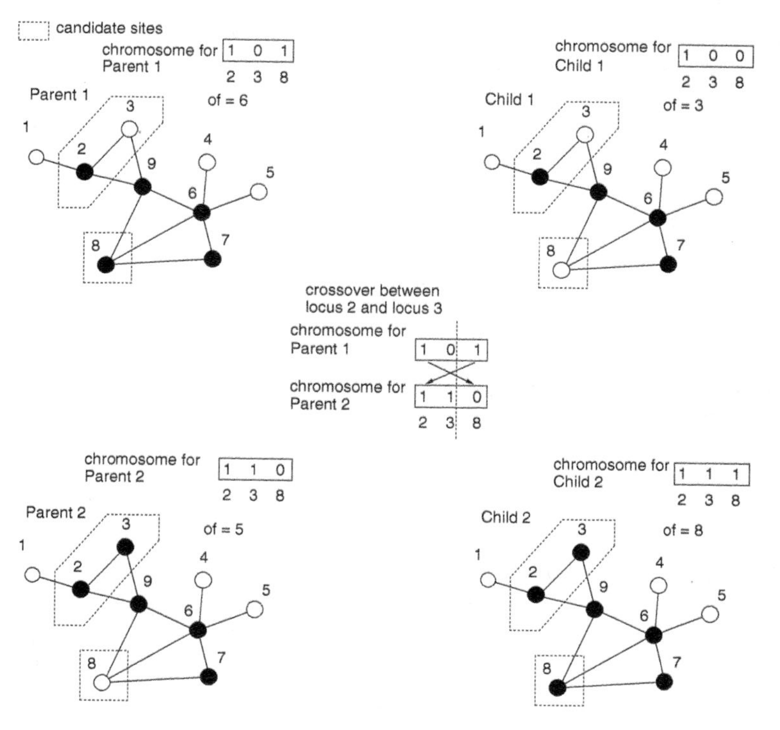

Figure 11.12 A simple example of coding a configuration optimization problem.

Problems

1. Use truth tables to show that $(A + B)' = A' \cdot B'$.
2. Generate a new raster from the input raster given below that indicates distance to nearest source raster cell, using a queen's neighbourhood to define nearest distance.

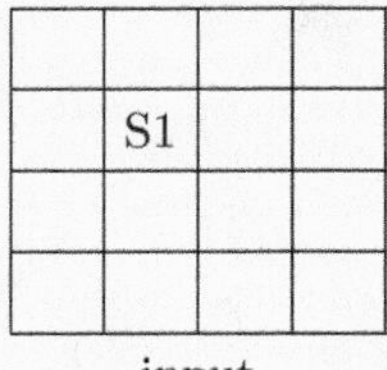

input

3. Implement Otsu segmentation, using the green band of the Victoria Landsat data. Segment the image into a binary image and interpret your results.
4. Use a membrane constraint stochastic interpolation on the Saskatchewan annual rainfall data. Compare the value estimated for Davidson with the actual value and the values calculated in this chapter using IDW interpolation and ordinary Kriging. Is a membrane prior model appropriate for rainfall data?
5. Think of three scenarios in which a geographic problem might be solved by considering a Pareto optimum solution. What are the disadvantages of the approach in each of these scenarios.

Chapter 12

Emerging Trends in Geographic Information

12.1 Introduction

The theories and technologies that collectively constitute GIS/GISci are rapidly changing, and predicting the future of the field is bound to be folly. A decade ago the practice of GIS was centred on the Desktop GIS Model, where GIS software was installed on a computer and a GIS analyst worked with the software and data to make maps, develop reports, and answer spatial questions. This paradigm is no longer, or unlikely to remain, the dominant mode of working with GIS. GIS-based technologies and applications have gone mainstream and have therefore become simplified and specialized to specific end-user applications (e.g., direction routing in Google Maps). As such, the role of classical GIS analysts who "do GIS" is receding, and taking their place are end-users, developers, and perhaps data scientists. The question remains: where does the GIScientist fit into this new paradigm—as theorist, as developer, as user? In this chapter we explore this question by highlighting areas of GIS we believe are on the cutting edge and show particular promise for new advancements in the acquisition, handling, and use of geographic information.

12.2 Data Acquisition

In 2003, K.C. Clarke [22] discussed the future of *geocomputation* and speculated that it would be driven by the extremes: in large-scale supercomputing and parallelized geospatial algorithms at one end, and nanoscale (i.e., thin) clients and sensors at the other. In the intervening decade or so, while the expansion of parallel (geo)computation has indeed continued, the nano trend has proven to be the more accurate forecast. Geosocial media streams, mobile phone-based sensors, low-cost location sensors, unmanned aerial vehicle technology, greater variety and access to satellite imagery, open data portals, and the like have created an unprecedented explosion in geographic data. This trend is likely to continue. For example, the introduction of the geolocation API in HTML5 provides the facility for every web page to have location-specific content. Two key developments will drive the use of geographic information to even more mainstream information technology than it is today. First, the increased networking of objects and peoples (i.e., the Internet of Things) will be geographically referenced. Coupled with location-aware web browsing and communication as a default, we can imagine virtually all information and communications technologies to have geographic representation. The underlying data modelling that supports these geographic representations and the move toward standardization of data technologies to support information interoperability then become paramount.

An important challenge for this to become reality is the mapping, modelling, and integration of indoor spaces within GIS. Applications of "indoor GIS" today include things like emergency routing, wayfinding, facilities management for large organizations, and tracking customer foot-traffic patterns within stores and shopping centres. OpenStreetMap also has several indoor mapping projects aiming to extend crowdsourced mapping to interiors of buildings. Taking this further, georeferenced and networked objects in the home could also be an area of future development as more and more devices become connected. The key research challenges for indoor GIS are the collection of indoor data and integration of existing indoor data (usually CAD) with GIS databases and applications.

What was perhaps not anticipated by previous forecasts of the geospatial landscape is the decidedly social element contributing to the geocoded social media and online interactions. While recent estimates of the percentage of Twitter activity that is geocoded put this at between 1% and 3% (see Morstatter et al. [81]), this is likely to increase as the benefits derived from "always on" geolocation begin to outweigh the costs associated with a loss of privacy and as the costs of mobile data usage continue to decline. The notion of individualized *augmented realities* (AR) delivering location- and individual-specific advertising through mobile and wearable computing is likely to further drive the geographic turn in modern society. For example, Layar is an augmented reality marketing company that has developed an app that allows users to scan magazines and newspapers with their smartphones in order to annotate the print content with dynamic real-time, location-specific content. This could be offering a lunch coupon at a nearby restaurant, good for the next 15 minutes, or providing a visual rendering of a news story based on features of the local environment. Location sensors also allow layering of geolocated context-aware content through the view of the phone's camera, such that the phone provides a mechanism to augment the real world with digital annotations (see Figure 12.1). While the idea of augmented reality have been around for some time (see Gronbaek et al. [54]), the widespread adoption of smartphones and apps suggests that the merging of digital and non-digital worlds is quickening. What will augmented realities mean for GIS?

We anticipate two key effects of this trend on GIS. The first is one of a further embedding and deepening of geographic information in society. The use of geographic information as part of location-sensitive applications that interact with digital content will create a GIS-by-default society. This is already beginning to happen. For example, search results from popular Internet search engines are now location-aware by default. As such, when a user searches for "Chinese food," the results set is restaurants located in the user's city or region. This occurs whether users are signed in to such services or not. Over time, Internet users will begin to expect this level of location awareness. Whether GIS—in its current form as a field of study interested in the technologies and theory underlying the handling of geographic information—is relevant to this transformation remains to be seen.

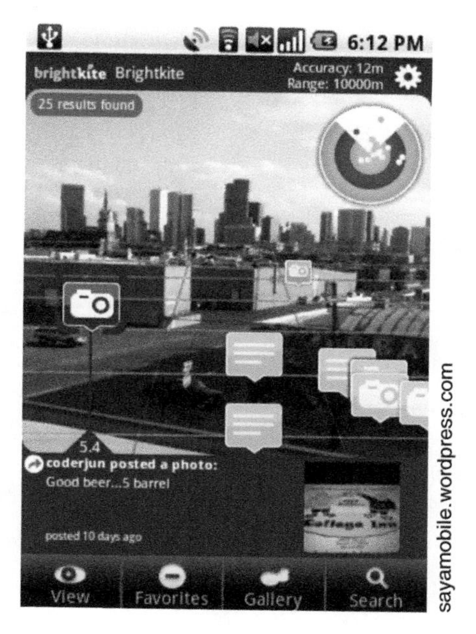

Figure 12.1 Augmented reality application.

The second key outcome of AR is that augmented realities will further blend digital and physical space through the *delivery* of digital annotations and content, and the same technologies will be used as *sensors* for geographic information. Mobile phones already track disease, spending, exercise, and health management, dating and storing these interactions in large distributed databases. The learning of human behaviours and interactions from such datasets via machine-learning algorithms has heralded extreme personalization of content (largely ads). We envisage that with an increasing share of such interactions being spatially referenced, there will be significant development of geospatial machine-learning algorithms. Machine-learning approaches have become very popular in prediction problems such as spam detection, recommendation engines, and topic modelling from web documents. Geospatial applications could include learning route optimization, landscape design, urban design, and place modelling and representation, and they could support artificial geospatial intelligence.

A related emerging challenge for GIS is the increasing phenomenon of machine-generated geographic information (MGI). An increasing share of the digital data being created today is created with little to no input from humans. Figure 12.2 presents the "big data pyramid," which represents the sources of data created in the world today as a pyramid, with human-generated data at the top, and machine-generated data making the largest (by size) share of data being created. If we were to extend this to the creation of GIS data (see Figure 12.3), we can envisage the vast majority of geographic information being created by sensors, mobile phones, and machine-generated processes. This implies that underlying data models will be increasingly important for handling these sources of information. The human-mediated production of geographic

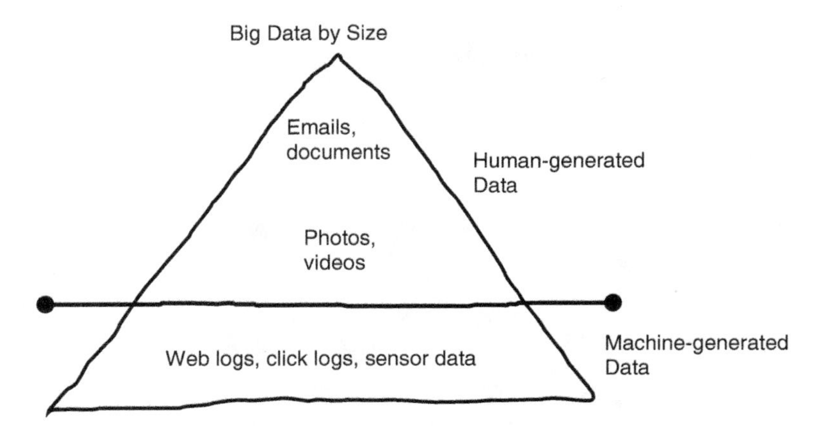

Figure 12.2 Big data pyramid.

Source: Adapted from Hadoop.org.

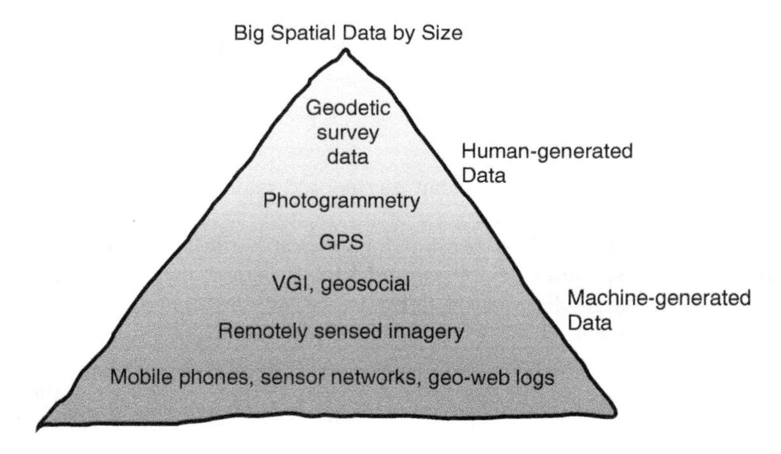

Figure 12.3 Big spatial data pyramid.

Source: Adapted from Hadoop.org.

information through geosocial activities represents a hybrid, human-machine source, as some portions of these data are generated by active participants and others by bots and automated processes. At the top of the pyramid is highly accurate survey-grade GIS data, which due to cost is likely to remain a small but important subset of the overall data distribution. The implications for GIS of Figure 12.3 are clear: greater need for integration of sensor data, development of robust spatial data models to handle and integrate human and machine-generated geographic information, and methods to link data obtained across the levels of the data hierarchy.

12.3 Data Modelling

The trends described above will pose important challenges for current approaches to handling geographic information. The first challenge already underway is the continued increase in the creation of geographic data. The idea of *big spatial data* has been articulated since the inception of the age of *big data*. While definitions of big data vary, many cite the five v's: volume, velocity, variety, veracity, and value. Each of these dimensions incurs challenges for geographic information (big spatial). What is common to most descriptions of big data, within the GIS community and elsewhere, is that data qualifies as *big* when traditional tools can no longer handle it. For GIS, this typically means relational database management systems, GIS file formats, spatial query and geoprocessing operations, and mapping. Each of these may need to be rethought to handle the large volumes of GIS data being created in the future.

Returning to Clarke's [22] forecast for parallel geocomputation, at this juncture, parallel geocomputation has failed to meet the demand of today's datasets commonly encountered. For example, Figure 12.4 is a map of the locations for all cargo vessels reported by satellite automatic identification system (AIS) in the North Pacific Ocean in one year. This dataset consists of over two million records. The total dataset for all ships obtained over the full year includes over a billion records and, maintained within a RDBMS stored on disk, occupies about 350 gigabytes. This type of dataset is typical for many GIScientists today, and the tools required for handling these datasets have

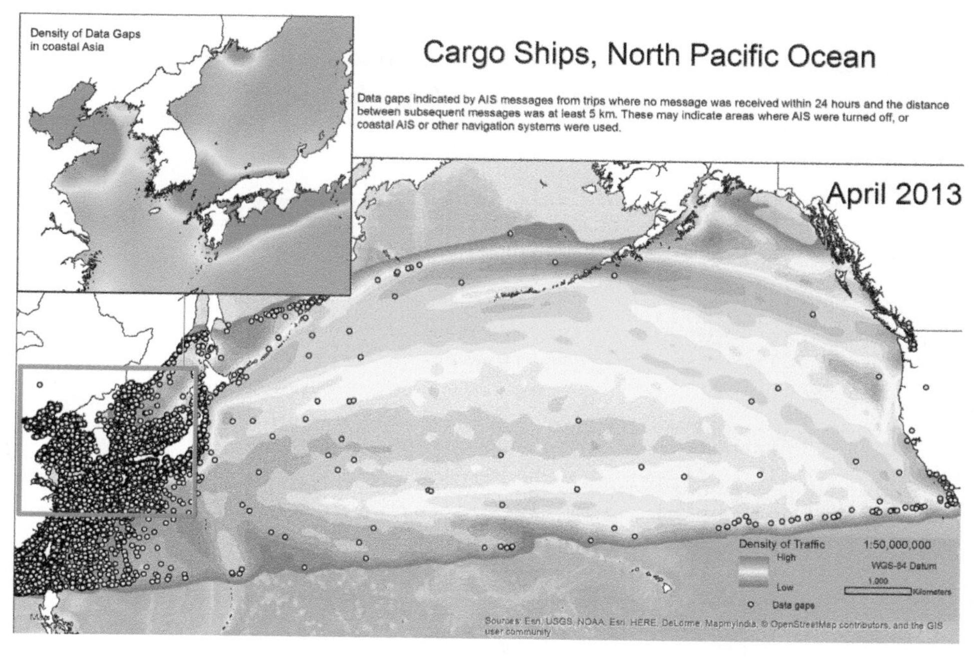

Figure 12.4 Cargo ship activity in the North Pacific Ocean.

not kept pace. Parallel and distributed file systems such as Hadoop have begun to solve this problem, but solving it for geospatial data remains a research frontier.

A key reason for the failure of parallel GIS to date has been a lack of a standardized environment for tackling parallelization problems. We'll highlight this situation through examples in both vector and raster, look at recent progress, and provide some possible avenues for future research. In the raster domain, parallel GIS development is commonly achieved through three types of computing architectures: graphical processing units (GPU), message passing interface (MPI), and what is called symmetrical multiprocessing (SMP). Each of these platforms differs with respect to how memory is shared between processors, how communication between processes occurs, and how input/output onto processors occurs. The lack of a standardized platform for parallel GIS computing means that algorithms are not portable between architectures and tend to be application-specific.

The degree to which the parallelization programming details can be hidden from the user/developer is an active area of research. Qin et al. [95] describe a set of operators for local, focal, and global raster operations in parallel that layer on top of any of the parallel architectures, thereby hiding some of the more idiosyncratic aspects of parallel programming. The details of the architecture are given in Figure 12.5. Further development of common middleware between parallel architectures and GIS representations is needed to fully realize the payoff in terms of computational speed.

An alternate approach is the trend toward cloud computing—centralizing data processing and delivering GIS analysis and data as a service over the Internet. The move toward delivery of data over the web has been underway for over a decade. For example, with recent releases of the open source software package Quantum GIS (www.qgis.org), data services from the cloud are delivered as base maps for use within the desktop software package. In the web, the technologies, frameworks, and processes giving rise to the explosion of geographic data delivered over the Internet—collectively termed the geoweb—have dramatically shifted how geographic information

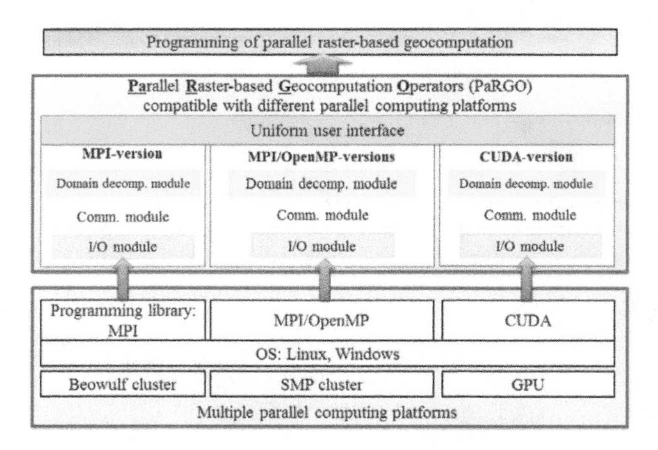

Figure 12.5 Middle tier layer for parallel raster GIS operations.

Source: From Qin et al. [95].

is consumed. There is a movement to delivering more-advanced GIS functionality from a central web-based tier to mobile, desktop, and even object clients over the Internet. The varying degrees of cloud-based services, generally suffixed with *as a service*, currently provide different models for future development (infrastructure as a service, software as a service, etc.). An existent service is Amazon's EC2, which provides monthly access to supercomputing power upon which geospatial applications and computations are built.

In addition to the pay-for-power model of EC2, many regional supercomputing clusters among universities and public sector organizations have been formed to fill this niche, echoing the early days of modern GIS. Tang and Feng [111] provide a case study of cloud computing for GIS by focusing on implementing map projection functions on a cloud-based GPU cluster. While experiments in [111] show significant speed-ups in the GPU-based cloud implementation over the CPU-based model, it is unclear whether this is the solution to the current big spatial data problem. We anticipate significant investment and development of GIS services and projects that fuse high-performance computing methods with cloud infrastructure and Internet-based delivery. However, at this point there remains little evidence that cloud-based GIS will emerge as more than a niche part of the field.

A danger amid the current hype over the size of one's data is that research into spatial representation and spatial data models will suffer. Recent history has shown us that there are many reasons technologies become predominant, not necessarily because they are superior to the alternatives. Consider the widespread use of shapefiles as the data format for storing and sharing GIS data. For many reasons, alternatives to the shapefile available at the time were superior—for example, ESRI's Coverage data model or the US Census TIGER/Line data model, which were topological data models. Yet the open specification and portability of the shapefile made it easy for developers and GIS researchers and practitioners to develop tools, customizations, or whole software packages to read and write shapefiles. As such, the format ascended to far greater use than was ever intended for it. While this is perhaps starting to change with the advent and widespread use of GeoJSON and its topological extension topoJSON as formats for interchange and representation of geographic information based on JavaScript, further development of core data models and representations is needed to better handle today's data-rich environment.

A simple example of this is given by Schneider [102], who develops an implementation of fuzzy spatial objects called Spatial Plateau Algebra, using existent spatial

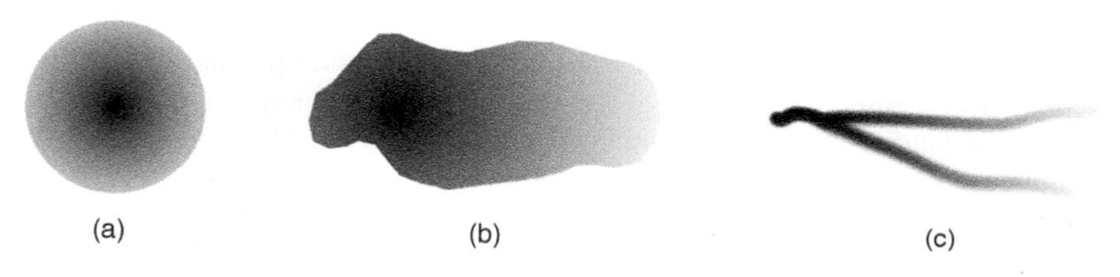

Figure 12.6 Visual representation of fuzzy spatial objects.
Source: Schneider [102].

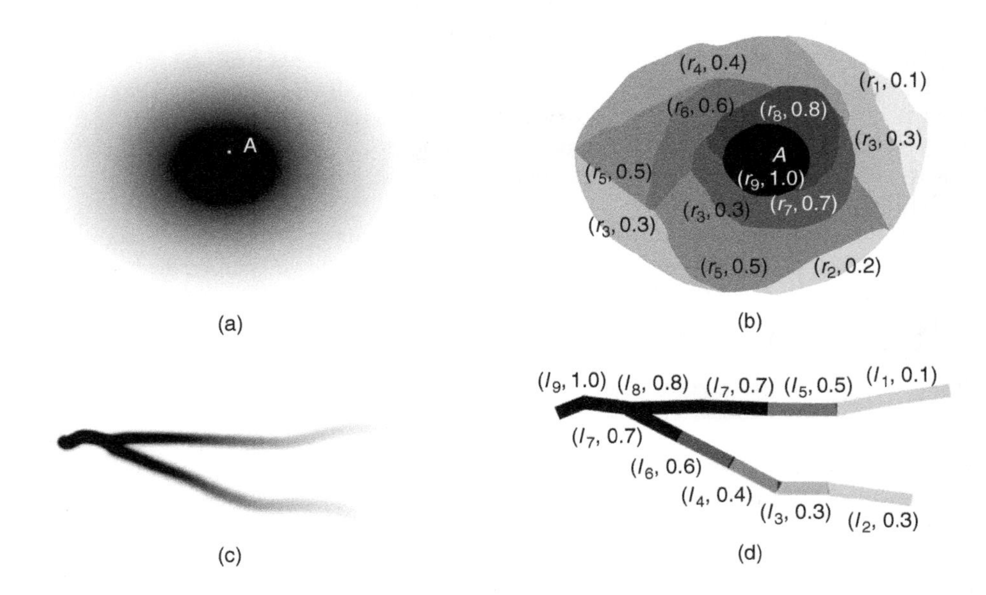

Figure 12.7 Conceptual representation and implementation of fuzzy spatial objects.
Source: Schneider [102].

data types. The data types and resultant objects yield fuzzy points, lines, and regions derived from their crisp counterparts (see Figure 12.7). Fuzzy objects have in them variable degrees of membership, which represent internal uncertainty within the objects; they therefore can represent *vague boundaries* and *blurred interiors*. While fuzzy spatial objects have been around for a while at a conceptual level, and fuzzy set theory has been employed widely in GIS, few examples of implementations of fuzzy spatial objects exist. Schneider defines each spatial plateau object with a finite set of membership values that are attached to component crisp objects, which in aggregate make up *plateau objects* (Figure 12.7). This differs from other fuzzy systems where membership values are made of fields (see Cova and Goodchild [28]) or points. Each of the component crisp objects is of variable size, and the precision in representation is determined by smaller and more numerous component objects. This system of types then supports fuzzy operations, such as intersect, union, range queries, and fuzzy spatial analysis, within available spatial data types and query languages.

Another interesting area of current research is the embedding of parallelism into GIS data structures to support data management operations such as spatially aware extract, transform, and load tools. The development of spatial indices in parallel architectures, for example, is an active research area (see, for example, Gao et al. [42]). The movement away from RDBMS as the basis for data storage in GIS would be a logical extension of the proliferation of distributed databases and the so-called NoSQL data storage options currently available. A relevant example today is the SpatialHadoop project based in the University of Minnesota (spatialhadoop.cs.umn.edu). Here, spatial indices are built as an extension to the popular Hadoop map/reduce framework. As data are increasingly becoming always-on, streaming, spatially-referenced feeds of information, with variable structure, the notion of normal forms will increasingly be eclipsed by real-time processing and what might be termed *ephemeral computing*. This does not indicate an end to the RDBMS, but rather that the frontier of GIS research and development will be in alternate data structures.

A promising example is the integration of parallelism at the database level, exemplified by a system called MapD (massively parallel database), which implements a relational database on a hybrid GPU/CPU architecture (see Mostak [82]) and supports querying, analysis, and rendering of big datasets (see Figure 12.8). The difficulty of a GPU database in the past was the high cost of transferring data over a relatively slow Peripheral Component Interface (PCI) bus to the CPU. This is achieved in MapD by a combination of architecture and procedures that are optimized for GPU/CPU integration. A key factor is that rather than using the CPU memory buffer / disk memory dichotomy of most RDBMS, MapD uses a three-tier model from GPU to CPU to disk, with each lower tier consisting of larger storage and slower access. Data are partitioned by tier according to frequency of use, so the most frequently used data are available directly from the GPU memory. In addition, at the schema level the data are stored in tables as columns rather than rows. Other advances include GPU-optimized query processing and novel spatial operations using GPU-accelerated rasterization-based algorithms for computing spatial intersections.

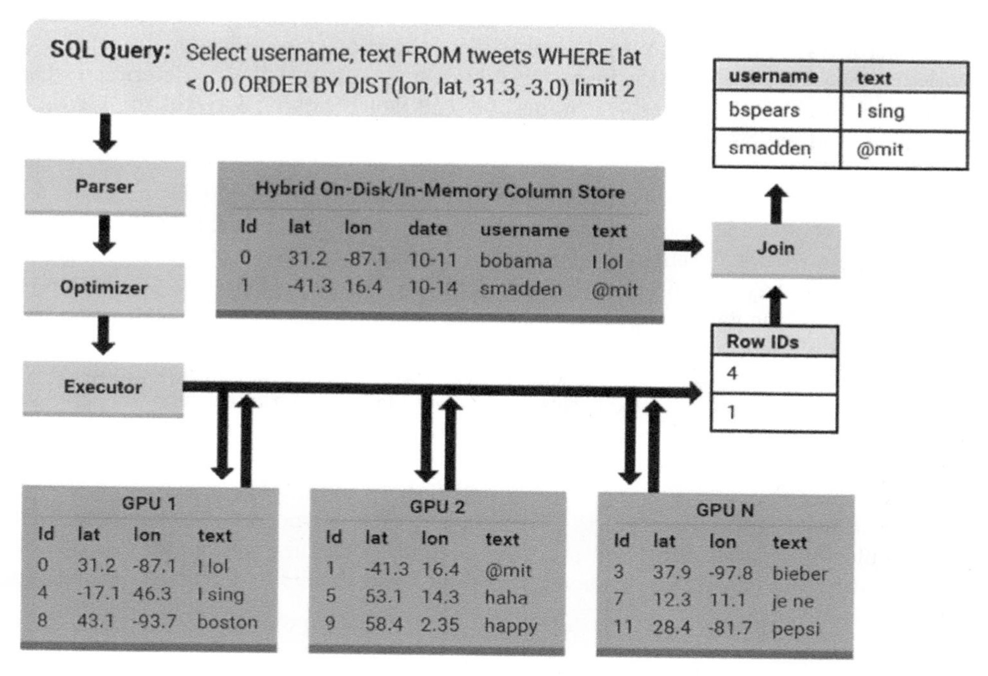

Figure 12.8 Query implementation in MapD GPU/CPU database.

Source: Mostak [82] © 2016 MapD Technologies, Inc.

12.4 GIS, Data, and Society

To date, the age of big data has been a decidedly corporate phenomenon. Whereas GIS first emerged in the fusion of university-based research and government agencies responsible for administering large environmental datasets, the big data evangelists of today are largely commercial in nature. And while the *applications* of big data are commercially oriented, many of the tools and communities supporting big data analysis are open-source. A good example is the R-Project for Statistical Computing. R developed as an experiment in statistical computing, based on Lisp-like functional programming (originally a language called Scheme). Today, R is a fully-fledged programming language that has full support for vector and raster spatial classes and has packages for standard GIS functionality in addition to numerous packages for spatial statistics, modelling, and visualization. Similar open source software initiatives in GIS such as QGIS, PostGIS, OpenStreetMap, and PySAL (in python) suggest that the trend toward collaborative development of tools for geospatial computing will continue and perhaps begin to eclipse the traditional role of commercial GIS vendors.

Barnes and Wilson [7] give an interesting interpretation of the big data movement as an extension of the *social physics* attempts throughout the 20th century to quantify human experience via generalizable laws and patterns that govern the social world as they do the physical. They recount the history of social physics and trace its lineage, ushered along by the spatial analysis and GIScience communities of geography and into modern big data. In a similar vein, the physicist Geoffery West famously reported by the New York Times to have "solved the city" by incorporating ever-larger datasets describing physical infrastructure, health, education, finance, weather, etc.; identifying statistical relationships between these variable sets; and generalizing to determine what makes certain cities thrive and others decline. GIS figures prominently in the big data movement, and critical approaches to big data are emerging in the literature. Graham et al. [53] map the uneven geography of user-generated geographic data and highlight the danger posed from over-reliance on unrepresentative data sources, and the potential for GIS analysis of these data distributions to reinforce social norms and perspectives that disadvantage under-represented groups in society. Similar criticisms by Boyd and Crawford [15] and Andrejevic [3] point to dangers in the way that emerging data-intensive society increases the gap between the data collectors (e.g., large corporations) and the those whom data are being collected about. The power disadvantage realized from this imbalance may have undesirable outcomes that deepen and codify social inequalities. A further issue relates to the implicit, rather than explicit, trading of services for personal information that permeates today's digital experiences. And while these are symptoms of much larger societal issues, it is likely that space, data, and privacy will be dominant themes in these debates. Echoing the GIS and Society critiques of the 1990s, there is now a push-back in the literature to incorporate critical perspectives into big data and in particular big spatial data. Spatial data handling is on the forefront of this debate as issues of locational privacy are being raised by governments and citizens around the globe.

Problems

1. Make a prediction about which of the emerging technologies discussed in this chapter will make the greatest impact on GIS practice over the next 10 years. Give an example.
2. Discuss the challenges of parallel computing for GIS data.

Appendix I

Mathematical Notation and Terminology

In this appendix we collect some mathematical ideas, notation, and definitions that will be useful throughout the text.

I.1 Sets

In this section we give a very brief and informal overview of some of the ideas of mathematical sets we use in the text.

{ } (curly brackets) indicate a collection of distinguishable objects.
We may also use capital letters to refer implicitly to a set of objects.

$$\text{e.g., } A = \{1, 3, 2, 7\}$$

Note that the order of the elements of the collection is ignored. If we need to consider the order of the collection of objects—in a tuple, for example, or in an ordered pair—we often use curved brackets to indicate this.

$$\text{e.g., } \underline{P}_i = (x_i, y_i)$$

The elements of sets may themselves be sets (see an example in Appendix II).

$x \in A$ indicates that object x is an element of set A.

$$\text{e.g., if } A \text{ as above, then } 3 \in A$$

$x \notin A$ indicates that object x is not an element of set A.

$$\text{e.g., if } A \text{ as above, then } 4 \notin A$$

$A \cup B$ indicates A union B and is defined as a new set such that $x \in A$ or $x \in B$.

$$\text{e.g., if } A \text{ as above and } B = \{2, 5\}, \text{ then } A \cup B = \{1, 2, 3, 5, 7\}$$

$A \cap B$ indicates A intersection B and is defined as a new set such that $x \in A$ and $x \in B$.

$$\text{e.g., if } A \text{ and } B \text{ as above, then } A \cap B = \{2\}$$

$A \setminus B$ indicates the set difference of A from B and is defined as a new set such that $x \in A$ and $x \notin B$.

$$\text{e.g., if } A \text{ and } B \text{ as above, then } A \setminus B = \{1, 3, 7\} \text{ and } B \setminus A = \{5\}$$

$|A|$ denotes the number of elements in set A.

$$\text{e.g., if } A \text{ as above, } |A| = 4$$

If A and B are sets, then $A = B$ means that A and B contain exactly the same elements.

$$\text{e.g., if } A \text{ and } B \text{ as above, then } A \neq B$$

If A and B are sets, then $A \subseteq B$ denotes that A is a subset of B and is defined as $x \in A$ implies $x \in B$.

$$\text{e.g., if } A \text{ as above, then } \{2, 7\} \subseteq A$$

If A and B are sets, then $A \subset B$ denotes that A is a proper subset of B and is defined as $x \in A$ implies $x \in B$ and $A \neq B$.

$$\text{e.g., if } B \text{ as above, then } \{2\} \subset B$$

$\emptyset$ indicates a set with no elements.

I.2 Summation

In this section we introduce the summation notation and list a few of its useful algebraic properties.

$$\sum_{i=1}^{n} a_i \text{ is shorthand notation for } a_1 + a_2 + \cdots + a_n$$

Note that in this notation i is an integer value and in the example above, 1 is the starting value (the first subscript on the variable being summed). Further, n is also an integer and is the ending subscript (the last subscript of the variable to be summed, in this case left unknown).

$$\text{e.g., if } n = 3,\ a_1 = 2,\ a_2 = 5 \text{ and } a_3 = 1 \text{, then}$$

$$\sum_{i=1}^{3} a_i = a_1 + a_2 + a_3 = 2 + 5 + 1 = 8$$

Products

$$\sum_{i=1}^{n} x_i^2 = x_1^2 + x_2^2 + \cdots + x_n^2$$

$$\sum_{i=1}^{n} x_i \cdot y_i = x_1 \cdot y_1 + x_2 \cdot y_2 + \cdots + x_n \cdot y_n$$

Associativity

$$\sum_{i=1}^{n} (x_i + y_i) = \sum_{i=1}^{n} x_i + \sum_{i=1}^{n} y_i$$

Distributivity

$$\sum_{i=1}^{n} c \cdot x_i = c \cdot \sum_{i=1}^{n} x_i \qquad \text{for } c \text{ a constant}$$

Sum of a constant

$$\sum_{i=1}^{n} c = n \cdot c \qquad \text{for } c \text{ a constant}$$

Partial sums

$$\sum_{i=1}^{n} x_i = \sum_{i=1}^{k} x_i + \sum_{i=k+1}^{n} x_i$$

$$x_1 + x_2 + \cdots + x_n = (x_1 + x_2 + \cdots + x_k) + (x_{k+1} + x_{k+2} + \cdots + x_n)$$

Double sums

$$\begin{aligned}\sum_{i=1}^{n} \sum_{j=1}^{m} x_{i,j} &= \sum_{i=1}^{n} (x_{i,1} + x_{i,2} + \cdots + x_{i,m}) \\ &= x_{1,1} + x_{1,2} + \cdots + x_{1,m} \\ &\quad + x_{2,1} + x_{2,2} + \cdots + x_{2,m} \\ &\quad \vdots \\ &\quad + x_{n,1} + x_{n,2} + \cdots + x_{n,m} \\ &= x_{1,1} + x_{1,2} + \cdots + x_{n,m}\end{aligned}$$

I.3 Vectors and Matrices

In this section we introduce the notation and basic algebra for vectors and matrices. This notation and algebra is extremely useful for multi-dimensional data.

Vector

A vector is commonly represented as a column of values indexed from 1 to n, where n is an integer. The dimension of the vector is $n \times 1$ meaning n rows by 1 column.

$$\underline{x} = \begin{bmatrix} x_1 \\ x_2 \\ \vdots \\ x_n \end{bmatrix}$$

Transpose of a vector

The transpose operation interchanges columns to rows. This changes the dimension of the column vector from $n \times 1$ to $1 \times n$.

$$\underline{x}^{\mathrm{T}} = [x_1, \quad x_2, \quad \cdots \quad , x_n]$$

Matrix

Note that the first integer in the double index indicates the row number of the entry in the matrix and the second integer indicates the column number.

$$\mathbf{X} = \begin{bmatrix} x_{1,1} & x_{1,2} & \cdots & x_{1,n} \\ x_{2,1} & x_{2,2} & \cdots & x_{2,n} \\ \vdots & \vdots & \ddots & \vdots \\ x_{n,1} & x_{n,2} & \cdots & x_{n,n} \end{bmatrix}$$

Transpose of a matrix

$$\mathbf{X}^{\mathrm{T}} = \begin{bmatrix} x_{1,1} & x_{2,1} & \cdots & x_{n,1} \\ x_{1,2} & x_{2,2} & \cdots & x_{n,2} \\ \vdots & \vdots & \ddots & \vdots \\ x_{1,n} & x_{2,n} & \cdots & x_{n,n} \end{bmatrix}$$

Vector addition

Note that the vectors must be the same dimension to be added.

$$\underline{x} + \underline{y} = \begin{bmatrix} x_1 \\ x_2 \\ \vdots \\ x_n \end{bmatrix} + \begin{bmatrix} y_1 \\ y_2 \\ \vdots \\ y_n \end{bmatrix} = \begin{bmatrix} x_1 + y_1 \\ x_2 + y_2 \\ \vdots \\ x_n + y_n \end{bmatrix}$$

Scalar vector multiplication

$$c \cdot \underline{x} = \begin{bmatrix} c \cdot x_1 \\ c \cdot x_2 \\ \vdots \\ c \cdot x_n \end{bmatrix}$$

Matrix addition

Note that the matrices must be the same dimension to be added.

$$\mathbf{X} + \mathbf{Y} = \begin{bmatrix} x_{1,1} + y_{1,1} & x_{1,2} + y_{1,2} & \cdots & x_{1,n} + y_{1,n} \\ x_{2,1} + y_{2,1} & x_{2,2} + y_{2,2} & \cdots & x_{2,n} + y_{2,n} \\ \vdots & \vdots & \ddots & \vdots \\ x_{n,1} + y_{n,1} & x_{n,2} + y_{n,2} & \cdots & x_{n,n} + y_{n,n} \end{bmatrix}$$

Scalar matrix multiplication

$$c \cdot \mathbf{X} = \begin{bmatrix} c \cdot x_{1,1} & c \cdot x_{1,2} & \cdots & c \cdot x_{1,n} \\ c \cdot x_{2,1} & c \cdot x_{2,2} & \cdots & c \cdot x_{2,n} \\ \vdots & \vdots & \ddots & \vdots \\ c \cdot x_{n,1} & c \cdot x_{n,2} & \cdots & c \cdot x_{n,n} \end{bmatrix}$$

Vector multiplication

Notice that vector and matrix multiplication requires the inner dimensions of the two inputs to match and the resultant product has the dimension of the outer input dimensions. In general, vector and matrix multiplication are not commutative.

$$\underbrace{\underline{x}^{\mathrm{T}} \cdot \underline{y}}_{(1\times n)(n\times 1)} = [x_1,\ x_2,\ \cdots,\ x_n] \cdot \begin{bmatrix} y_1 \\ y_2 \\ \vdots \\ y_n \end{bmatrix} = \underbrace{\sum_{i=1}^{n} x_i \cdot y_i}_{(1\times 1)}$$

$$\underbrace{\underline{x} \cdot \underline{y}^{\mathrm{T}}}_{(n\times 1)(1\times n)} = \begin{bmatrix} x_1 \\ x_2 \\ \vdots \\ x_n \end{bmatrix} \cdot [y_1,\ y_2,\ \cdots,\ y_n] = \underbrace{\begin{bmatrix} x_1 \cdot y_1 & x_1 \cdot y_2 & \cdots & x_1 \cdot y_n \\ x_2 \cdot y_1 & x_2 \cdot y_2 & \cdots & x_2 \cdot y_n \\ \vdots & \vdots & \ddots & \vdots \\ x_n \cdot y_1 & x_n \cdot y_2 & \cdots & x_n \cdot y_n \end{bmatrix}}_{(n\times n)}$$

Matrix multiplication

$$\underbrace{\mathbf{X} \cdot \mathbf{Y}}_{(n\times m)(m\times p)} = \begin{bmatrix} x_{1,1} & x_{1,2} & \cdots & x_{1,m} \\ x_{2,1} & x_{2,2} & \cdots & x_{2,m} \\ \vdots & \vdots & \ddots & \vdots \\ x_{n,1} & x_{n,2} & \cdots & x_{n,m} \end{bmatrix} \cdot \begin{bmatrix} y_{1,1} & y_{1,2} & \cdots & y_{1,p} \\ y_{2,1} & y_{2,2} & \cdots & y_{2,p} \\ \vdots & \vdots & \ddots & \vdots \\ y_{m,1} & y_{m,2} & \cdots & y_{m,p} \end{bmatrix}$$

$$= \underbrace{\begin{bmatrix} \sum_{i=1}^{m} x_{1,i} \cdot y_{i,1} & \sum_{i=1}^{m} x_{1,i} \cdot y_{i,2} & \cdots & \sum_{i=1}^{m} x_{1,i} \cdot y_{i,p} \\ \sum_{i=1}^{m} x_{2,i} \cdot y_{i,1} & \sum_{i=1}^{m} x_{2,i} \cdot y_{i,2} & \cdots & \sum_{i=1}^{m} x_{2,i} \cdot y_{i,p} \\ \vdots & \vdots & \ddots & \vdots \\ \sum_{i=1}^{m} x_{n,i} \cdot y_{i,1} & \sum_{i=1}^{m} x_{n,i} \cdot y_{i,2} & \cdots & \sum_{i=1}^{m} x_{n,i} \cdot y_{i,p} \end{bmatrix}}_{(n\times p)}$$

I.4 Mathematical Expectation

In this section we give basic definitions and properties, with minimal explanation, of mathematical expectation and variance. In the expressions below, X and Y are random variables, $p(x)$ is the probability of $X = x$ and a, b, and c are constants. These results are used in some of the derivations in the text.

Mathematical Expectation—discrete case

$$E[X] = \sum_{i=1}^{n} x_i \cdot p(x_i) \quad , \qquad \text{where} \quad \sum_{i=1}^{n} p(x_i) = 1$$

Mathematical Expectation—continuous case

$$E[X] = \int_{-\infty}^{\infty} x \cdot p(x)dx \quad , \qquad \text{where} \quad \int_{-\infty}^{\infty} p(x)dx = 1$$

$$E[a \cdot X + b] = a \cdot E[X] + b$$

$$E[a \cdot X + b \cdot Y + c] = a \cdot E[X] + b \cdot E[Y] + c$$

Mathematical Variance—discrete case

$$\text{var}[X] = \sum_{i=1}^{n} (x_i - E[X])^2 \cdot p(x_i) \quad , \qquad \text{where} \quad \sum_{i=1}^{n} p(x_i) = 1$$

Mathematical Variance—continuous case

$$\text{var}[X] = \int_{-\infty}^{\infty} (x - E[X])^2 \cdot p(x)dx \quad , \qquad \text{where} \quad \int_{-\infty}^{\infty} p(x)dx = 1$$

$$\text{var}[X] = E[X^2] - (E[X])^2$$

$$\text{var}[a \cdot X + b] = a^2 \cdot \text{var}[X]$$

$$\text{var}[a \cdot X + b \cdot Y + c] = a^2 \cdot \text{var}[X] + b^2 \cdot \text{var}[Y] + 2 \cdot a \cdot b \cdot cov[X, Y]$$

Appendix II

Mathematical Graphs

In this appendix we introduce some formal definitions and theoretical results for a series of mathematical objects collectively referred to as mathematical graphs.

A *mathematical graph*, G, is defined as a set $\{V, E\}$ of *vertices* and *edges*.

The vertex set of G, $V\{G\}$, must be nonempty.

An edge joins 2 vertices. The edge set of G, $E\{G\}$ may be empty.

If an edge $e \in E$ joins vertices $u \in V$ and $v \in V$, then u and v are called *incident* to edge e, and u and v are considered *adjacent* vertices.

If $e_1 \in E$ and $e_2 \in E$ share a common incident vertex v, then they are referred to as *adjacent* edges.

A *complete graph*, K, has every vertex adjacent to every other vertex.

A *planar graph* is a graph that is embeddable in the plane, with no edges intersecting except at their incident vertices.

Parallel edges are edges that join the same pair of vertices. Graphs that allow parallel edges are called *multi-graphs*.

A *loop* is an edge that joins a vertex to itself. Graphs that allow loops are called *pseudo-graphs*.

A *directed graph*, G^{D}, is a set of vertices, $V\{G^{\mathrm{D}}\}$, and a set of directed edges, $E\{G^{\mathrm{D}}\}$. The directed edges are represented via a set of ordered pairs of vertices, (u, v), that

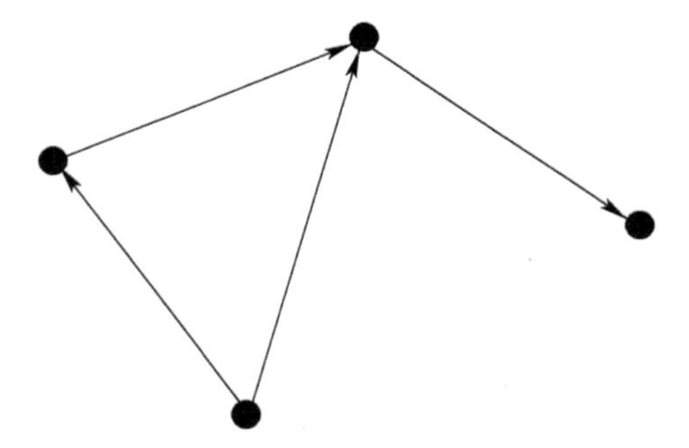

Figure II.1 An example of a directed, planar graph.

induce an orientation on their incident edge, $e \in E\{G^{\mathrm{D}}\}$. That is, a directed graph is a graph where the edges are assigned a direction or orientation. Figure II.1 is an example of a directed, planar graph.

A *dual graph*, G^{d}, is a graph created from a planar graph G by the following process:

1. A vertex is created for each enclosed region in G.
2. A universe vertex is created for the area outside all enclosed regions in G.
3. Edges are added between region vertices that share a common boundary in the underlying planar graph G (including the universe vertex).
4. All parallel edges are deleted.

Figure II.2 is an example of a dual graph with its underlying planar graph.

A *walk* from vertex u to vertex v in a graph G is a finite alternating sequence,

$$u = u_0, e_1, u_1, e_2, \ldots, u_{k-1}, e_k, u_k = v \quad ,$$

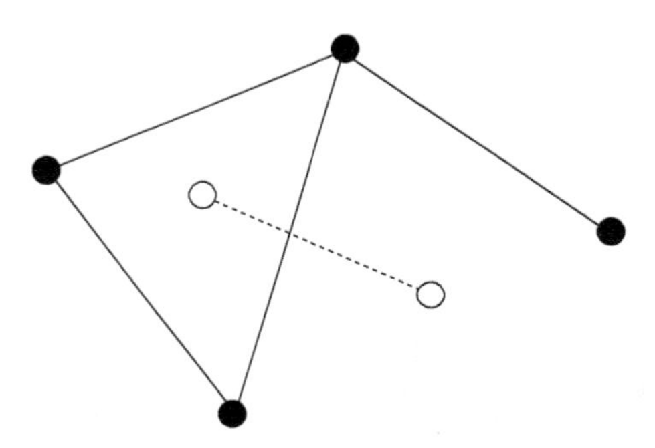

Figure II.2 An example of a dual graph with its underlying planar graph.

of vertices and edges, where u and v need not be distinct. If $u = v$ the walk is closed; otherwise it is open.

A u to v *trail* is a u to v walk with no edges repeated.

A u to v *path* is a u to v trail with no vertices repeated.

A vertex u is *connected* to vertex v if there is a u to v path in G.

A graph G is called *connected*, if every pair of its vertices is connected.

A *circuit* is a closed trail with at least one edge.

A *cycle* is a circuit,

$$v_1, v_2, \ldots, v_n, v_1 \ (n \geq 3) \quad ,$$

whose n vertices are distinct.

A *tree* is a connected, acyclic (without cycles) graph. The number of edges in a tree is the minimum number for a given number of vertices to make the graph connected and is equal to $v - 1$ (where v is the number of vertices).

A *weighted graph* is a graph whose edges or vertices are assigned numeric values.

A *minimum spanning tree* (*MST*) is a tree with the additional property that the sum of weights on the edges is minimal.

II.1 Prim's Algorithm for Finding a Minimum Spanning Tree (MST)

input: A set of points, $X = \{\underline{x}_1, \ldots, \underline{x}_n\}$
output: An MST for X, consisting of a vertex set S and an edge set T

initialize: $S = \{\underline{v}\}$ for any vertex $\underline{v}$, $T = \emptyset$

while $|T| < n - 1$ do

$\quad (\underline{w} \in X \backslash S \ , \ \underline{v}^* \in S) = \underset{\underline{z} \in X \backslash S}{\operatorname{argmin}}\{d(\underline{z}, S)\}$

$\quad S = S \cup \{\underline{w}\}$

$\quad T = T \cup \{\text{edge from } \underline{v}^* \text{ to } \underline{w}\}$

end

II.2 Euler's Equation

We now present a sketch proof that Euler's equation, $v - e + f = 2$, holds for all connected planar (finite) graphs.

Steps:

1. Choose any connected planar (finite) graph.
2. Systematically reduce the size of the graph by either
   a. removing one outside edge (removes one edge and one face)
   or
   b. removing a dangling edge (removes one edge and one vertex).
3. Calculate Euler's equation.
4. Continue until left with one vertex (the minimum size graph).

Note that at every iteration Step 3, Euler's equation is satisfied and that since we are free to choose *any* connected planar (finite) graph as a starting point, the result holds for *all* connected planar (finite) graphs.

QED

Appendix III

Why $(n-1)$ and not n in s^2?

In this appendix we collect some mathematical ideas, notation, and definitions that will be useful for further study, using an explanation of the form of the equation of s^2 as a structuring theme.

Maximum likelihood estimation

Consider a family of probability density functions (pdfs) given by

$$f(x \mid \theta)\ ,\ \theta \in \Omega \quad ,$$

where Ω is the parameter space.

The probability that $X_1 = x_1,\ X_2 = x_2,\ \ldots\ ,\ X_n = x_n$ is the joint pdf

$$f(x_1 \mid \theta) \cdot f(x_2 \mid \theta) \cdot \ \ldots\ \cdot f(x_n \mid \theta)\ ,\ \theta \in \Omega$$

or, regarded as a function of θ,

$$L(\theta) = f(x_1 \mid \theta) \cdot f(x_2 \mid \theta) \cdot \ \ldots\ \cdot f(x_n \mid \theta)\ ,\ \theta \in \Omega \quad .$$

If we find a function $u(\ x_1\ ,\ x_2\ ,\ \ldots\ ,\ x_n) = \hat{\theta}$ such that L is maximized, then u is the maximum likelihood estimate (MLE) of θ.

For example, consider the normal pdf,

$$N(\theta_1,\ \theta_2) = \frac{1}{\sqrt{2 \cdot \pi \cdot \theta_2}} \cdot \exp\left[-\frac{(x-\theta_1)^2}{2 \cdot \theta_2}\right] \quad .$$

This gives

$$L(\theta_1\ ,\ \theta_2\ ,\ x_1\ ,\ x_2\ ,\ \ldots\ ,\ x_n) = \left(\frac{1}{\sqrt{2 \cdot \pi \cdot \theta_2}}\right)^n \cdot \exp\left[-\frac{\sum_{i=1}^{n}(x_i - \theta_1)^2}{2 \cdot \theta_2}\right] \quad .$$

It will make further calculations easier if we take the natural logarithm of both sides, giving

$$\ln\ L(\theta_1\ ,\ \theta_2\ ,\ x_1\ ,\ x_2\ ,\ \ldots\ ,\ x_n) = -\frac{\sum_{i=1}^{n}(x_i - \theta_1)^2}{2 \cdot \theta_2} - \frac{n \cdot \ln(2 \cdot \pi \cdot \theta_2)}{2} \quad .$$

To maximize L we proceed as usual and take partial derivatives and set these equal to zero; that is,

$$\frac{\partial\ \ln L}{\partial \theta_1} = 0 \qquad \frac{\partial\ \ln L}{\partial \theta_2} = 0 \quad .$$

First, with respect to θ_1,

$$\frac{\partial\ \ln L}{\partial \theta_1} = 0$$

$$\frac{\partial}{\partial \theta_1}\left(-\frac{\sum_{i=1}^{n}(x_i - \theta_1)^2}{2 \cdot \theta_2} - \frac{n}{2} \cdot \ln(2 \cdot \pi \cdot \theta_2) \right) = 0$$

$$\frac{-\sum_{i=1}^{n} 2 \cdot (x_i - \theta_1)^{(2-1)} \cdot (-1)}{2 \cdot \theta_2} - 0 = 0$$

$$\frac{\sum_{i=1}^{n}(x_i - \theta_1)}{\theta_2} = 0$$

$$\sum_{i=1}^{n}(x_i - \theta_1) = 0$$

$$\sum_{i=1}^{n} x_i - \sum_{i=1}^{n} \theta_1 = 0$$

$$\sum_{i=1}^{n} x_i = \sum_{i=1}^{n} \theta_1$$

$$\sum_{i=1}^{n} x_i = n \cdot \theta_1$$

$$\theta_1 = \frac{\sum_{i=1}^{n} x_i}{n} = \bar{X} \quad .$$

So the sample mean $\bar{X}$ is the MLE of θ_1 ($\hat{\theta}_1 = \bar{X}$) for a normal distribution. Next, with respect to θ_2, we have

$$\frac{\partial \ \ln L}{\partial \theta_2} = 0$$

$$\frac{\partial}{\partial \theta_2}\left(-\frac{\sum_{i=1}^{n}(x_i - \theta_1)^2}{2 \cdot \theta_2} - \frac{n}{2} \cdot (\ln(2 \cdot \pi) + \ln(\theta_2)) \right) = 0$$

$$\frac{\sum_{i=1}^{n}(x_i - \theta_1)^2}{2 \cdot {\theta_2}^2} - \frac{n}{2 \cdot \theta_2} = 0$$

$$\frac{\sum_{i=1}^{n}(x_i - \theta_1)^2}{2 \cdot {\theta_2}^2} = \frac{n}{2 \cdot \theta_2}$$

$$\frac{\sum_{i=1}^{n}(x_i - \theta_1)^2}{n} = \theta_2 = s_*^2 \quad .$$

So s_*^2 is the MLE of θ_2 ($\hat{\theta}_2 = s_*^2$) for a normal distribution.

Note that s_*^2 is not quite the same as our familiar sample variance s^2. To explore the reason for this difference we must introduce the idea of biased and unbiased estimators. An estimator is said to be unbiased if

$$\text{E[estimate of the statistic]} = \text{the statistic} \quad .$$

In words, an estimator of a statistic is unbiased if its expected value equals the statistic. We now need to recall a few facts about mathematical expectation and variance (see also Appendix II).

Mathematical expectation and variance—discrete case

$$\mu = \text{E}[X] = \sum_{i=1}^{n} x_i \cdot p(x_i) \quad \text{and}$$

$$\sigma^2 = \text{var}[X] = \sum_{i=1}^{n}(x_i - \text{E}[X])^2 \cdot p(x_i) , \quad \text{where}$$

$$\sum_{i=1}^{n} p(x_i) = 1 \quad .$$

These definitions lead to the following useful identities:

$$\mathrm{E}\left[a \cdot X + b\right] = a \cdot \mathrm{E}\left[X\right] + b$$

$$\mathrm{E}\left[\sum_{i=1}^{n} X_i\right] = \sum_{i=1}^{n} \mathrm{E}\left[X_i\right]$$

$$\mathrm{var}\left[X\right] = \mathrm{E}\left[X^2\right] - \left(\mathrm{E}\left[X\right]\right)^2$$

$$\Rightarrow \mathrm{E}\left[X^2\right] = \mathrm{var}\left[X\right] + \left(\mathrm{E}[X]\right)^2$$

$$\Rightarrow \mathrm{E}\left[X^2\right] = \sigma^2 + \mu^2$$

and

$$\mathrm{var}\left[a \cdot X + b\right] = a^2 \cdot \mathrm{var}\left[X\right] \quad ,$$

where a and b are constants.

We will use the above identities in the following derivations.

The MLE $\hat{\theta}_1 = \bar{X}$ *is an unbiased estimator of* μ

$$\mathrm{E}\left[\bar{X}\right] = \mathrm{E}\left[\frac{\sum_{i=1}^{n} x_i}{n}\right] = \frac{1}{n} \cdot \sum_{i=1}^{n} \mathrm{E}\left[x_i\right] = \frac{1}{n} \cdot \sum_{i=1}^{n} \mu = \frac{n \cdot \mu}{n} = \mu \quad .$$

The MLE $\hat{\theta}_2 = s_*^2$ *is a biased estimator of* σ^2

$$\begin{aligned}
\mathrm{E}\left[s_*^2\right] &= \mathrm{E}\left[\frac{\sum_{i=1}^{n} (x_i - \theta_1)^2}{n}\right] \\
&= \mathrm{E}\left[\frac{\sum_{i=1}^{n} (x_i - \bar{X})^2}{n}\right] \\
&= \frac{1}{n} \cdot \mathrm{E}\left[\sum_{i=1}^{n} (x_i - \bar{X})^2\right] \\
&= \frac{1}{n} \cdot \mathrm{E}\left[\sum_{i=1}^{n} \left(x_i^2 - 2 \cdot x_i \cdot \bar{X} + \bar{X}^2\right)\right]
\end{aligned}$$

$$
\begin{aligned}
&= \frac{1}{n} \cdot \mathrm{E}\left[\sum_{i=1}^{n} x_i^2 - \sum_{i=1}^{n} 2 \cdot x_i \cdot \bar{X} + \sum_{i=1}^{n} \bar{X}^2\right] \\
&= \frac{1}{n} \cdot \left(\mathrm{E}\left[\sum_{i=1}^{n} x_i^2\right] - \mathrm{E}\left[\sum_{i=1}^{n} 2 \cdot x_i \cdot \bar{X}\right] + \mathrm{E}\left[\sum_{i=1}^{n} \bar{X}^2\right]\right) \\
&= \frac{1}{n} \cdot \left(\mathrm{E}\left[\sum_{i=1}^{n} x_i^2\right] - \mathrm{E}\left[2 \cdot \bar{X} \cdot \sum_{i=1}^{n} x_i\right] + \mathrm{E}\left[\bar{X}^2 \cdot \sum_{i=1}^{n} 1\right]\right) \\
&= \frac{1}{n} \cdot \left(\mathrm{E}\left[\sum_{i=1}^{n} x_i^2\right] - \mathrm{E}\left[2 \cdot \bar{X} \cdot n \cdot \bar{X}\right] + \mathrm{E}\left[n \cdot \bar{X}^2\right]\right) \\
&= \frac{1}{n} \cdot \left(\sum_{i=1}^{n} \mathrm{E}\left[x_i^2\right] - \mathrm{E}\left[2 \cdot \bar{X} \cdot n \cdot \bar{X}\right] + \mathrm{E}\left[n \cdot \bar{X}^2\right]\right) \\
&= \frac{1}{n} \cdot \left(n \cdot \mathrm{E}\left[x_i^2\right] - 2 \cdot n \cdot \mathrm{E}\left[\bar{X}^2\right] + n \cdot \mathrm{E}\left[\bar{X}^2\right]\right) \\
&= \mathrm{E}\left[x_i^2\right] - \mathrm{E}\left[\bar{X}^2\right] \\
&= \left(\sigma^2 + \mu^2\right) - \left(\mathrm{var}\left[\bar{X}\right] + \left(\mathrm{E}\left[\bar{X}\right]\right)^2\right) \\
&= \sigma^2 + \mu^2 - \mathrm{var}\left[\bar{X}\right] - \mu^2 \\
&= \sigma^2 - \mathrm{var}\left[\bar{X}\right] \\
&= \sigma^2 - \mathrm{var}\left[\frac{\sum_{i=1}^{n} x_i}{n}\right] \\
&= \sigma^2 - \frac{1}{n^2} \cdot \sum_{i=1}^{n} \mathrm{var}\left[x_i\right] \\
&= \sigma^2 - \frac{n \cdot \sigma^2}{n^2} \\
&= \sigma^2 - \frac{\sigma^2}{n} \\
&= \frac{(n-1)}{n} \cdot \sigma^2 \neq \sigma^2 \quad .
\end{aligned}
$$

So $\hat{\theta}_2 = s_*^2$ is a biased estimator of σ^2. Now further note that replacing n in the denominator of s_*^2 with $(n-1)$ (that is, creating s^2) makes an unbiased estimator of σ^2. You can see this quite easily by doing this substitution yourself in the derivation above.

Appendix IV

Details on Information Measure

Proof of equivalence of the two forms of H

$$H = -\sum_{j=1}^{m}\left(\frac{f_j}{n}\right)\cdot\ln\left(\frac{f_j}{n}\right)$$

$$= -\frac{1}{n}\cdot\sum_{j=1}^{m} f_j\cdot\ln\left(\frac{f_j}{n}\right)$$

$$= -\frac{1}{n}\cdot\sum_{j=1}^{m} f_j\cdot(\ln(f_j)-\ln(n))$$

$$= -\frac{1}{n}\cdot\sum_{j=1}^{m} f_j\cdot\ln(f_j)+\frac{1}{n}\cdot\sum_{j=1}^{m} f_j\cdot\ln(n)$$

$$= \frac{1}{n}\cdot\sum_{j=1}^{m} f_j\cdot\ln(n)-\frac{1}{n}\cdot\sum_{j=1}^{m} f_j\cdot\ln(f_j)$$

$$= \frac{1}{n}\cdot\ln(n)\cdot\sum_{j=1}^{m} f_j-\frac{1}{n}\cdot\sum_{j=1}^{m} f_j\cdot\ln(f_j)$$

$$= \frac{1}{n}\cdot\ln(n)\cdot n-\frac{1}{n}\cdot\sum_{j=1}^{m} f_j\cdot\ln(f_j)$$

$$= \ln(n)-\frac{1}{n}\cdot\sum_{j=1}^{m} f_j\cdot\ln(f_j)$$

We now show why it is reasonable to take $0 \cdot \ln(0) = 0$ in the calculation of information measure. We show that

$$\lim_{p \to 0+} p \cdot \ln(p) = 0 \quad .$$

We use L'Hospital's Rule as the basis for the argument (after Stewart [107], p. 311).

The conditions needed to apply L'Hospital's Rule are as follows:

If both

$$\lim_{x \to 0} f(x) = 0 \qquad \text{and} \qquad \lim_{x \to 0} g(x) = 0$$

or

$$\lim_{x \to 0} f(x) = \pm\infty \qquad \text{and} \qquad \lim_{x \to 0} g(x) = \pm\infty \quad ,$$

then

$$\lim_{x \to 0} \frac{f(x)}{g(x)} = \lim_{x \to 0} \frac{f'(x)}{g'(x)} \quad .$$

In our case,

$$\lim_{p \to 0+} p = 0 \qquad \text{and} \qquad \lim_{p \to 0+} \ln(p) = -\infty \quad .$$

And to get the fractional form we can write

$$p = \frac{1}{\frac{1}{p}}$$

and note that

$$\lim_{p \to 0+} \frac{1}{p} = \infty \quad .$$

Thus we can use L'Hospital's Rule:

$$\lim_{p \to 0+} p \cdot \ln(p) = \lim_{p \to 0+} \frac{\ln(p)}{\frac{1}{p}} = \lim_{p \to 0+} \frac{\frac{1}{p}}{-\frac{1}{p^2}} = \lim_{p \to 0+} \frac{-p^2}{p} = \lim_{p \to 0+} (-p) = 0 \quad .$$

Appendix V

Rotation Transformation

Here we explain the form of the rotation transformation. We will present the results in hierarchical manner, working back to more-basic results so readers can decide how much detail they need to understand the ideas, based on their background knowledge.

V.1 The Basic Result

Please see Figure V.1 to help follow along with the argument. Observe that the following relationships hold:

$$x = r \cdot \cos \alpha \ , \ \ y = r \cdot \sin \alpha$$

and

$$x' = r \cdot \cos(\alpha + \beta) \ , \ \ y' = r \cdot \sin(\alpha + \beta) \quad .$$

Combining these results and using the double angle formulas (see next section), we get

$$\begin{aligned} x' &= r \cdot \cos(\alpha + \beta) \\ &= r \cdot (\cos \alpha \cdot \cos \beta - \sin \alpha \cdot \sin \beta) \\ &= r \cdot \cos \alpha \cdot \cos \beta - r \cdot \sin \alpha \cdot \sin \beta \\ &= x \cdot \cos \beta - y \cdot \sin \beta \end{aligned}$$

and

$$\begin{aligned} y' &= r \cdot \sin(\alpha + \beta) \\ &= r \cdot (\sin \alpha \cdot \cos \beta + \cos \alpha \cdot \sin \beta) \\ &= r \cdot \sin \alpha \cdot \cos \beta + r \cdot \cos \alpha \cdot \sin \beta) \\ &= y \cdot \cos \beta + x \cdot \sin \beta \\ &= x \cdot \sin \beta + y \cdot \cos \beta \quad . \end{aligned}$$

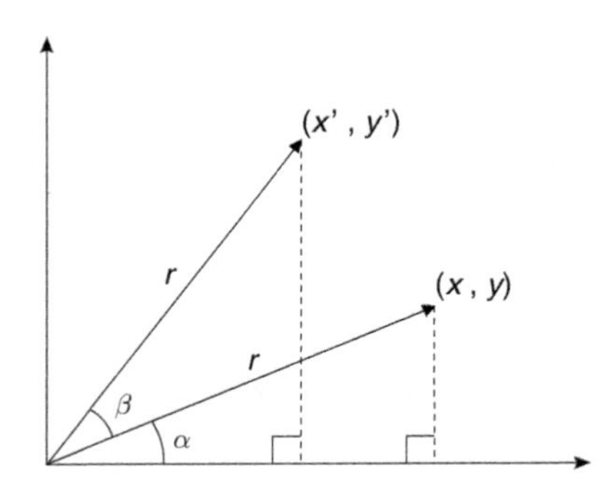

Figure V.1 Rotation transformation.

In matrix notation this is given as

$$\begin{bmatrix} x' \\ y' \end{bmatrix} = \begin{bmatrix} \cos\beta & -\sin\beta \\ \sin\beta & \cos\beta \end{bmatrix} \cdot \begin{bmatrix} x \\ y \end{bmatrix} .$$

V.2 The Double Angle Formulas

The following equations for the sine and cosine double angle formulas can be observed directly from Figure V.2.

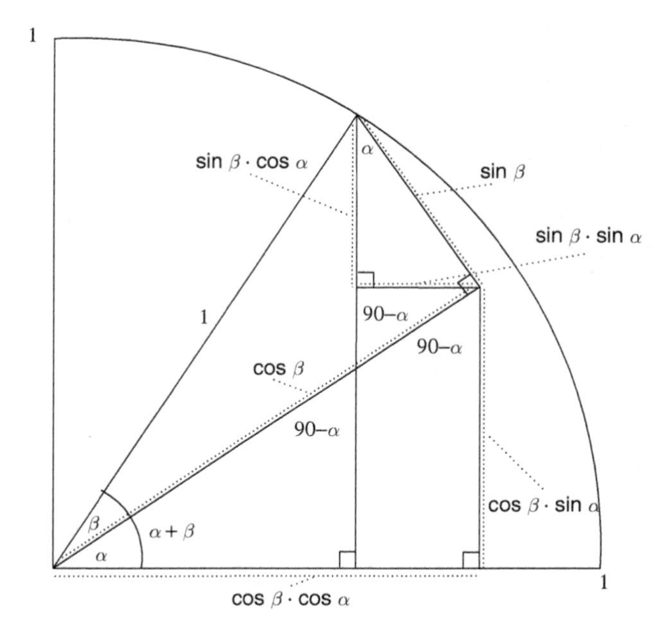

Figure V.2 Sine and cosine double angle formulas.

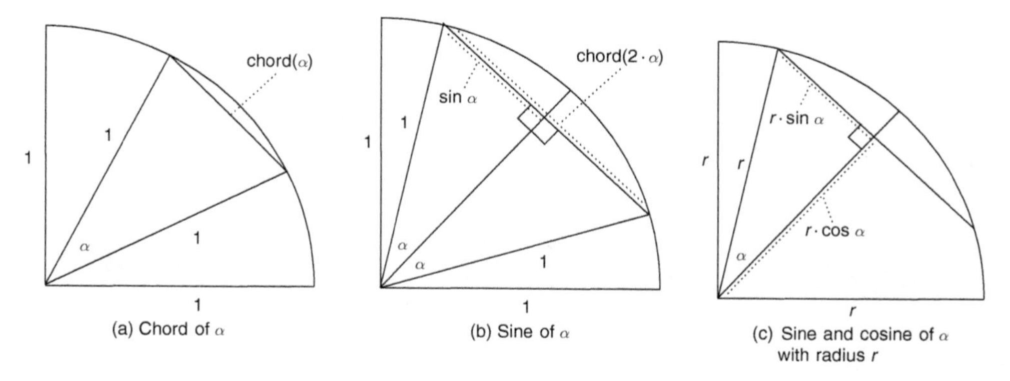

Figure V.3 Some details for the sine function.

$$\begin{aligned}\sin(\alpha+\beta) &= \sin\beta\cdot\cos\alpha+\cos\beta\cdot\sin\alpha\\ &= \sin\alpha\cdot\cos\beta+\cos\alpha\cdot\sin\beta\\ \cos(\alpha+\beta) &= \cos\beta\cdot\cos\alpha-\sin\beta\cdot\sin\alpha\\ &= \cos\alpha\cdot\cos\beta-\sin\alpha\cdot\sin\beta \quad .\end{aligned}$$

V.3 Sine and Cosine Definitions

The length of the arc subtended by the angle α in radians is α radians. A chord of a circle is a straight line connecting two points on the circle. Chord(α) is the chord between the two points prescribed by the angle α. The sine function is formally defined as

$$\sin\alpha = \frac{1}{2}\cdot\text{chord}(2\cdot\alpha) \quad .$$

These relationships are illustrated in Figures V.2 and V.3.

Appendix VI

Projections and Transformations

Here we explain the form of the spatial ellipsoidal (or geodetic) coordinate system. The following notes were adapted from pages 47 to 52 of Torge [113].

VI.1 The Rotational Ellipsoid

The earth's surface can be approximated quite well by an oblate (flattened at the poles) rotational ellipsoid; that is, an ellipsoid obtained by rotating a meridinal ellipse about its minor axis. First consider the meridinal ellipse, which is just an ellipse as shown in Figure VI.1, where a is the length of the semimajor axis and b is the length of the semiminor axis. The following relationships are defined for an ellipse, flattening f and first eccentricity e:

$$f = \frac{a-b}{a}$$

and

$$e = \frac{\sqrt{a^2-b^2}}{a} \qquad \text{or} \qquad e^2 = \frac{a^2-b^2}{a^2} \quad .$$

From the above definitions we get

$$\frac{b}{a} = 1 - f = \sqrt{1-e^2}$$

or

$$\frac{b^2}{a^2} = 1 - e^2 \quad . \tag{VI.1}$$

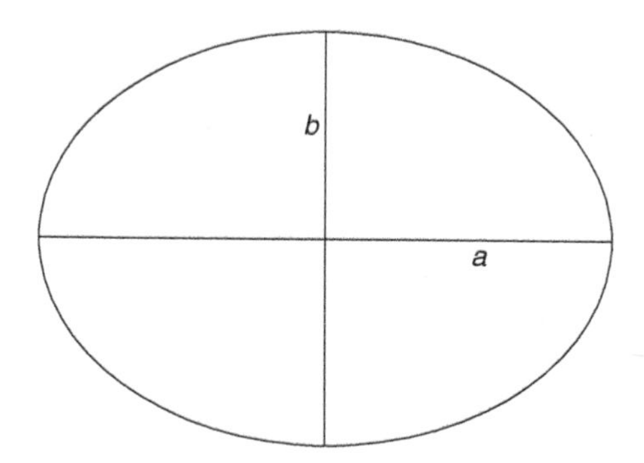

Figure VI.1 An ellipse.

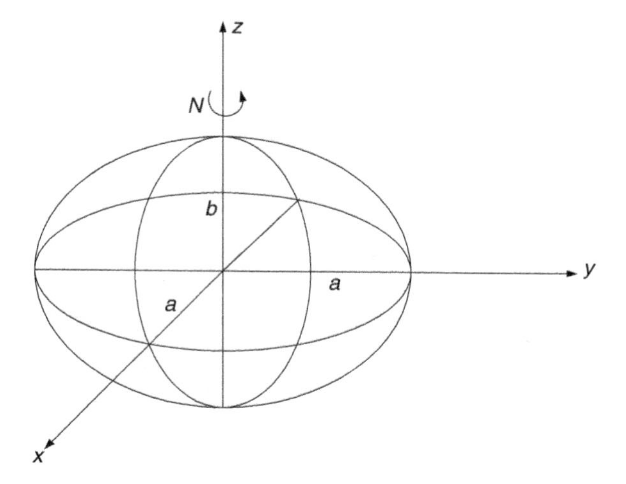

Figure VI.2 The rotational ellipsoid, Euclidean coordinates.

If the origin of the rotational ellipsoid is set to (0,0,0) and the z-axis set to the minor axis of the ellipsoid, the equation for the surface of the ellipsoid is given by

$$\frac{x^2}{a^2} + \frac{y^2}{a^2} + \frac{z^2}{b^2} = 1 \quad . \tag{VI.2}$$

See Figure VI.2.

VI.2 Ellipsoidal Geographic Coordinates

Further, we can define ellipsoidal geographic coordinates using geographic latitude ϕ and geographic longitude λ, where ϕ is the angle measured in the meridian plane between the equatorial plane (the x, y plane) of the ellipsoid and the surface normal

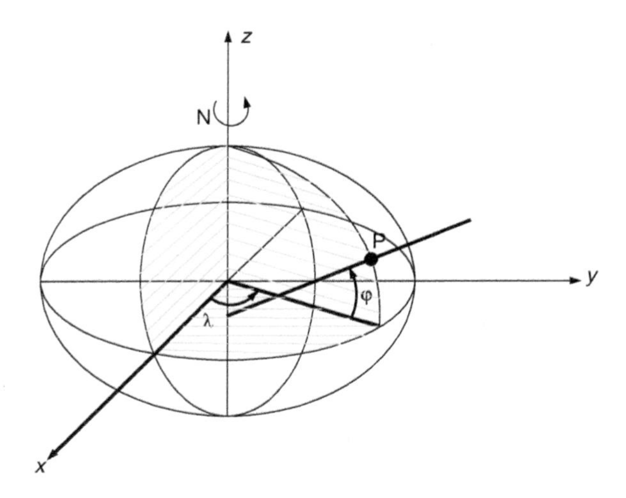

Figure VI.3 The rotational ellipsoid, geographic coordinates.

at P. Note that the surface normal at P is the direction perpendicular to the plane tangent to the ellipsoid at P. Geographic longitude λ is the angle measured in the equatorial plane between the zero meridian (the x-axis) and the meridian plane of P (see Figure VI.3). Note that ϕ is positive northward and negative southward, and λ is positive eastward.

Now we introduce a polar coordinate representation of the x, y coordinates. We have

$$x = p \cdot \cos(\lambda) \qquad \text{and} \qquad y = p \cdot \sin(\lambda) \quad , \tag{VI.3}$$

where p is the radius of a circle of latitude and

$$p = \sqrt{x^2 + y^2}$$

or

$$p^2 = x^2 + y^2 \quad .$$

Next we substitute p into equation VI.2:

$$\frac{x^2}{a^2} + \frac{y^2}{a^2} + \frac{z^2}{b^2} = 1$$

$$\Rightarrow \qquad \frac{x^2 + y^2}{a^2} + \frac{z^2}{b^2} = 1$$

$$\Rightarrow \qquad \frac{p^2}{a^2} + \frac{z^2}{b^2} = 1 \quad .$$

Differentiating (implicitly) with respect to p, we get

$$\frac{d}{dp}\left(\frac{p^2}{a^2}+\frac{z^2}{b^2}=1\right)$$

$$\Rightarrow \quad \frac{d}{dp}\left(\frac{p^2}{a^2}\right)+\frac{d}{dp}\left(\frac{z^2}{b^2}\right)=\frac{d}{dp}(1)$$

$$\Rightarrow \quad \frac{1}{a^2}\cdot\left(\frac{d}{dp}(p^2)\right)+\frac{1}{b^2}\cdot\left(\frac{d}{dp}(z^2)\right)=\frac{d}{dp}(1)$$

$$\Rightarrow \quad \frac{1}{a^2}\cdot\left(\frac{d}{dp}(p^2)\right)+\frac{1}{b^2}\cdot\left(\left(\frac{dz}{dz}\right)\frac{d}{dp}(z^2)\right)=\frac{d}{dp}(1)$$

$$\Rightarrow \quad \frac{1}{a^2}\cdot\left(\frac{d}{dp}(p^2)\right)+\frac{1}{b^2}\cdot\left(\left(\frac{dz}{dp}\right)\frac{d}{dz}(z^2)\right)=\frac{d}{dp}(1)$$

$$\Rightarrow \quad \frac{2\cdot p}{a^2}+\frac{2\cdot z}{b^2}\cdot\left(\frac{dz}{dp}\right)=0$$

$$\Rightarrow \quad \frac{2\cdot z}{b^2}\cdot\left(\frac{dz}{dp}\right)=-\frac{2\cdot p}{a^2}$$

$$\Rightarrow \quad \frac{z}{b^2}\cdot\left(\frac{dz}{dp}\right)=-\frac{p}{a^2}$$

$$\Rightarrow \quad \frac{dz}{dp}=-\frac{b^2}{a^2}\cdot\left(\frac{p}{z}\right) \quad .$$

Note that $\frac{dz}{dp}$ is the slope of the tangent to the ellipsoid at P. Thus, referring to Figure VI.4, we have

$$\begin{aligned}\frac{dz}{dp} &= -\tan(90-\phi)\\ &= -\frac{\sin(90-\phi)}{\cos(90-\phi)}\\ &= -\frac{\cos(\phi)}{\sin(\phi)}\\ &= -\cot(\phi) \quad .\end{aligned}$$

So thus far we have established the relationships

$$\frac{dz}{dp}=-\frac{b^2}{a^2}\cdot\left(\frac{p}{z}\right)=-\cot(\phi) \quad . \qquad \text{(VI.4)}$$

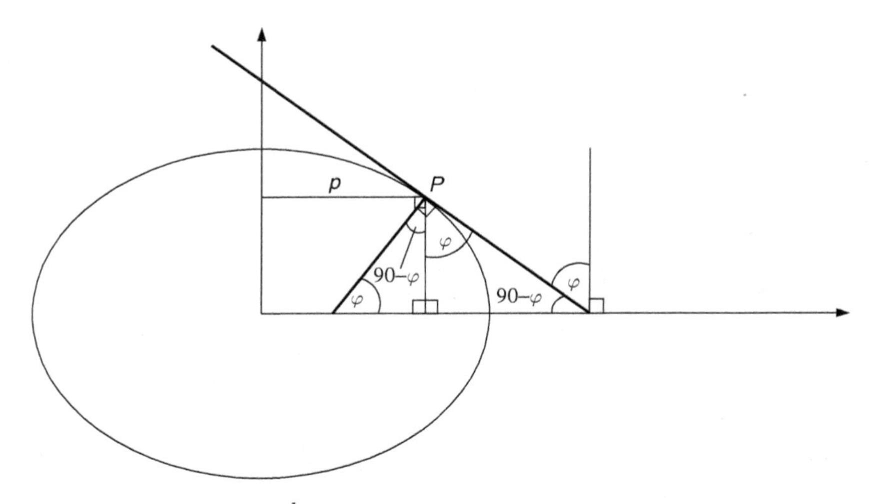

Figure VI.4 Ellipsoid tangent $\frac{dz}{dp}$.

Having these relationships in hand we can now move on to a parametric representation of p and z in terms of only ϕ and the constants a and b. Now, starting from VI.4, we have

$$\begin{aligned} & -\frac{b^2}{a^2}\cdot\left(\frac{p}{z}\right) = -\cot(\phi) \\ \Rightarrow \quad & \frac{b^2}{a^2}\cdot\left(\frac{p}{z}\right) = \cot(\phi) \\ \Rightarrow \quad & \frac{b^2}{a^2}\cdot\frac{p}{z} = \frac{\cos(\phi)}{\sin(\phi)} \\ \Rightarrow \quad & \frac{b^4}{a^4}\cdot\frac{p^2}{z^2} = \frac{\cos^2(\phi)}{\sin^2(\phi)} \quad . \end{aligned}$$

Now recall that from VI.2 we have

$$\begin{aligned} & \frac{x^2}{a^2}+\frac{y^2}{a^2}+\frac{z^2}{b^2} = 1 \\ \Rightarrow \quad & \frac{p^2}{a^2}+\frac{z^2}{b^2} = 1 \\ \Rightarrow \quad & \frac{z^2}{b^2} = 1-\frac{p^2}{a^2} \\ \Rightarrow \quad & \frac{z^2}{b^2} = \frac{a^2}{a^2}-\frac{p^2}{a^2} \\ \Rightarrow \quad & \frac{z^2}{b^2} = \frac{a^2-p^2}{a^2} \\ \Rightarrow \quad & z^2 = b^2\cdot\frac{(a^2-p^2)}{a^2} \quad . \end{aligned}$$

Substituting this result for z^2 back where we left off, we get

$$\Rightarrow \quad \frac{b^4}{a^4} \cdot p^2 \cdot \left(\frac{a^2}{b^2 \cdot (a^2 - p^2)} \right) = \frac{\cos^2(\phi)}{\sin^2(\phi)}$$

$$\Rightarrow \quad \frac{b^2}{a^2} \cdot \frac{p^2}{(a^2 - p^2)} = \frac{\cos^2(\phi)}{\sin^2(\phi)}$$

$$\Rightarrow \quad \frac{p^2}{(a^2 - p^2)} = \frac{a^2 \cdot \cos^2(\phi)}{b^2 \cdot \sin^2(\phi)}$$

$$\Rightarrow \quad p^2 = (a^2 - p^2) \cdot \left(\frac{a^2 \cdot \cos^2(\phi)}{b^2 \cdot \sin^2(\phi)} \right)$$

$$\Rightarrow \quad p^2 = a^2 \cdot \left(\frac{a^2 \cdot \cos^2(\phi)}{b^2 \cdot \sin^2(\phi)} \right) - p^2 \cdot \left(\frac{a^2 \cdot \cos^2(\phi)}{b^2 \cdot \sin^2(\phi)} \right)$$

$$\Rightarrow \quad p^2 \cdot \left(1 + \frac{a^2 \cdot \cos^2(\phi)}{b^2 \cdot \sin^2(\phi)} \right) = \frac{a^4 \cdot \cos^2(\phi)}{b^2 \cdot \sin^2(\phi)}$$

$$\Rightarrow \quad p^2 \cdot \left(\frac{b^2 \cdot \sin^2(\phi) + a^2 \cdot \cos^2(\phi)}{b^2 \cdot \sin^2(\phi)} \right) = \frac{a^4 \cdot \cos^2(\phi)}{b^2 \cdot \sin^2(\phi)}$$

$$\Rightarrow \quad p^2 \cdot (b^2 \cdot \sin^2(\phi) + a^2 \cdot \cos^2(\phi)) = a^4 \cdot \cos^2(\phi)$$

$$\Rightarrow \quad p^2 = \frac{a^4 \cdot \cos^2(\phi)}{b^2 \cdot \sin^2(\phi) + a^2 \cdot \cos^2(\phi)}$$

$$\Rightarrow \quad p = \frac{a^2 \cdot \cos(\phi)}{\sqrt{b^2 \cdot \sin^2(\phi) + a^2 \cdot \cos^2(\phi)}}$$

$$\Rightarrow \quad p = \frac{a^2 \cdot \cos(\phi)}{\sqrt{a^2 \cdot \cos^2(\phi) + b^2 \cdot \sin^2(\phi)}} \quad . \tag{VI.5}$$

Similarly, we can derive the following expression for z:

$$z = \frac{b^2 \cdot \sin(\phi)}{\sqrt{a^2 \cdot \cos^2(\phi) + b^2 \cdot \sin^2(\phi)}} \quad .$$

Next we need an expression for the radius of curvature in the prime vertical. Referring to Figure VI.5, we have

$$p = v(\phi) \cdot \cos(\phi) \tag{VI.6}$$

$$\Rightarrow \quad v(\phi) = \frac{p}{\cos(\phi)} \quad .$$

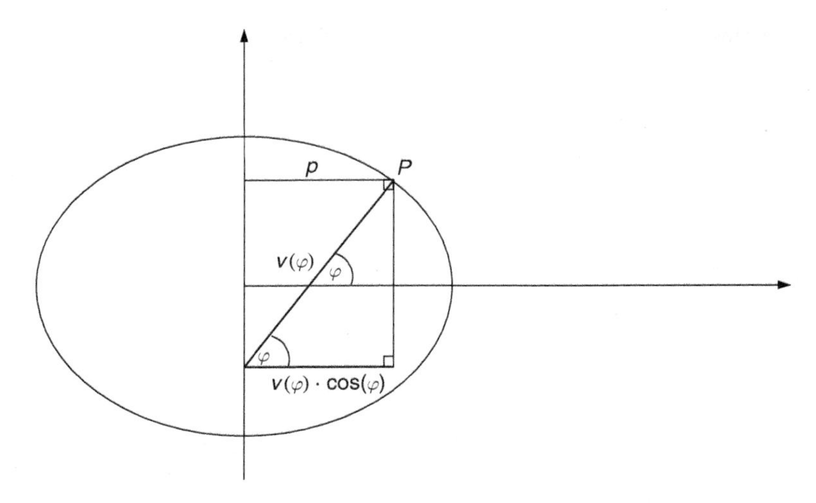

Figure VI.5 The radius of curvature in the prime vertical.

Substituting our expression for p from VI.5, we get

$$v(\phi) = \frac{1}{\cos(\phi)} \cdot \frac{a^2 \cdot \cos(\phi)}{\sqrt{a^2 \cdot \cos^2(\phi) + b^2 \cdot \sin^2(\phi)}}$$

$$= \frac{a^2}{\sqrt{a^2 \cdot \cos^2(\phi) + b^2 \cdot \sin^2(\phi)}}$$

$$= \frac{a^2}{a \cdot \sqrt{\frac{a^2}{a^2} \cdot \cos^2(\phi) + \frac{b^2}{a^2} \cdot \sin^2(\phi)}}$$

$$= \frac{a}{\sqrt{\cos^2(\phi) + \left(\frac{b^2}{a^2}\right) \cdot \sin^2(\phi)}}$$

$$= \frac{a}{\sqrt{(1 - \sin^2(\phi)) + \left(\frac{b^2}{a^2}\right) \cdot \sin^2(\phi)}}$$

$$= \frac{a}{\sqrt{1 + \left(\frac{b^2}{a^2} - 1\right) \cdot \sin^2(\phi)}}$$

$$= \frac{a}{\sqrt{1 - \left(1 - \frac{b^2}{a^2}\right) \cdot \sin^2(\phi)}}$$

$$= \frac{a}{\sqrt{1 - \left(\frac{a^2 - b^2}{a^2}\right) \cdot \sin^2(\phi)}}$$

$$v(\phi) = \frac{a}{\sqrt{1 - e^2 \cdot \sin^2(\phi)}} \quad . \tag{VI.7}$$

After the above preliminary results we are now ready to derive the final form of the spatial ellipsoidal coordinate system, or geodetic coordinates. Until now we have created the machinery for specifying a location on the geoide (or roughly the ellipsoid of rotation that approximates mean sea level). We need to add in height above the geoide, h, to specify the complete coordinate system for a position on the surface of the earth (see Figure VI.6).

Now we use VI.3 and VI.6 to get

$$x = p \cdot \cos(\lambda)$$
$$y = p \cdot \sin(\lambda)$$

$$\Rightarrow$$

$$x = v(\phi) \cdot \cos(\phi) \cdot \cos(\lambda)$$
$$y = v(\phi) \cdot \cos(\phi) \cdot \sin(\lambda) \quad .$$

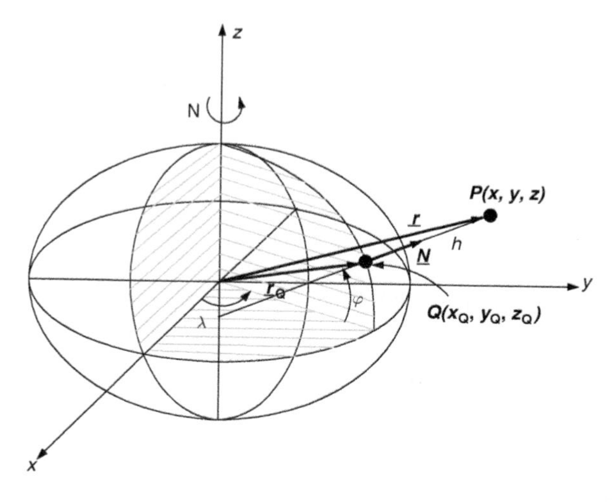

Figure VI.6 Spatial ellipsoidal coordinates.

Further, we use VI.1, VI.4, and VI.6 to give

$$\frac{b^2}{a^2} \cdot \frac{p}{z} = \frac{\cos(\phi)}{\sin(\phi)}$$

$$\Rightarrow \qquad \left(1 - e^2\right) \cdot \frac{p}{z} = \frac{\cos(\phi)}{\sin(\phi)}$$

$$\Rightarrow \qquad z = p \cdot \left(1 - e^2\right) \cdot \frac{\sin(\phi)}{\cos(\phi)}$$

$$\Rightarrow \qquad z = v(\phi) \cdot \cos(\phi) \cdot \left(1 - e^2\right) \cdot \frac{\sin(\phi)}{\cos(\phi)}$$

$$\Rightarrow \qquad z = v(\phi) \cdot \left(1 - e^2\right) \cdot \sin(\phi) \quad .$$

In summary, for a point on the geoide,

$$\underline{r}_Q = \begin{bmatrix} x_Q \\ y_Q \\ z_Q \end{bmatrix} = \begin{bmatrix} v(\phi) \cdot \cos(\phi) \cdot \cos(\lambda) \\ v(\phi) \cdot \cos(\phi) \cdot \sin(\lambda) \\ v(\phi) \cdot \left(1 - e^2\right) \cdot \sin(\phi) \end{bmatrix} \quad .$$

To include height above the geoide, h, consider $\underline{r}$ (referring to Figure VI.6),

$$\underline{r} = \underline{r}_Q + h \cdot \underline{n} \quad ,$$

where

$$\underline{n} = \begin{bmatrix} \cos(\phi) \cdot \cos(\lambda) \\ \cos(\phi) \cdot \sin(\lambda) \\ \sin(\phi) \end{bmatrix} \quad .$$

Thus in summary,

$$\underline{r} = \begin{bmatrix} x \\ y \\ z \end{bmatrix} = \begin{bmatrix} (v(\phi) + h) \cdot \cos(\phi) \cdot \cos(\lambda) \\ (v(\phi) + h) \cdot \cos(\phi) \cdot \sin(\lambda) \\ \left(v(\phi) \cdot \left(1 - e^2\right) + h\right) \cdot \sin(\phi) \end{bmatrix} \quad .$$

References

[1] E.H.L. Aarts and J.K. Lenstra. Introduction. In E.H.L. Aarts and J.K. Lenstra, eds, *Local Search in Combinatorial Optimization.* John Wiley & Sons Ltd, 1997.

[2] P.G.V. Abeyratne, W.E. Featherstone, and D.A. Tantrigoda. On the geodetic datums in Sri Lanka. *Survey Review* 42(317): 229–39, 2010.

[3] M. Andrejevic. Big data, big questions the big data divide. *International Journal of Communication* 8(0): 17, June 2014.

[4] L. Anselin. What is special about spatial data? Alternative perspectives on spatial data analysis (89–4). *eScholarship,* April 1989.

[5] R.H. Atkin. *Mathematical Structure in Human Affairs.* Heinemann Educational Books, 1974.

[6] T.C. Bailey and A.C. Gatrell. *Interactive Spatial Data Analysis.* Prentice Hall, 1995.

[7] T.J. Barnes and M.W. Wilson. Big data, social physics, and spatial analysis: the early years. *Big Data & Society* 1(1), April, 2014.

[8] M. Batty, A.S. Fotheringham, and P. Longley. Fractal geometry and urban morphology. In S.N. Lam and L. De Cola, eds, *Fractals in Geography.* Prentice Hall, 1993.

[9] B. Baumgart. A polyhedron representation for computer vision. *Proceedings AFIPS National Conference* 44: 589–96, 1975.

[10] Y. Bedard and E. Bernier. Supporting multiple representations with spatial databases views management and the concept of VUEL. In ISPRS/ICA *Joint Workshop on Multi-Scale Representations of Spatial Data,* Ottawa, Canada, 2002.

[11] E. Bodansky, A.R. Gribov, and M. Pilouk. Smoothing and compression of lines obtained by raster-to-vector conversion. In D. Blostein and Y.-B. Kwon, eds, *Graphics Recognition Algorithms and Applications*, number 2390 in *Lecture Notes in Computer Science*, pp. 256–65. Berlin, Heidelberg: Springer, 2002.

[12] B. Boots, R. Feick, N. Shiode, and S.A. Roberts. Investigating recursive point voronoi diagrams, in geographic information science. In *Proceedings of GIScience 2002, Second International Conference on Geographic Information Science, Boulder, Colorado*, volume 2478 of *Springer Lecture Notes in Computer Science*, pp. 1–22, 2002.

[13] B. Boots and N. Shiode. Preliminary observations on recursive Voronoi diagrams. In *Proceedings of the GIS Research UK 9th Annual Conference* GISRUK 2001, pp. 452–5, 2001.

[14] F. Bouillé. Structuring cartographic data and spatial processes with the hypergraph-based data structure. In G. Dutton, [ed]., *First International Advanced Study Symposium on Topological Data Structures for Geographic Information Systems,* Volume 5 – *Data Structures: Surficial and Multi-Dimensional*, Harvard Papers on Geographic Information Systems. Harvard, 1978.

[15] D. Boyd and K. Crawford. Critical questions for big data. *Information, Communication & Society* 15(5): 662–79, 2012.

[16] P.A. Burrough and A.U. Frank. Concepts and paradigms in spatial information: are current geographic information systems truly generic? *International Journal of Geographical Information Systems* 9(2): 101–16, 1995.

[17] P.A. Burrough and R.A. McDonnell. *Principles of Geographical Information Systems* (2nd edn). Oxford, New York: Oxford University Press, April 1998.

[18] M.L. Casado. Some basic mathematical constraints for the geometric conflation problem. In *Proceedings of the 7th International Symposium on Spatial Accuracy Assessment in Natural Resources and Environmental Sciences*, pp. 264–74, 2006.

[19] G. Chander, B.L. Markham, and D.L. Helder. Summary of current radiometric calibration coefficients for Landsat MSS, TM, ETM+, and EO-1 ALI sensors. *Remote Sensing of Environment* 113(5): 893–903, 2009.

[20] P.P.-S. Chen. The entity-relationship model: toward a unified view of data. *ACM Trans. Database Syst.* 1(1): 9–36, March 1976.

[21] K.C. Clarke. *Getting Started with Geographic Information Systems.* Prentice Hall Series in Geographic Information Science (3rd edn). Prentice Hall, 2001.

[22] ———. Geocomputation's future at the extremes: high performance computing and nanoclients. *Parallel Computing* 29(10): 1281–95, October 2003.

[23] E. Clementini and P. Di Felice. An algebraic model for spatial objects with indeterminate boundaries. In P.A. Burrough and A.U. Frank, eds, *Geographic Objects with Indeterminate Boundaries*, volume 2 of GISDATA, pp. 155–69. Taylor & Francais, 1996.

[24] E.F. Codd. A relational model of data for large shared data banks. *Communications of the ACM* 13(6): 377–87, 1970.

[25] J. Cohen. A coefficient of agreement for nominal scales. *Educational and Psychological Measurement* 20: 37–46, 1960.

[26] J.T. Coppock and D.W. Rhind. The history of GIS. *Geographical Information Systems: Principles and Applications* 1(1): 21–43, 1991.

[27] T.H. Cormen, R.L. Rivest, C.E. Leiserson, and C. Stein. *Introduction to Algorithms* (3rd edn). MIT Press, 2009.

[28] T. Cova and M. Goodchild. Extending geographical representation to include fields of spatial objects. *International Journal of Geographical Information Science* 16(6): 509–32, 2002.

[29] D. Crandall and N. Snavely. Modeling people and places with Internet photo collections. *Communications of the ACM* 55(6): 52–60, 2012.

[30] R. Crane. *A Simplified Approach to Image Processing: Classical and Modern Techniques in C.* Hewlett-Packard Professional Books. Prentice Hall PTR, 1997.

[31] N. Cressie. *Statistics for Spatial Data.* Wiley, 1993.

[32] R.E. Deakin. *A Guide to the Mathematics of Map Projections.* Technical report, School of Mathematical and Geospatial Sciences, RMIT University, 2004.

[33] K. Deb, A. Pratap, S. Agarak, and T. Meyarivan. A fast and elitist multiobjective genetic algorithm: NSGA-II. *IEEE Transactions on Evolutionary Computing* 8(2): 182–97, 2002.

[34] D.H. Douglas and T.K. Poiker. Algorithms for the reduction of the number of points required to represent a digitized line or its caricature. *Cartographica: The International Journal for Geographic Information and Geovisualization* 10(2): 112–22, 1973.

[35] W.E. Dramstad, J.D. Olson, and R.T.T. Forman. *Landscape Ecology Principles in Landscape Architecture and Land-Use Planning.* Harvard Graduate School of Design: Island Press, American Society of Landscape Architects, 1996.

[36] I.L. Dryden and K.V. Mardia. *Statistical Shape Analysis.* Wiley, 1998.

[37] K.R. Ferreira, G. Camara, and A.M.V. Monteiro. An algebra for spatiotemporal data: from observations to events. *Transactions in GIS* 18(2): 253–69, April 2014.

[38] P. Fieguth. *Statistical Image Processing and Multidimensional Modeling.* Springer, 2011.

[39] P. Fisher. Concepts and paradigms of spatial data. In M. Craglia and H. Couclelis, eds, *Geographic Information Research: Bridging the Atlantic, National Science Foundation and European Science Foundation*, pp. 297–307. Taylor & Francis, 1997.

[40] G.M. Foody. Thematic map comparison: evaluating the statistical significance of differences in classification accuracy. *Photogrammetric Engineering and Remote Sensing* 70(5): 627–33, 2004.

[41] R.T.T. Forman. *Land Mosaics: The Ecology of Landscapes and Regions.* Cambridge University Press, 1997.

[42] S. Gao, C. Ji, C. Xu, and N. Yang. Parallel spatial nearest neighbour query based on grid index. In *IEEE Workshop on Electronics, Computer and Applications*, pp. 273–6, May 2014.

[43] A. Getis and B. Boots. *Models of Spatial Processes.* Cambridge: Cambridge University Press, 1978.

[44] H.R. Gimblett. *Integrating Geographic Information Systems and Agent-Based Modeling Techniques for Simulating Social and Ecological Processes.* Oxford University Press, 2001.

[45] C.M. Gold. PAN graphs: an aid to GIS analysis. *International Journal of Geographic Information Systems* 2(1): 29–41, 1988.

[46] G.H. Golub and C.F. Van Loan. *Matrix Computations* (3rd edn). Johns Hopkins University Press, 1996.

[47] M.F. Goodchild. A general framework for error analysis in measurement-based GIS. *Journal of Geographical Systems* 6(4): 323–4, 2004.

[48] ———. GIScience geography, form, and process. *Annals of the Association of American Geographers* 94(4): 709–14, December 2004.

[49] ———. Citizens as sensors: the world of volunteered geography. *GeoJournal* 69(4): 211–21, 2007.

[50] M.F. Goodchild and J.A. Glennon. Crowdsourcing geographic information for disaster response: a research frontier. *International Journal of Digital Earth* 3(3): 231–41, 2010.

[51] M.F. Goodchild and G.J. Hunter. A simple positional accuracy measure for linear features. *International Journal of Geographical Information Science* 11(3): 299–306, April 1997.

[52] P. Gould. Is statistix inferens the geographical name for a wild goose? *Economic Geography*, 46: 439–48, June 1970.

[53] M. Graham, B. Hogan, R.K. Straumann, and A. Medhat. Uneven geographies of user-generated information: patterns of increasing informational poverty. *Annals of the Association of American Geographers* 104(4): 746–64, 2014.

[54] K. Gronbaek, P.P. Vestergaard, and P. Orbaek. Towards geo-spatial hypermedia: concepts and

prototype implementation. In *Proceedings of the Thirteenth ACM Conference on Hypertext and Hypermedia*, HYPERTEXT '02, pp. 117–26, New York: ACM, 2002.

[55] L. Guibas and J. Stolfi. Primitives for the manipulation of general subdivisions and the computation of Voronoi diagrams. *Association for Computing Machinery Transactions on Graphics* 4(2): 74–123, 1985.

[56] A. Guttman. R-trees: a dynamic index structure for spatial searching. *SIGMOD Rec.* 14(2): 47–57, June 1984.

[57] T. Hadzilacos and N. Tryfona. Evaluation of database modeling methods for geographic information systems. *Australasian Journal of Information Systems* 6(1), July 1998.

[58] A. Hagen. Fuzzy set approach to assessing similarity of categorical maps. *International Journal of Geographical Information Science* 17(3): 235–49, 2003.

[59] M. Haklay. How good is volunteered geographical information? A comparative study of OpenStreetMap and ordnance survey datasets. *Environment and Planning B: Planning and Design* 37(4): 682–703, 2010.

[60] M. Harrower and M. Bloch. Mapshaper.org: a map generalization web service. *IEEE Computer Graphics and Applications* 26(4): 22–7, 2006.

[61] R. Healey, S. Dowers, B. Gittings, and M. Mineter, eds. *Parallel Processing Algorithms for GIS.* Taylor & Francis, 1998.

[62] J. Herring. *OpenGIS Implementation Specification for Geographic information—Simple Feature Access—Part 2: SQL Option.* Technical report, Open Geospatial Consortium, Inc., 2006. http://www.opengeospatial.org.

[63] G. Heuvelink. Propagation of error in spatial modelling with GIS. In *Geographical Information Systems: Principles, Techniques, Applications, and Management*, vol. 2, pp. 207–17. John Wiley & Sons, 1999.

[64] C.S. Holling. Cross-scale morphology, geometry, and dynamics of ecosystems. *Ecological Monographs* 62(4): 447–502, 1992.

[65] S.P. Jackson, W. Mullen, P. Agouris, A. Crooks, A. Croitoru, and A. Stefanidis. Assessing completeness and spatial error of features in volunteered geographic information. *ISPRS International Journal of Geo-Information* 2(2): 507–30, 2013.

[66] B. Kanefsky, N.G. Barlow, and V.C. Gulick. Can distributed volunteers accomplish massive data analysis tasks? *Lunar and Planetary Science* 32: 1272, 2001.

[67] D. Kaplan. *Statistical Modeling: A Fresh Approach.* Project MOSAIC, 2011.

[68] L. Kettner, K. Mehlhorn, S. Pion, S. Schirra, and C. Yap. Classroom examples of robustness problems in geometric computations. *Computational Geometry: Theory and Applications* 40: 61–78, 2008.

[69] B. Klinkenberg. The true cost of spatial data in Canada. *The Canadian Geographer* 47(1): 37–49, 2003.

[70] D.E. Knuth. *Lecture Notes in Computer Science*, vol. 606. Springer-Verlag, 1991.

[71] W. Kuhn. Geospatial semantics: Why, of what, and how? In S. Spaccapietra and E. Ziminyi, eds, *Journal on Data Semantics III*, no. 3534 in *Lecture Notes in Computer Science*, pp. 1–24. Berlin, Heidelberg: Springer, January 2005.

[72] G. Langran. *Time in Geographic Information Systems.* Taylor & Francis, New York, 1992.

[73] G. Langran and N.R. Chrisman. A framework for temporal geographic information. *Cartographica: The International Journal for Geographic Information and Geovisualization* 25(3): 1–14, 1988.

[74] R. Laurini and D. Thompson. *Fundamentals of Spatial Information Systems.* APIC Series, no. 37. Academic Press, 1992.

[75] J.-G. Leu and L. Chen. Polygonal approximation of 2-D shapes through boundary merging. *Pattern Recognition Letters* 7(4): 231–8, April 1988.

[76] M. Ligas and P. Banasik. Conversion between Cartesian and geodetic coordinates on a rotational ellipsoid by solving a system of nonlinear equations. *Geodesy and Cartography* 60(2): 145–59, 2012.

[77] C.P. Lo and A.K.W. Yeung. *Concepts and Techniques of Geographic Information Systems* (2nd edn). Upper Saddle River, NJ: Prentice Hall, 2007.

[78] K.V. Mardia and P.E. Jupp. *Directional Statistics.* Wiley, 1999.

[79] M. Mitchell. *An Introduction to Genetic Algorithms.* MIT Press, 1996.

[80] M. Molenaar. *An Introduction to the Theory of Spatial Object Modelling for GIS.* Research Monographs in GIS. Taylor & Francis, 1998.

[81] F. Morstatter, J. Pfeffer, H. Liu, and K.M. Carley. *Is the Sample Good Enough? Comparing Data from Twitter Streaming API with Twitter Firehose.* 2013.

[82] T. Mostak. *An overview of MapD (Massively Parallel Database).* MapD whitepaper, 2014. http://www.map-d.com/docs/mapd-whitepaper.pdf.

[83] D.E. Muller and F.P. Preparata. Finding the intersection of two convex polyhedra. *Theoretical Computer Science* 7(2): 217–336, 1978.

[84] T. Nelson, B. Boots, and M.A. Wulder. Large-area mountain pine beetle infestations: spatial data representation and accuracy. *The Forestry Chronicle* 82(2): 243–52, 2006.

[85] A. Okabe, B. Boots, K. Sugihara, and S.N. Chui. *Spatial Tessellations: Concepts and Applications of Voronoi Diagrams* (2nd edn). Wiley, 2000.

[86] Open GIS Consortium Inc. *OpenGIS Implementation Standard for Geographic Information–Simple Feature Access–Part 1: Common Architecture.* Technical report, 2011. http://www.opengeospatial.org.

[87] S. Openshaw and P.J. Taylor. A million or so correlation coefficients: three experiments on the modifiable areal unit problem. *Statistical Applications in the Spatial Sciences* 127: 127–44, 1979.

[88] J. O'Rourke. *Computational Geometry in C* (2nd edn). Cambridge University Press, 1998.

[89] D. O'Sullivan and D. Unwin. *Geographic Information Analysis* (2nd edn). Wiley, 2010.

[90] D. Pantazis. CON.G.O.O.: A conceptual formalism for geographic database design. In M. Craglia and H. Couclelis, eds, *Geographic Information Research: Bridging the Atlantic.* National Science Foundation and European Science Foundation, pp. 348–67. Taylor & Francis, 1997.

[91] E. Pebesma. Spacetime: spatio-temporal data in R. *Journal of Statistical Software* 51(7): 1–30, 2012.

[92] D. Peuquet. Making space for time: issues in space-time data representation. *GeoInformatica* 5: 11–32, 2001.

[93] S. Pope, L. Copland, and D. Mueller. Loss of multiyear landfast sea ice from Yelverton Bay, Ellesmere Island, Nunavut, Canada. *Arctic, Antarctic, and Alpine Research* 44(2): 210–21, 2012.

[94] A. Pourabdollah, J. Morley, S. Feldman, and M. Jackson. Towards an authoritative OpenStreetMap: conflating OSM and OS OpenData national maps road network. *ISPRS International Journal of Geo-Information* 2(3): 704–28, 2013.

[95] C.-Z. Qin, L.-J. Zhan, A. Zhu, and C.-H. Zhou. A strategy for raster-based geocomputation under different parallel computing platforms. *International Journal of Geographical Information Science* (April 2014): 1–18.

[96] D.B. Richardson. Real-time space-time integration in GIScience and geography. *Annals of the Association of American Geographers* 103(5): 1062–71, September 2013.

[97] S.A. Roberts, G.B. Hall, and P.H. Calamai. Analyzing forest fragmentation using spatial autocorrelation, graphs and GIS. *International Journal of Geographic Information Science* 14(2): 185–204, March 2000a.

[98] ———. A pre-categorical spatial-data metamodel. *Environment and Planning B: Planning and Design* 33(6): 881–901, 2006.

[99] C. Robertson, T. Nelson, B. Boots, and M.A. Wulder. STAMP: spatialtemporal analysis of moving polygons. *Journal of Geographical Systems* 9(3): 207–27, 2007.

[100] Y. Sadahiro and M. Umemura. A computational approach for the analysis of changes in polygon distributions. *Journal of Geographical Systems* 3: 137–54, 2001.

[101] M. Schneider. Finite resolution crisp and fuzzy objects. In P. Forer, A.G.O. Yeh, and J. He, eds, *Proceedings 9th International Symposium on Spatial Data Handling*, pp. 5a.3–5a.17, August 2000.

[102] ———. Spatial plateau algebra for implementing fuzzy spatial objects in databases and GIS: spatial plateau data types and operations. *Applied Soft Computing* 16, March 2014.

[103] J. Serra. *Image Analysis and Mathematical Morphology.* Academic Press Inc., 1982.

[104] D. Sinton. The inherent structure of information as a constraint to analysis: mapped thematic data as a case study. *Harvard Papers on Geographic Information Systems,* vol. 6: 1–17, 1978.

[105] Smallworld. Smallworld home page. Technical report, 2015. http://www.gedigitalenergy.com/gis.htm.

[106] J. Stell and J. Webster. Oriented matroids as the foundation for space in GIS. *Computers, Environment and Urban Systems* 31: 379–92, 2007.

[107] J. Stewart. *Calculus: Early Trancendentals* (5th edn). Thompson, 2003.

[108] D. Sui, S. Elwood, and M.F. Goodchild, eds. *Crowdsourcing Geographic Knowledge.* Dordrecht, Netherlands: Springer, 2013.

[109] M.A. Summerfield. Populations, samples and statistical inference in geography. *The Professional Geographer* 35(2): 143–9, 1983.

[110] A.Y. Tang, T.M. Adams, and E.L. Usery. A spatial data model design for feature-based geographical information systems. *International Journal of Geographic Information Systems* 10(5): 643–59, 1996.

[111] W. Tang and W. Feng. Parallel map projection of vector-based big spatial data: coupling cloud computing with graphics processing units. *Computers, Environment and Urban Systems*, February 2014.

[112] D.M. Theobald and M.D. Gross. EML: A modeling environment for exploring landscape dynamics. *Computers, Environment and Urban Systems* 18(3): 193–204, 1994.

[113] W. Torge. *Geodesy*. de Gruyter, 1981.

[114] D.C. Tsichritizis and A. Klug, eds. *The ANSI/X3/SPARC DBMS Framework Report of the Study Group on Database Management Systems*. Technical Report 3, American National Standards Institute, 1978.

[115] H. Tveite and S. Langaas. An accuracy assessment method for geographical line data sets based on buffering. *International Journal of Geographical Information Science* 13(1): 27–47, 1999.

[116] T. Ubeda and M.J. Egenhofer. Topological error correcting in GIS. In M. Scholl and A. Voisard, eds, *Advances in Spatial Databases*, no. 1262 in *Lecture Notes in Computer Science*, pp. 281–97. Berlin, Heidelberg: Springer, January 1997.

[117] M. Visvalingam and J.D. Whyatt. Line generalisation by repeated elimination of points. *The Cartographic Journal* 30(1): 46–51, 1993.

[118] R.B. Voight. The New Haven Census Use Study: an experience in the practical applications of statistical data. *Revue de l'Institut International de Statistique / Review of the International Statistical Institute* 38(3): 369, 1970.

[119] J. Webster. Cell complexes, oriented matroids and digital geometry. *Theoretical Computer Science* 305: 491–502, 2003.

[120] K. Weiler. Edge-based structures for solid modeling in curved-surface environments. *Computer Graphics Applications* 5(1): 21–40, 1985.

[121] G. Weisbuch. *Complex Systems Dynamics: An Introduction to Automata Networks*, vol. 2 of Santa Fe Institute Studies in the Sciences of Complexity, Lecture Notes. Addison-Wesley, 1991.

[122] S. Wolfram. *A New Kind of Science*. Wolfram Media, Inc., 2002.

[123] M.F. Worboys. *GIS: A Computing Perspective*. Taylor & Francis, 1995.

[124] F. Wu. Simland: a prototype to simulate land conversion through the integrated GIS and CA with AHP-derived transition rules. *International Journal of Geographic Information Science* 12(1): 63–82, 1998.

[125] T. Yasseri, A. Spoerri, M. Graham, and J. Kertesz. *The Most Controversial Topics in Wikipedia: A Multilingual and Geographical Analysis*. Technical Report ArXiv:1305.5566, May 2013.

[126] A. Zabala and X. Pons. Effects of lossy compression on remote sensing image classification of forest areas. *International Journal of Applied Earth Observation Geoinformation*, 13: 43–51, 2011.

Index